Power System Operation, Utilization, and Control

This book presents power system analysis methods that cover all aspects of power systems operation, utilization, control, and system management.

At the beginning of each chapter, an introduction is given describing the objectives of the chapter. The authors present power system parameters in a lucid, logical, step-by-step approach.

In recognition of requirements by the Accreditation Board for Engineering and Technology (ABET) on integration of engineering computer tools, the authors demonstrate the use of MATLAB® programming in obtaining solutions to engineering power problems. MATLAB is introduced in a student-friendly manner and follow up is given in Appendix A. The use of MATLAB and power system applications are presented throughout the book.

Practice problems immediately follow each illustrative example. Students can follow the example step-by-step to solve the practice problems. These practice problems test students' comprehension and reinforce key concepts before moving on to the next chapter.

In each chapter, the authors discuss specific application aspects of the chapter's concepts using computer programming. The material covered in the chapter applies to at least one or two practical problems to students see how the concepts are used in real-life situations.

Thoroughly worked examples are provided at the end of every section. These examples give students a solid grasp of the solutions and the confidence to solve similar problems themselves.

Designed for a three-hour semester course on **Power System Operation, Utilization, and Control**, this book is intended as a textbook for a senior-level undergraduate student in electrical and computer engineering. The prerequisites for a course based on this book are knowledge of standard mathematics, including calculus and complex numbers, and basic undergraduate engineering courses.

Power System Operation, Utilization, and Control

John Fuller
Pamela Obiomon
and
Samir I. Abood

CRC Press is an imprint of the
Taylor & Francis Group, an **informa** business

First edition published 2023
by CRC Press
6000 Broken Sound Parkway NW, Suite 300, Boca Raton, FL 33487-2742

and by CRC Press
4 Park Square, Milton Park, Abingdon, Oxon, OX14 4RN

CRC Press is an imprint of Taylor & Francis Group, LLC

ISBN: 978-1-032-27745-5 (hbk)
ISBN: 978-1-032-27764-6 (pbk)
ISBN: 978-1-003-29396-5 (ebk)

DOI: 10.1201/9781003293965

Typeset in Times
by SPi Technologies India Pvt Ltd (Straive)

Dedicated to

Dedication is given to my family that has been beside me in reaching this point in life and especially to my mother, Bernice Fuller, who sacrificed so that this period in time is a reality.

John Fuller

To my husband and children with all my love. I owe you an unbiased world. To all my family, the symbol of love and giving.

Pamela Obiomon

To great parents, who never stop giving of themselves in countless ways. To my beloved brothers and sisters.

To my dearest wife, who leads me through the valley of darkness with the light of hope and support.

To beloved kids are Daniah and Mustafa, whom I can't force myself to stop loving.

Samir I. Abood

Contents

Preface

Power system operation, utilization, and control present a detailed analysis of ways to control electrical power systems and clarify severe outages due to the sustained growth of signal oscillations in modern interconnected power systems. The identification branch for controlling the electrical power system is called power system control and operation. This book aims to better understand the control of power systems, their applications, and optimization.

The book begins with the study of the concept of basic power generation. It then presents these applications in different configurations with detailed diagrams. It optimizes the stability scheme's location in power and the application of power electronic devices and control on the power system. This book is intended for college students, community colleges, and universities. The book is also intended for researchers, technicians, technologists, and skilled specialists in power generation and control of power systems. It presents the relationship between the power system's quantities of control and management. The book's major goal is to give a concise introduction to controlling power systems covered in two semesters. The book is appropriate for Juniors, Senior Undergraduate Students, Graduate Students, Industry Professionals, Researchers, and Academics.

This book focuses on modeling power system components with attention to control equipment. The one possible exception reflects the concern of the time the book came into being, namely analysis of the linear system model for detection and mitigation of possible poorly damped operating conditions.

Typical topics in the electric power field and allied areas such as hydro and thermal power plants are treated in this book using a system approach to apply the most modern theories and methods.

The ultimate product of the steam plant is electricity that is exported to the grid or steam or hot water sent to a nearby process plant, industry, or housing complex, but the control systems will be relatively similar in both cases. Nonetheless, designing these systems is a specialized effort, as much an art as a science, and the goal in this introduction is to highlight the breadth and depth of knowledge that it necessitates.

Power system operators are facing greater challenges than in the past. Such challenges as scheduling and handling generation resources due to the electricity market operation of transmission networks being close to their technical limits due to difficulties in building new transmission facilities. Also there are generation uncertainties due to renewable energy sources' intermittency and less accurate forecasts, or even natural forces such as earthquakes.

This book is organized into nine chapters, with a short review of the basic electric power systems concept in Chapter 1, it discusses synchronous machines and simplified cylindrical rotor (non-salient) synchronous machines for the steady-state condition. A power angle characteristic for salient-pole synchronous machines condition is explained in this chapter. Modeling of synchronous generators is included in Chapter 2. The chapter also introduces system modeling types, introduces the basic operating principles of the generator, turbine, system identification, and describes the mathematical model of these devices. Chapter 3 deals with power and frequency control and system behavior in a single area and multi areas. It also presents the power-frequency characteristic of an interconnected system and the effect of governor characteristics.

Chapter 4 presents the concept of voltage and reactive power control. The voltage control methods and the relation between voltage, power, and reactive power at a node are presented in this chapter. Also, the reactive power compensation and harmonic effect on load balancing are presented in this chapter.

Chapter 5 presents some aspects of power system optimization and familiarization with principles of power system utilization, especially from the commercial viewpoint. This chapter introduces optimization fundamentals, linear programming, and optimization problem formulation.

The description of economic dispatch in power systems, types, and specifications is discussed in Chapter 6 under the title of economic dispatch. Also, it includes mathematic formulation analysis procedures of electric power system optimization techniques. Besides, it identifies strategies that can be applied to evaluate power loss and power flow control.

Chapter 7 deals with unit commitment, concept, operation, and Problem Formulation. The description of the Dynamic Programming Method and Unit commitment time consideration is discussed in this chapter.

Chapter 8 presents the state estimation of the power system. The chapter includes the Weighted Least Square algorithms, error characteristics, Newton iteration, bad data detection, Supervisory Control, and Data Acquisition, and Phasor Measurement Units.

Finally, Chapter 9 focuses on load management and forecasting. This chapter introduces load forecasting solution techniques, load curves and factors, and forecasting modeling.

In recognition of requirements by the Accreditation Board for Engineering and Technology (ABET) on integrating computer tools, the use of MATLAB® is encouraged in a student-friendly manner. The reader does not need to have previous knowledge of MATLAB. The material of this text can be learned without MATLAB. However, the authors highly recommend that the reader studies this material in conjunction with the MATLAB Student Version. Appendix A of this text provides a practical introduction to MATLAB.

MATLAB® is a registered trademark of The Math Works, Inc. For product information, please contact:

The Math Works, Inc.
3 Apple Hill Drive
Natick, MA 01760-2098
Tel: 508-647-7000
Fax: 508-647-7001
E-mail: info@mathworks.com
Web: www.mathworks.com

Acknowledgments

Acknowledgment has to be given to Prairie View A&M University for providing a platform where Faculty, administrators, Ph.D. students, private, and governmental entities can contribute to a book that will be used in the education of future power engineers, and also for providing a resource to contribute to advancing the knowledge of power systems and continuing the technical foundation building for the production of future engineers.

It is our pleasure to acknowledge the outstanding help and support of the team at CRC Press in preparing this book, especially from Nora Konopka and Prachi Mishra,

The authors appreciate the suggestions and comments from several reviewers, including prof. Zainab Ibrahim/ University of Baghdad/Electrical Engineering Department, and Dr. Muna Fayyadh / American InterContinental University, their frank and positive criticisms improved this work considerably.

Finally, we express our profound gratitude to our families, without whose cooperation this project would have been difficult, if not impossible. We appreciate feedback from professors and other users of this book. We can be reached at jhfuller@pvamu.edu, phobiomon@pvamu.edu, and siabood@pvamu.edu.

Authors

John Fuller is a professor of Electrical and Computer Engineering at Prairie View A&M University in Prairie View, Texas. Dr. Fuller received a BSEE degree from Prairie View A&M University and a master's and Ph.D. degree from the University of Missouri, Columbia. He has researched several funded projects over a 44-year teaching career in Higher Education. His research efforts' major projects are as follows: Hybrid energy systems, Stepper Motor Control, design, and building of a solar-powered car, nuclear survivability, characterization on non-volatile memory devices, and nuclear detection/sensor evaluation, and some other electrical and computer-related projects. Dr. Fuller is presently the coordinator of Title III funding to the Department of Electrical and Computer Engineering in developing a solar-powered home. He is also associate director of the Center for Big Data Management located in the ECE Department. In addition to teaching and research duties with college-level students, he is also active in the PVAMU summer programs for middle and high school students. Dr. Fuller has also held administrative positions of department head of Electrical Engineering and interim dean of the College of Engineering at Prairie View A&M University. In 2018, he was recognized as the Texas A&M System Regents Professor.

Pamela Obiomon is the dean of the Roy G. Perry College of Engineering at Prairie View A&M University. She is the seventh dean of the college and the first female to serve in this role. Pamela served as the department head of the Department of Electrical and Computer Engineering at PVAMU. She earned a B.S. in Electrical Engineering from the University of Texas at Arlington, an M.S. in Engineering from Prairie View A&M University, and a Ph.D. in Electrical Engineering from Texas A&M University.

Samir I. Abood received his B.S. and M.S. from the University of Technology, Baghdad, Iraq, in 1996 and 2001, respectively. He got his Ph.D. in the Electrical and Computer Engineering Department at Prairie View A & M University. From 1997 to 2001, he worked as an engineer at the University of Technology. From 2001 to 2003, he was a professor at the University of Baghdad and Al-Nahrain University. From 2003 to 2016, Mr. Abood was a professor at Middle Technical University / Baghdad-Iraq. From 2018 to the present, he has worked at Prairie View A&M University/ Electrical and Computer Engineering Department. He is the author of 30 papers and ten books. His main research interests are sustainable power and energy systems, microgrids, power electronics and motor drives, digital PID Controllers, digital methods for electrical measurements, digital signal processing, and control systems.

1 Synchronous Machines

As we know that early man survived by building shelter for security from the weather and carnivorous animals. In building this shelter, he lifted a stone to become a part of a protective wall and realized that the larger stones were quite heavier than the smaller stones with no knowledge that gravity created the difference. As man advanced in technological knowledge, he discovered that the earth had a gravitational force that made the stones heavier and allowed him to walk upright. As man progressed, he discovered that the earth contained a mineral known as magnetite that could be processed to become a permanent magnet. A naturally occurring form of magnetic known as loadstone was constructed as a magnetite compass that used the earth's magnetic field.

Progression of technology led to the scientific discovery that a wire carrying a current would produce a magnetic field, and if this wire is wound on a spool, it will produce a concentrated magnetic field. The knowledge of producing and controlling magnetic fields leads to the creation of electrical machines.

The term Synchronous Machine is used to broadly define synchronous generators and synchronous motors. An AC system has some advantages over a DC system. Therefore, the AC system is exclusively used to generate, transmit, and distribute electric power. The machine which converts mechanical power into AC electrical power is called a Synchronous Generator or Alternator. However, the same machine can be operated as a motor and is known as a Synchronous Motor. This chapter will explain the concept of synchronous machines, characteristics of the power angle curve, and per unit quantity application in the power system. Synchronous machines rely on three-phase currents applied to a stator to produce three separate magnetic fields. The three magnetic fields combine to produce one main rotating field, and the machine operates as a motor. When an external source of rotational power is applied to turn a permanent or temporary magnet within three-phase coils, the machine operates as a generator. This chapter explains the concept of synchronous machines, characteristics of power angle curve, and per unit quantity applications in power systems.

1.1 SIMPLIFIED MODELS OF CYLINDRICAL ROTOR (NON-SALIENT) SYNCHRONOUS MACHINES FOR THE STEADY-STATE CONDITION

The flux linkage for an N-turn concentrated coil (aa') will be maximum ($N\varphi$) at $\omega t = 0$ and zero at $\omega t = \pi/2$. Assuming distributed winding, the flux linkage θ_a will vary as the cosine of the angle ωt as shown in Figure 1.1. The current flowing from a^+ to a^- produces a magnetic field θ. Also an applied magnetic field will produce a voltage e at the terminal. Thus, the flux linkage with the coil (a) is

$$\theta_a = N\varphi \cos \omega t \tag{1.1}$$

DOI: 10.1201/9781003293965-1

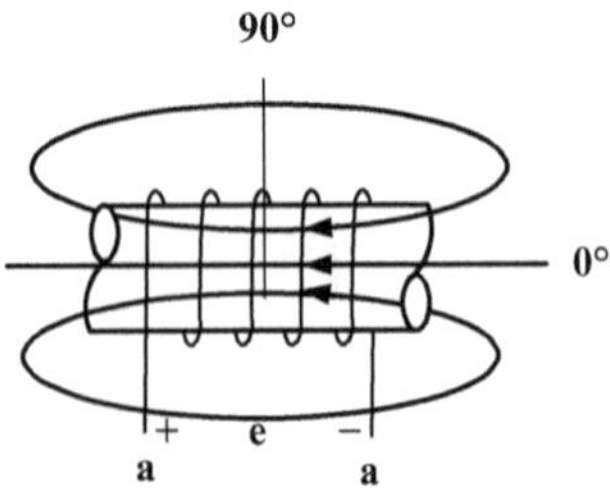

FIGURE 1.1 Flux linkage around a coil and induced voltage.

The voltage induced in coil aa' is obtained from Faraday's law as

$$e_a = \frac{d\theta}{dt} = \omega N\varphi \sin \omega t$$

$$= E_{\max} \sin \omega t \tag{1.2}$$

$$= E_{\max} \cos\left(\omega t - \frac{\pi}{3}\right) \tag{1.3}$$

where

$$E_{\max} = \omega N\varphi = 2\pi fN\varphi \tag{1.4}$$

The r.m.s. value of the generated voltage is:

$$E_{rms} = \frac{2\pi}{\sqrt{2}} fN\varphi \tag{1.5}$$

$$E_{rms} = 4.44 fN\varphi \tag{1.6}$$

K_w – winding factor

$$E_{rms} = 4.44 K_w fN\varphi \tag{1.7}$$

Figure 1.2 shows a two-pole three-phase synchronous generator, the frequency of the induced armature voltages depends on the rotor's speed and the number of poles. For which the machine is wound.

$$f = \frac{Pn}{120} \tag{1.8}$$

where n is the rotor speed in r.p.m., referred to as synchronous speed.

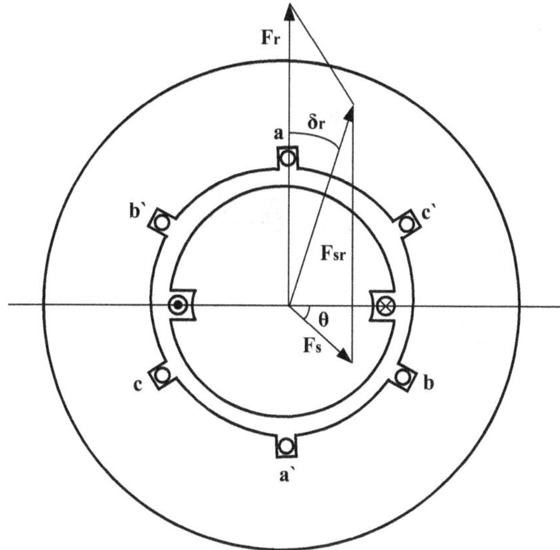

FIGURE 1.2 Two-poles three-phase synchronous generator.

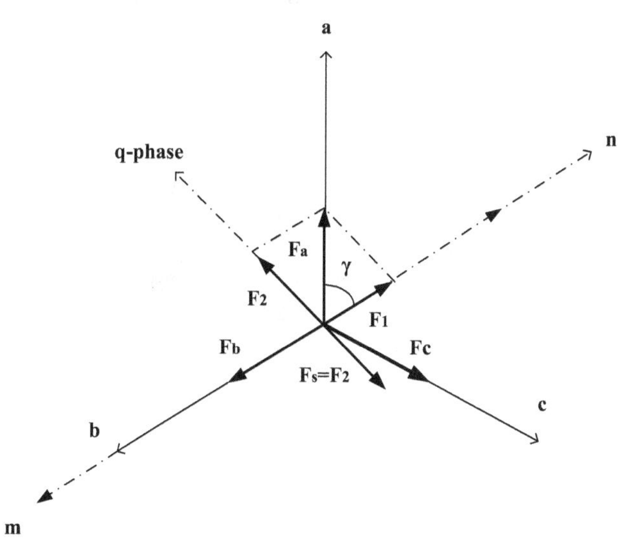

FIGURE 1.3 The mmf phasor diagram of the three-phase salient pole synchronous generator.

During normal conditions, the generator operates synchronously with the power grid. This results in three-phase balanced currents in the armature. Assuming the current in phase (a) is lagging the generated e.m.f. in that phase by an angle θ, which is indicated by line mn in Figure 1.3.

$$i_a = I_{\max} \sin\left(\omega t - \theta\right) \tag{1.9}$$

$$i_b = I_{\max} \sin\left(\omega t - \frac{2\pi}{3} - \theta\right) \tag{1.10}$$

$$i_c = I_{\max} \sin\left(\omega t - \frac{4\pi}{3} - \theta\right) \tag{1.11}$$

At any instant, each phase winding produces a sinusoidally distributed magnetic motive force (mmf) wave with its peak (maximum) along the phase winding axis.

These sinusoidally distributed fields can be represented by vectors referred to as space phasors.

The amplitude of the mmf $\{f_a(\theta)\}$ is represented by the vector Fa along the axis of phase (a). Similarly, the amplitude of the $f_b(\theta)$ and $f_c(\theta)$ are vectors F_b and F_c along their respective axis.

The mmf amplitudes are proportional to the instantaneous value of the phase current, i.e.:

$$F_a = Ki_a = K\,I_{\max} \sin\left(\omega t - \theta\right) = F_{\max} \sin\left(\omega t - \theta\right) \tag{1.12}$$

$$F_b = Ki_b = K\,I_{\max} \sin\left(\omega t - \frac{2\pi}{3} - \theta\right) = F_{\max} \sin\left(\omega t - \frac{2\pi}{3} - \theta\right) \tag{1.13}$$

$$F_c = Ki_c = K\,I_{\max} \sin\left(\omega t - \frac{4\pi}{3} - \theta\right) = F_{\max} \sin\left(\omega t - \frac{4\pi}{3} - \theta\right) \tag{1.14}$$

K is proportional to the number of armature turns per phase and a function of the winding type. The resultant armature mmf is the vector sum of the above mmfs. A suitable method for finding the resultant mmf is to project these mmfs on line *nm* and obtain the resultant in-phase and quadrature-phase components.

The resultant in-phase components are:

$$F_1 = F_m \sin\left(\omega t - \theta\right)\cos\left(\omega t - \theta\right) + F_m \sin\left(\omega t - \frac{2\pi}{3} - \theta\right) \times \cos\left(\omega t - \frac{2\pi}{3} - \theta\right) + F_m \times \sin\left(\omega t - \frac{4\pi}{3} - \theta\right)\cos\left(\omega t - \frac{4\pi}{3} - \theta\right) \tag{1.15}$$

We know, $\sin\alpha\cos\alpha = \frac{1}{2}\sin 2\alpha$

$$\therefore F_1 = \frac{F_m}{2}\left[\sin 2\left(\omega t - \theta\right) + \sin 2\left(\omega t - \frac{2\pi}{3} - \theta\right) + \sin 2\left(\omega t - \frac{4\pi}{3} - \theta\right)\right] \tag{1.16}$$

Equation 1.16 is the sum of three sinusoidal functions displaced by $\frac{2\pi}{3}$ radians (120°), which add up to zero, i.e.

$$F_1 = 0 \tag{1.17}$$

The resultant in quadrature components are:

$$F_2 = F_m \sin(\omega t - \theta)\sin(\omega t - \theta) + F_m \sin\left(\omega t - \frac{2\pi}{3} - \theta\right) \times \sin\left(\omega t - \frac{2\pi}{3} - \theta\right) + F_m \times \sin\left(\omega t - \frac{4\pi}{3} - \theta\right)\sin\left(\omega t - \frac{4\pi}{3} - \theta\right) \tag{1.18}$$

We know,

$$\sin^2 \alpha = \frac{1 - \cos 2\alpha}{2}$$

$$F_2 = \frac{F_m}{2}\left[3 - \cos 2(\omega t - \theta) + \cos 2\left(\omega t - \frac{2\pi}{3} - \theta\right) + \cos 2\left(\omega t - \frac{4\pi}{3} - \theta\right)\right] \times \cos 2(\omega t - \theta) + \cos 2 \times \left(\omega t - \frac{2\pi}{3} - \theta\right) + \cos 2\left(\omega t - \frac{4\pi}{3} - \theta\right) = 0 \tag{1.19}$$

$$\therefore F_2 = \frac{3F_m}{2} \tag{1.20}$$

Thus, the amplitude of the resultant armature mmf or (stator mmf) becomes:

$$F_S = \frac{3}{2} F_m \tag{1.21}$$

where

$$F_m = KI_{\max} \tag{1.22}$$

K – proportional to no. of turns and type of the winding.

Therefore, the resultant armature mmf (F_S) has a constant amplitude perpendicular to the line mn, and rotates at a constant speed an in synchronism with the field mmf (F_r).

From the phasor diagram shown in Figure 1.4

$$E = E_{sr} + jX_{ar}I_a \tag{1.23}$$

where:

E excitation voltage, when $I_a = 0$
E_{sr} e.m.f generated due to φ_{sr}
X_{ar} the reactance of the armature reaction

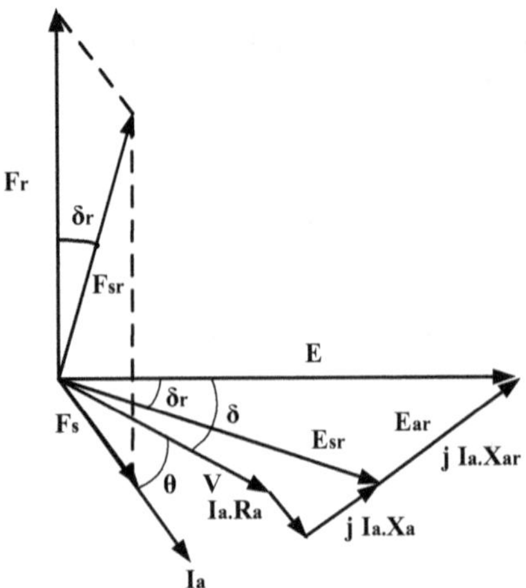

FIGURE 1.4 Cylindrical rotor generator phasor diagram.

The terminal voltage (V) is less than E_{sr} by the amount of resistive voltage drop $R_a I_a$ and leakage reactance voltage drop $X_l I_a$, thus:

$$E = V + \left[R_a + j\left(X_l + X_{ar}\right)\right] I_a \tag{1.24}$$

where X_l - Leakage reactance of armature or

$$E = V + \left[R_a + jX_s\right)\right] I_a \tag{1.25}$$

where $X_s = X_l + X_{ar}$ is known as *Synchronous Reactance.*

According to Equation 1.25, a simple per phase model for a cylindrical rotor generator for the steady-state condition is shown in Figure 1.5.

The armature resistance R_a generally is much smaller than the synchronous reactance X_s and is often neglected for evaluation purposes. The two buses circuit connected to the infinite bus becomes that shown in Figure 1.6.

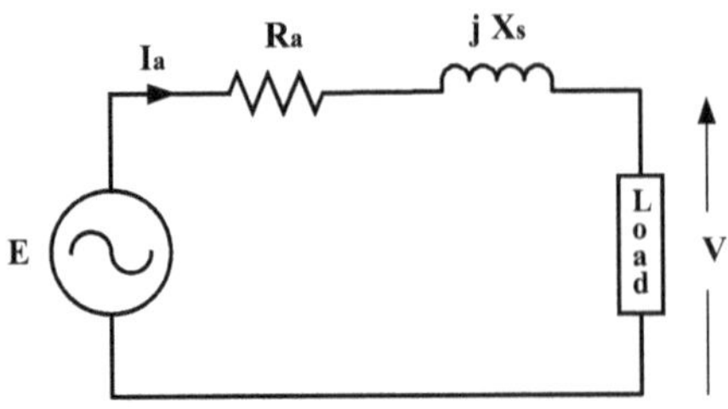

FIGURE 1.5 Equivalent circuit of one phase diagram of three-phase synchronous generator.

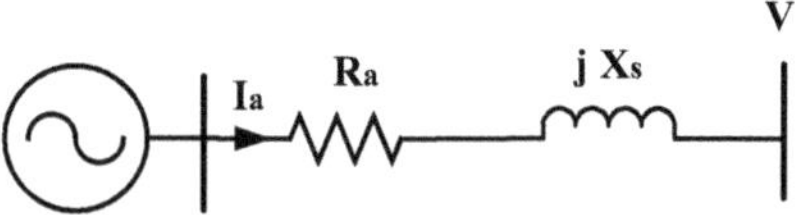

FIGURE 1.6 Synchronous generator connected to an infinite bus.

1.2 POWER ANGLE CHARACTERISTICS

The three-phase complex power at the generator terminal is

$$S_{3-\varphi} = 3VI_a^* \tag{1.26}$$

where V is the phase-to-neutral terminal voltage. If the phasor voltages are in polar form, the armature current for a single-phase is

$$I_a = \frac{|E|\angle\delta - |V|\angle 0}{|Z_S|\angle\gamma} \tag{1.27a}$$

And by the conjugate of the armature current

$$I_a^* = \frac{|E|\angle -\delta - |V|\angle 0}{|Z_S|\angle -\gamma} \tag{1.27b}$$

where the synchronous impedance is $|Z_S| = \sqrt{R_a^2 + X_s^2}$

$$S_{3-\varphi} = 3V.I_a^* \tag{1.28}$$

$$\therefore \quad S_{3-\varphi} = 3\frac{|E||V|}{|Z_S|}\angle(\gamma-\delta) - 3\frac{|V|^2}{|Z_S|}\angle\gamma \tag{1.29}$$

So the active power is given by

$$P_{3-\varphi} = 3\frac{|E||V|}{|Z_S|}\cos(\gamma-\delta) - 3\frac{|V|^2}{|Z_S|}\cos\gamma \tag{1.30a}$$

And the reactive power is given by

$$Q_{3-\varphi} = 3\frac{|E||V|}{|Z_S|}\sin(\gamma-\delta) - 3\frac{|V|^2}{|Z_S|}\sin\gamma \tag{1.30b}$$

Since X_s is much greater than R_a, then R_a is neglected, giving $Z_S = jX_S$ and $\gamma = 90°$

$$\therefore\ P_{3-\varphi} = 3\frac{|E||V|}{X_S} \tag{1.31}$$

$$Q_{3-\varphi} = 3\frac{|V|}{X_S}\left[|E|\cos(\delta) - |V|\right] \tag{1.32}$$

$$\therefore\ P_{\max(3-\varphi)} = 3\frac{|E||V|}{X_S} \tag{1.33}$$

$P_{\max(3-\varphi)}$ is called steady-state stability limit or static stability limit. Equation 1.32 shows that for small amount of δ, cosδ is nearly unity, then

$$Q_{3-\varphi} \approx 3\frac{|V|}{X_S}\left[|E| - |V|\right]. \tag{1.34}$$

1.3 POWER ANGLE CHARACTERISTICS FOR SALIENT POLE SYNCHRONOUS MACHINES FOR THE STEADY-STATE CONDITION

The model developed in the previous section is only valid for cylindrical rotor generators with uniform air gaps. The salient pole rotor results in non-uniformity of the magnetic reluctance of the air gap. The reluctance along the d-axis is less than that along the q-axis. Therefore, the reactance has a high value X_d along the d-axis and a low value X_q along q-axis.

The voltage drop in the armature due to these reactances can be taken into account by resolving the armature current I_a into two components I_d and I_q. The phasor diagram for the circuit of Figure 1.6 with R_a neglected is shown in Figure 1.7.

The excitation voltage magnitude is

$$|E| = |V|\cos\delta + X_d I_d \tag{1.35}$$

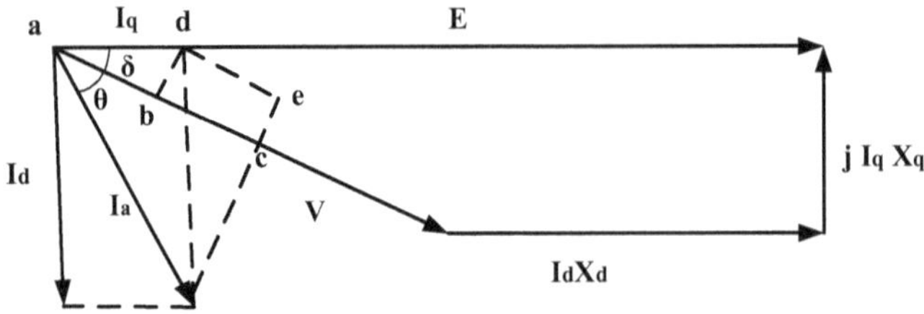

FIGURE 1.7 Phasor diagram of a salient pole generator.

$$P_{3-\varphi} = 3|V||I|\cos\theta \tag{1.36}$$

From the phasor diagram

$$|I_a|\cos\theta = ab + de(bc)$$

$$= I_q\cos\delta + I_d\sin\delta \tag{1.37}$$

$$\therefore \quad P_{3-\varphi} = 3|V|\left(I_q\cos\delta + I_d\sin\delta\right) \tag{1.38}$$

From phasor diagram:

$$|V|\sin\delta = X_q I_q$$

$$I_q = \frac{|V|\sin\delta}{X_q} \tag{1.39}$$

From Equation 1.35,

$$I_d = \frac{|E| - |V|\cos\delta}{X_d} \tag{1.40}$$

Put Equations 1.39 and 1.40 in Equation 1.38 and consider

$$\cos\delta\sin\delta = \frac{\sin 2\delta}{2}$$

$$\therefore \quad P_{3-\varphi} = 3\frac{|E||V|}{X_d}\sin\delta + 3|V|^2\frac{X_d - X_q}{2X_dX_q}\sin 2\delta \tag{1.41}$$

Similarly:

$$\therefore \quad Q_{3-\varphi} = 3\frac{|E||V|}{X_d}\cos\delta - 3|V|^2\frac{X_d + X_q}{2X_dX_q} + 3|V|^2\frac{X_d - X_q}{2X_dX_q}\cos 2\delta \tag{1.42}$$

For the non-salient (cylindrical rotor) machine,

$$X_d = X_q = X_S$$

Therefore, Equations 1.41 and 1.42 become:

$$\therefore\ P_{3-\varphi} = 3\frac{|E||V|}{X_d}\sin\delta \tag{1.43}$$

The same Equation 1.31 is applicable, if X_d is replaced by X_S

$$Q_{3-\varphi} = 3\frac{|V|}{X_d}\left[|E|\cos(\delta) - |V|\right] \tag{1.44}$$

The first term in Equation 1.41 is the same as the expression obtained for cylindrical rotor machine Equation 1.43. The second term introduces the effect of saliency. This term is the power corresponding to the reluctance torque and is independent of the field current because it does not contain the excitation voltage (E).

Power angle characteristics for salient and non-salient pole machines are shown in Figure 1.8; these characteristics are drawn using Equation 1.41, where the salient power is added to non-salient power to give the resultant power.

Figure 1.9 shows power angle characteristics at a constant torque angle, where the power for salient pole machine is greater than the power of cylindrical type at the same torque angle.

Figure 1.10 shows power angle characteristics at a constant value of power. The torque angle for the salient pole machine is less than the torque angle of the cylindrical type at the same power. Therefore, the salient pole rotor machine is more stable than the cylindrical rotor machine.

Figure 1.11 shows a cylindrical (non-salient) power-rotor angle generator curve. The maximum curve is represented as steady-state stability limit or P_{max} as described in Equation 1.33.

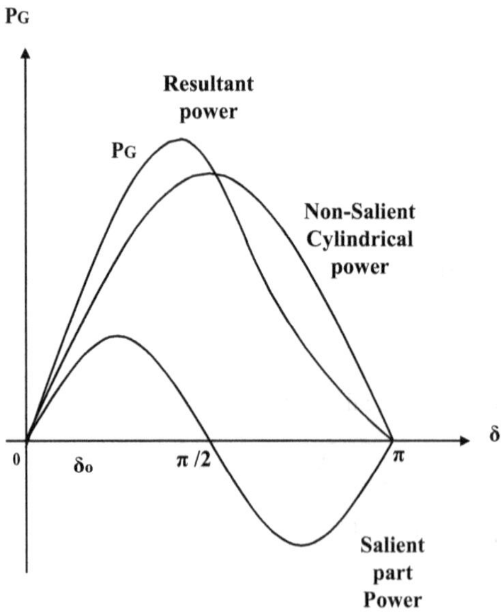

FIGURE 1.8 Power angle characteristics for salient and non-salient pole machines.

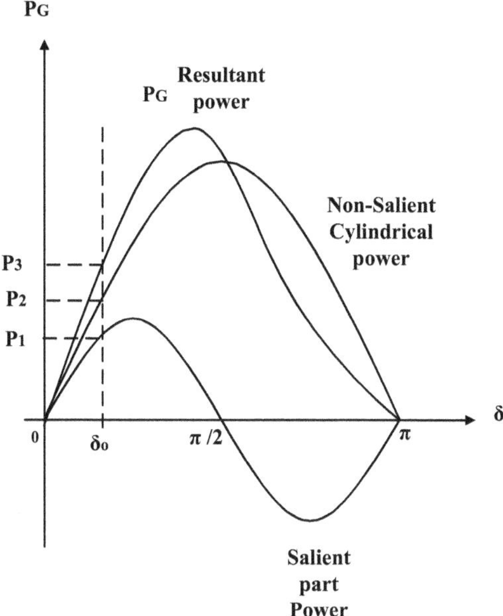

FIGURE 1.9 Power angle characteristics at a constant value of torque angle.

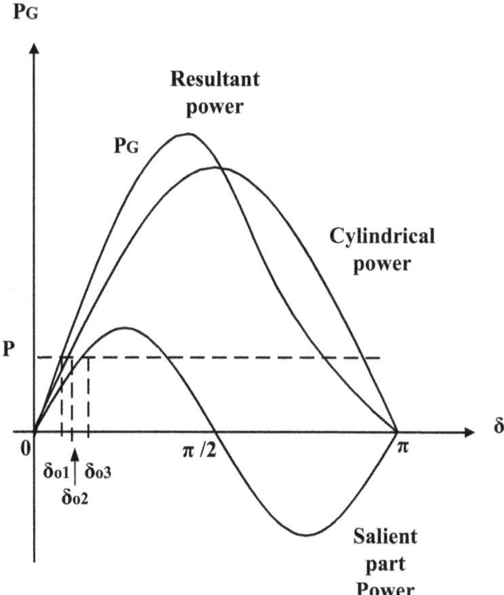

FIGURE 1.10 Power angle characteristics at a constant value of power.

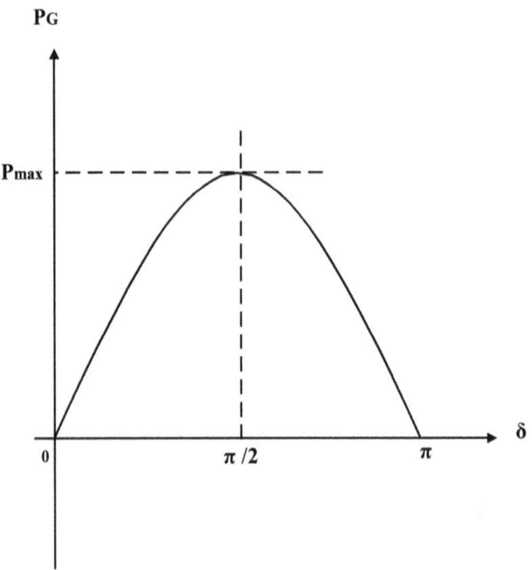

FIGURE 1.11 Cylindrical (non-salient) power-rotor angle generator curve.

1.4 PER UNIT QUANTITY

When calculating a power system network with two or more voltage levels, it is cumbersome to convert currents to different voltage levels at each point where they flow through a transformer. The change in the current is inversely proportional to the transformer turns ratio. To simplify these calculations, we can use a per-unit system.

In this system, base quantities are assumed for each voltage level, and the per-unit quantities are calculated as per the below-given equation:

$$\text{Per unit quantity} = \frac{Actual\ quantity}{Base\ quantity} \tag{1.45}$$

The four electrical quantities (voltages, current, power, and impedance) are so related that the selection of base values for any two of them determines the remaining two base values.

Usually, base apparent power in MVA and base voltage in kV are quantities selected to specify the base values.

For a single-phase system or three-phase on per phase basis, the following relationships hold:

$$Base\ current = \frac{Base\ VA}{Base\ votage} \tag{1.46}$$

$$Base\ impedance = \frac{Base\ voltage}{Base\ current} \tag{1.47}$$

$$Per\,unit\ voltage = \frac{Actual\ voltage}{Base\ voltage} \tag{1.48}$$

$$Per\,unit\ current = \frac{Actual\ current}{Base\ current} \tag{1.49}$$

$$Per\,unit\,impedance = \frac{Actual\,impedance}{Base\,impedance} \tag{1.50}$$

$$Base\ impedance = \frac{(Base\ voltage\,kV)^2}{Base\,kVA} \times 1000 \tag{1.51}$$

If we choose base kVA and base voltage in kV to mean kVA for the total of the three-phase base voltage for the line to line, we find

$$Base\ current = \frac{Base\,kVA\,3\ phase}{\sqrt{3} \times Base\,kV} \tag{1.52}$$

In a three-phase system, the per-unit three-phase kVA and voltage on the three-phase basis is equal to the per unit per phase kVA and voltage on the per-phase basis.

Sometimes it is necessary to convert per-unit quantities from one base to another. The conversion formula for the impedance can be written as:

$$z_{pu\,new} = z_{pu\,old}\left(\frac{v_{old}}{v_{new}}\right)^2 \frac{S_{new}}{S_{old}} \tag{1.53}$$

Example 1.1

A three-phase synchronous generator characterized by the following parameters X_d = 0.8 p.u, X_q = 0.3 p.u,and negligible armature resistance. The machine is connected directly to the infinite bus bar, as shown in Figure 1.12, with an infinite bus voltage of 1.0 per unit. The generator delivers a real power of 1.0 p.u at 0.707 lagging power factor.

Determine

i. The EMF voltage and torque angle.
ii. The power angle equation.

For the cases

a. Neglecting the saliency effect.
b. Including the effect of saliency.

Solution

a. Neglecting the saliency effect.
   i. The voltage behind transient reactance.

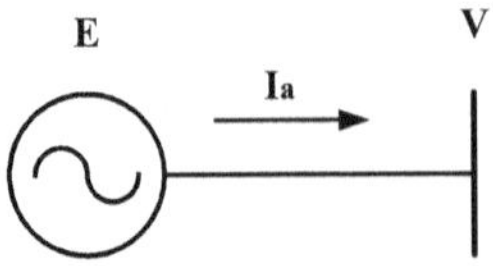

FIGURE 1.12 Power system configuration of Example 1.1.

$$I_a = \frac{P}{V.\cos\theta}\angle -\cos^{-1}(p.f) = \frac{1.0}{1.0\times 0.707}\angle -\cos^{-1}0.707$$

$$I_a = 1.414\angle -45° p.u$$

$$E = V + I_a.X_d$$

$$E = 1.0\angle 0 + 1.414\angle -45° \times j0.8$$

$$= 1.97\angle 23.96° p.u$$

ii. The transient power angle equation

$$P_e = \frac{|V||E|}{X_d}\sin\delta = \frac{1.0\times 1.97}{0.8}\sin\delta$$

$$= 2.46\sin\delta \; p.u$$

b. Including the effect of saliency.
   i. The voltage behind transient reactance

$$\delta = \tan^{-1}\left[\frac{X_q.|I_a|\cos\theta}{|V| + X_q.|I_a|\sin\theta}\right]$$

$$= \tan^{-1}\left[\frac{0.3\times 1.414\times 0.707}{1.0 + 0.3\times 1.414\times 0.707}\right]$$

$$= 13°$$

$$|E| = |V|\cos\delta + X_d|I_a|\sin(\delta + \theta)$$

$$= 1.0\times 0.707 + 0.8\times 1.414\times\sin(13° + 45°)$$

$$= 1.666\text{p.u}$$

$$|E_q| = |E| + \frac{(X_q - X_d).|V|\cos\theta}{X_d}$$

$$= 1.666 + \frac{(0.3 - 0.8) \times 1.0 \times 0.707}{0.8} = 1.224\text{p.u}$$

ii. The transient power angle equation

$$P_e = \frac{|V||E_q'|}{X_d}\sin\delta + \frac{(|V|)^2 (X_d - X_q)}{2X_d.X_q}\sin 2\delta$$

$$= \frac{1.0 \times 1.224}{0.8}\sin\delta + \frac{(1.0)^2 (0.8 - 0.3)}{2 \times 0.8 \times 0.3}\sin 2\delta$$

$$= 1.53\sin\delta + 1.666\sin 2\delta\,\text{p.u}$$

Example 1.2

A three-phase synchronous generator is characterized by the following parameters $X_d = 1.1\text{p.u}, X_q = 0.8\text{p.u}, X_d' = 0.35\text{p.u},$ and negligible armature resistance. The machine is connected directly to the infinite bus bar as shown in figure 1.13, with an infinite bus voltage of 1.0 per unit. The generator delivers a real power of 0.8 P.U at a 0.8 Lagging power factor.

Determine

i. The voltage behind transient reactance.
ii. The transient power angle equation.

For the cases

a. Neglecting the saliency effect.
b. Including the effect of saliency.

Solution

a. Neglecting the saliency effect.
   i. The voltage behind transient reactance.

$$I_a = \frac{P}{V.\cos\theta}\angle -\cos^{-1}(p.f) = \frac{0.8}{1.0 \times 0.8}\angle -\cos^{-1}0.8$$

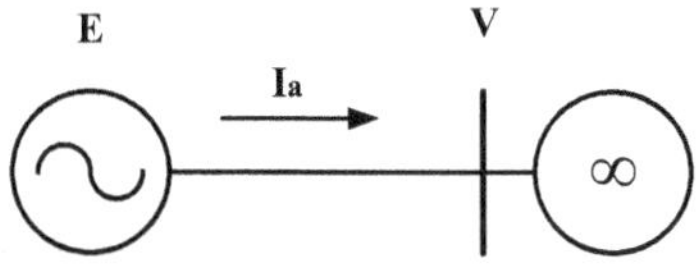

E V Ia ∞

FIGURE 1.13 Power system configuration of Example 1.2.

$$=1.0\angle-36.87°\text{p.u}$$

$$E' = V + I_a.X'_d$$

$$E' = 1.0\angle 0 + 1.0\angle -36.87° \times j0.35$$

$$=1.242\angle 13.03°\text{p.u}$$

ii. The transient power angle equation

$$P_e = \frac{|V||E|}{X'_d}\sin\delta = \frac{1.0\times1.242}{0.35}\sin\delta$$

$$=3.548\sin\delta\,\text{p.u}$$

b. Including the effect of saliency.
   i. The voltage behind transient reactance

$$\delta = \tan^{-1}\left[\frac{X_q.|I_a|\cos\theta}{|V|+X_q.|I_a|\sin\theta}\right]$$

$$=\tan^{-1}\left[\frac{0.8\times1.0\times0.8}{1.0+0.8\times1.0\times0.6}\right]$$

$$=23.38°$$

$$|E| = |V|\cos\delta + X_d|I_a|\sin(\delta+\theta)$$

$$=1.0\times\cos(23.38°)+1.1\times1.0\times\sin(23.38°+36.87°)$$

$$=1.786\,p.u$$

$$|E'_q| = \frac{X'_d|E|+(X_d-X'_d).|V|\cos\theta}{X_d}$$

$$|E'_q| = \frac{0.35\times1.786+(1.1-0.35)\times1.0\times0.8}{1.1} = 1.113\text{p.u}$$

ii. The transient power angle equation

$$P_e = \frac{|V||E'_q|}{X'_d}\sin\delta + \frac{(|V|)^2(X'_d-X_q)}{2X'_d.X_q}\sin2\delta$$

$$=\frac{1.0\times1.113}{0.35}\sin\delta+\frac{(1.0)^2(0.35-0.8)}{2\times0.35\times0.8}\sin2\delta$$

$$=3.182\sin\delta-0.8\sin2\delta\,p.u$$

Example 1.3

A three-phase synchronous generator is characterized by the following parameters $X_g = 0.8$ p.u,and negligible armature resistance. The machine is connected directly to the infinite bus bar as shown in figure 1.14, With an infinite bus voltage of 1.0 Per unit. The generator delivers a real power of 0.9 P.U at a 0.6 Lagging power factor and neglects the saliency effect.

Determine

i. The EMF voltage and torque angle.
ii. The power angle equation.
iii. If the EMF voltage calculated in (i) above is increased by 10% at the same active power, determine the new value of Q, δ, and power factor.

Solution

i. The voltage behind transient reactance.

$$I_a = \frac{P}{V.\cos\theta}\angle -\cos^{-1}(p.f) = \frac{0.6}{1.0\times 0.9}\angle -\cos^{-1}0.9$$

$$= 0.667\angle -25.84°\text{p.u}$$

$$E = V + I_a.X_d$$

$$E = 1.0\angle 0 + 0.667\angle -25.84° \times j0.8$$

$$= 1.323\angle 21.28°\text{p.u}$$

ii. The transient power angle equation

$$P_e = \frac{|V||E|}{X_d}\sin\delta = \frac{1.0\times 1.323}{0.8}\sin\delta$$

$$= 1.653\sin\delta\ \text{p.u}$$

iii. If the EMF voltage calculated in (i) above is increased by 10% at the same active power, determine the δ, Q, and power factor.

$$|E| = 1.323 + 10\%\times 1.323 = 1.4553\text{p.u}$$

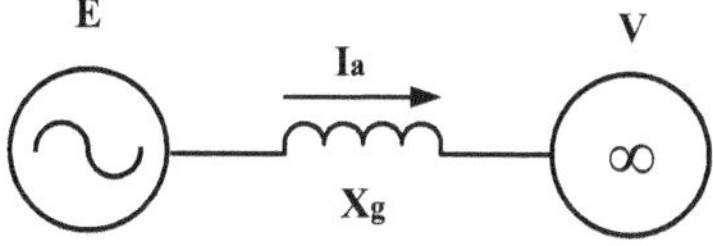

FIGURE 1.14 Power system configuration of Example 1.3.

$$0.6 = \frac{1.0 \times 1.4553}{0.8} \sin\delta$$

$$\delta = 13.83°$$

$$Q = \frac{|V|}{X_g}\left[|E|\cos\delta - |V|\right]$$

$$= \frac{1.0}{0.8}\left[1.4553\cos 13.83° - 1.0\right] = 0.516\text{p.u}$$

$$I_a = \frac{|E|\angle\delta - |V|\angle 0°}{jX_g} = \frac{1.4553\angle 13.83° - 1.0\angle 0°}{j0.8} = 0.675\angle -49.9°\text{p.u}$$

$$p.f = \cos 49.9° = 0.644\,lagging.$$

1.5 PARALLEL OPERATION OF SYNCHRONOUS GENERATORS

Most synchronous generators operate in parallel with other synchronous generators to supply power to the same power network. Clear advantages of this arrangement are as follows:

1. Some generators can supply a bigger load.
2. A failure of a single generator does not result in a total power loss to the load, increasing the reliability of the power system.
3. Individual generators may be removed from the power system for maintenance without shutting down the load.
4. A single generator not operating at near full load might be quite inefficient. While having several generators in parallel, turning off some of them when operating the rest at the near full-load condition is possible.

1.5.1 Conditions Required for Paralleling

A diagram in Figure 1.15 shows that Generator two (oncoming generator) will be connected in parallel when the switch *S* is closed. However, closing the switch at an arbitrary moment can severely damage both generators.

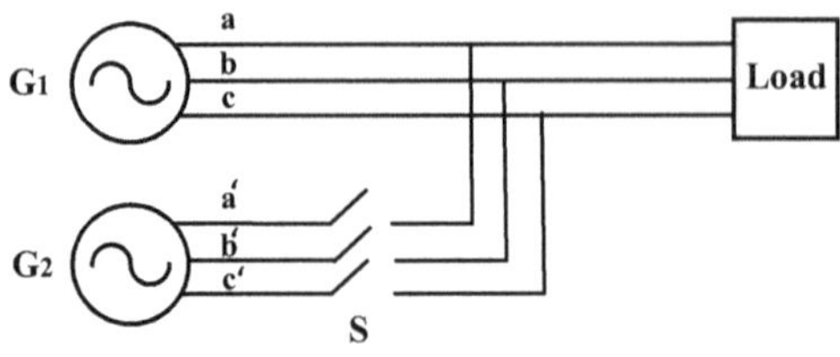

FIGURE 1.15 Two synchronous generators are connected in parallel.

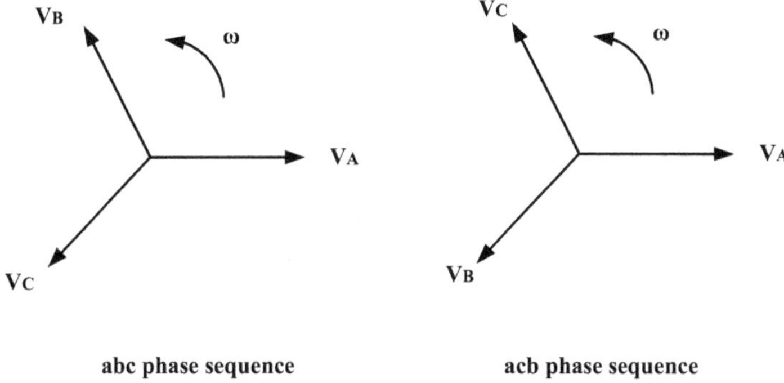

FIGURE 1.16 The phase voltages at different sequences.

If voltages are not the same in both lines (i.e., in a and a', b and b', etc.), a very large current will flow when the switch is closed. Therefore, to avoid this, voltages coming from both generators must be the same. Thus, the following conditions must be met:

1. The RMS line voltages of the two generators must be equal.
2. The two generators must have the same phase sequence.
3. The phase angles of the two phases must be identical.
4. The frequency of the oncoming generator must be slightly higher than the frequency of the running system.

If the phase sequences are different, as shown in Figure 1.16, then if one pair of voltages (phases a) are in phase, the other two pairs will be 120° out of phase creating huge currents in these phases.

If the generators' frequencies are different, a large power transient may occur until the generators stabilize at a common frequency. The frequencies of the two machines must be very close to each other but not exactly equal. If frequencies differ by a small amount, the oncoming generator's phase angles will change slowly concerning the running system's phase angles. If the angles between the voltages can be observed, it is possible to close the switch S when the machines are in phase.

PROBLEMS

1.1 A Three-phase synchronous generator characterized by the following parameters $X_d = 0.6$ p.u, $X_q = 0.35$ p.u, and negligible armature resistance. The machine is connected directly to an infinite bus bar. With the infinite bus voltage of 1.05 per unit, the generator delivers a real power of 1.2 p.u at a 0.67 lagging power factor.

Determine

i. The EMF voltage and torque angle.
ii. The power angle equation.

For the cases

a. Neglecting the saliency effect.
b. Including the effect of saliency.

1.2 A three-phase synchronous generator characterized by the following parameters $X_d = 0.96\ p.u, X_q = 0.65\ p.u, X_d^{'} = 0.45\ p.u,$ and negligible armature resistance. The machine is connected directly to an infinite bus bar with a bus voltage of 1.0 per unit. The generator delivers a real power of 0.95 p.u at a 0.85 lagging power factor.

Determine

i. The voltage behind transient reactance.
ii. The transient power angle equation.

For the cases

a. Neglecting the saliency effect.
b. Including the effect of saliency.

1.3 A Three-phase synchronous generator characterized by the following parameters X_g = 1.1 p.u, and negligible armature resistance. The machine is connected directly to an infinite bus bar, with an infinite bus voltage of 1.0 per unit. The generator delivers a real power of 0.98 p.u at a 0.8 lagging power factor and neglects the saliency effect.

Determine

i. The EMF voltage and torque angle.
ii. The power angle equation.
iii. If the EMF voltage calculated in (i) above decreased by 10% at the same active power, determine the new value of Q, δ, and power factor.

2 Modeling of Synchronous Generator

Power systems are increasingly expected to transmit power for long distances due to economic considerations and hence the need for advanced system identification techniques for efficient multimachine power system control. Turbogenerators are at the heart of power generation work under a wide variety of conditions. They are extremely non-linear, time-varying, fast-acting, multivariable (MIMO) devices, normally connected to the power system through a transmission system. Their dynamic characteristics differ as conditions change, but outputs must be synchronized to meet power system operation requirements.

This chapter aims to find the dynamic response (system behavior modeling) of the turbogenerator stages in a power station and discuss load division between generators.

2.1 IMPORTANCE OF MODELING

Recent innovations in power systems, resulting from technical advancements, generation, load patterns improvements, and economic considerations, have revealed numerous problems and requirements. Large power stations were installed in locations away from load centers, with the consequent danger of dynamic instability under some circumstances. Modern generators have high per-unit reactance values and rely on increasingly complex control systems for efficient operation. Successful computer controllers for turbogenerators have become possible with advanced computer technology, mathematical tools, and modern control theory.

Due to many modeling applications in different fields, such as chemical processes, biomedical systems transport, electrical power systems, etc., system modeling and recognition importance has attracted considerable attention. A mathematical model can explain the system's actions and predict and control the present and future situation. Modeling plays a crucial role in researching electrical power systems output and designing new turbogenerator controllers. Numerous mathematical models were used to explain this large and complex structure. Representations have become increasingly precise, keeping pace with modern technology and new problems in the operation and control of power systems.

Most of the previous modeling literature was designed for linearized station analytical models. A turbogenerator can be represented by a non-linear set of equations in which coefficients depend on calculated parameter values. Designing a controller is generally appropriate for reducing the non-linear model's order and linearizing the equations by considering minor deviations around the selected operating state.

System identification means constructing a model from experimental input–output data, i.e., identification is used to evaluate system characteristics to build a

DOI: 10.1201/9781003293965-2

mathematical model from the input–output sets. An alternative approach is to apply device identification methods to obtain models directly from the station, either in the transfer function, state-variable, or input–output format.

An essential issue in each model structure is whether the framework being examined can be adequately described within that structure.

2.2 TURBOGENERATOR IDENTIFICATION

A model must accurately describe the system's behavior in designing a predictive control for a dynamic system. Due to complex phenomena, physics laws are often not adequate to satisfactorily explain system dynamics in thermal station application.

In these cases, the designer turns to field data to test methodology. Experiments can be performed to excite the station and directly evaluate its response. This modeling and estimating unknown data parameters of the station are generally called device identification. The identification may also be made "online," thus depending on exciting station activity.

Device recognition is a core component of the adaptive scheme. Over a decade, adaptive control techniques were extensively studied, and numerous successful implementations were published.

A truth well appreciated by those involved in applications is that any practical design problem is typically complex, involving several factors such as technical constraints, economic trade-offs, human considerations, and model uncertainty; although well understood in principle, the application of adaptive control to stations and processes does not guarantee the good results expected by analysis and unmodeled dynamics, colored noise, and inadequate signal excitation degrade device detection and thus control in operation. Adaptive control applications applied to turbogenerator station models have produced favorable results.

2.3 THERMAL STATION

Nearly all major power systems operate through a series of processes based on the theoretical thermodynamic cycle called the Rankine cycle since the Rankine cycle is the basis for steam power cycles. A steam turbine consists of four stages (high pressure, intermediate pressure, reheater, and low pressure). Driven generators (turbogenerators) work at high speed. The rotor carries an exciting field. A schematic diagram of a thermal power station is shown in Figure 2.1.

Two elements critical for generator and motor operation are: the mechanical and the electrical torque. The mechanical torque T_m and the electrical torque T_e are considered positive for the synchronous generator. This means that T_m is the resultant shaft torque, which tends to accelerate the rotor in the positive θ_m the direction of rotation, as shown in Figure 2.2(a). Under the steady-state operation of the generator T_m and T_e are equal, and the accelerating torque T_a is zero. There is no acceleration or deceleration of the rotor masses, and the resultant constant speed is the synchronous speed. The rotating masses, which include the rotor of the generator and the prime mover, are said to be in synchronism with the other machines operating at synchronous speed in the power system.

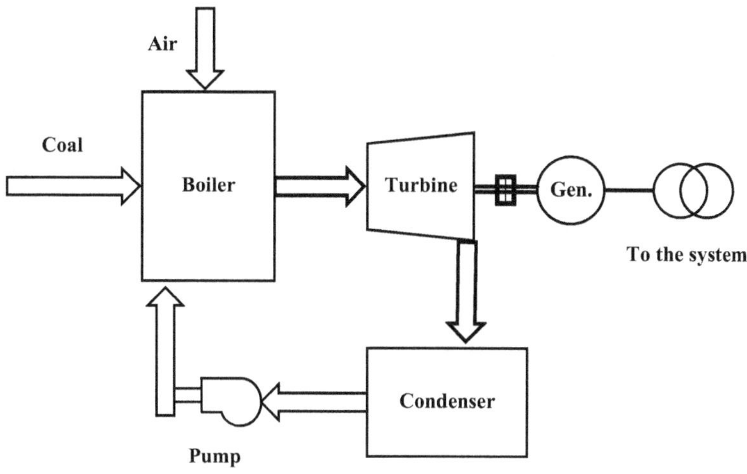

FIGURE 2.1 The thermal power station.

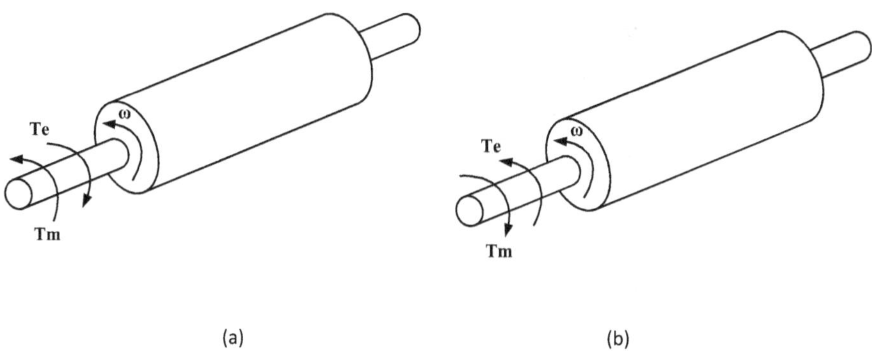

FIGURE 2.2 Machine rotor mechanical and electrical torque and speed direction for (a) a generator and (b) a motor.

The synchronous motor's direction of power flow is opposite to that in the generator, as shown in Figure 2.2(b).

Equation 2.1 governing a synchronous machine's rotor's motion is based on the elementary principle of dynamics. It states that accelerating torque is the product of the moment of inertia of the rotor times its angular acceleration. Applying a swing equation to a synchronous machine to small perturbation, we have

$$\frac{2H}{\omega_S}\frac{d^2\Delta\delta}{dt^2} = \Delta P_m - \Delta P_e \tag{2.1}$$

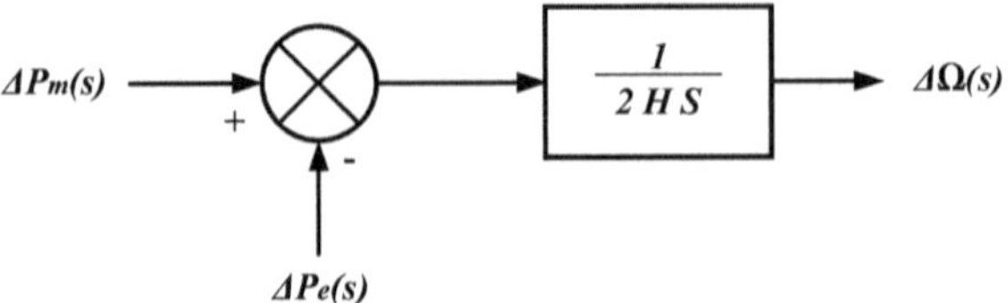

FIGURE 2.3 Generator block diagram.

or in terms of small deviation in speed:

$$\frac{d\Delta\frac{\omega}{\omega_S}}{dt} = \frac{1}{2H}\left(\Delta P_m - \Delta P_e\right) \tag{2.2}$$

With speed expressed in per unit, without explicit per unit notation, we have

$$\frac{d\Delta\omega}{dt} = \frac{1}{2H}\left(\Delta P_m - \Delta P_e\right) \tag{2.3}$$

Taking the Laplace transform of Equation (2.3), we obtain

$$\Delta\Omega(s) = \frac{1}{2HS}\left[\Delta P_m(s) - \Delta P_e(s)\right] \tag{2.4}$$

where:

$\Delta\Omega(s)$: The deviation in speed in Laplace form
H: inertia constant (M joul/MVA)
δ: torque angle (degree)
ω: rotor speed (radian/sec)
Δ: deviation from the steady-state
P_m: mechanical power is applied to the rotor (per unit)
P_e: electrical output power from the generator (per unit)
ω_S: synchronous speed in electrical units.

A block diagram that summarizes Equation 2.4 is given in Figure 2.3.

2.4 TURBINE MODEL

The mechanical power source, the prime mover (e.g., steam turbine), operates from coal or fossil fuel burning. The turbine is a complex device with many non-linearities and was initially proposed to introduce a relatively simple model containing only the critical features. The vapor pressure at the turbine inlet valves is assumed to be constant, and the vapor mass flow is proportional to the valve position. Each stage is

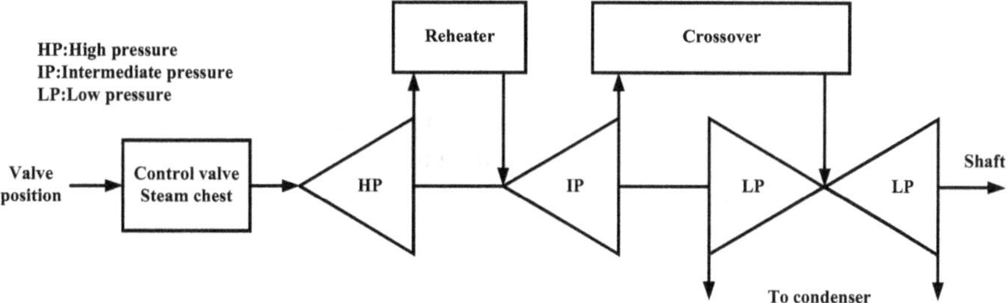

FIGURE 2.4 General form of the turbine with reheater.

defined by a first-order transfer function, weighing and summing outputs according to their relative contributions to the total shaft torque. Figure 2.4 shows a general turbine with a reheater.

The prime mover is a three-stage turbine with a reheater, and a fourth-order differential equation is used to describe the inertia of steam mass flow in the individual stages and the time delay of the reheater. The corresponding equations are

$$\frac{dY_H}{dt} = \frac{G_M P_0 - Y_H}{\tau_H} \tag{2.5}$$

$$\frac{dY_R}{dt} = \frac{Y_H - Y_R}{\tau_R} \tag{2.6}$$

$$\frac{dY_I}{dt} = \frac{G_I Y_R - Y_I}{\tau_I} \tag{2.7}$$

$$\frac{dY_L}{dt} = \frac{Y_I - Y_L}{\tau_L} \tag{2.8}$$

A transfer function is assumed to represent the time delay of the turbine valves, with appropriate limits on valve positions and opening and closing rates:

$$\frac{dG_M}{dt} = \frac{U_M - G_M}{\tau_M} \tag{2.9}$$

$$\frac{dG_I}{dt} = \frac{U_I - G_I}{\tau_{IV}} \tag{2.10}$$

$$T_M = F_H Y_H + F_I Y_I + F_L Y_L \tag{2.11}$$

where:

Y_H: the output of the high-pressure stage per unit
Y_I: the output of the intermediate-pressure stage per unit
Y_L: the output of the low-pressure stage per unit
G_M: the position of intercept valves
G_I: the position of inlet valves
Po: internal boiler steam pressure
τ_H: time constant associated with the high-pressure stage of the turbine
τ_I: time constant associated with the intermediate-pressure stage of the turbine
τ_L: time constant associated with the low-pressure stage of the turbine
τ_M: time constant of intercept valves
τ_{IV}: time constant of inlet valves
T_M: $F_H Y_H$ is the contribution of the high-pressure stage to the total shaft torque.

It can be concluded that there is a merger between the turbine and generator from prior knowledge of the station; turbogenerator dynamics for off-line is known. This knowledge indicates a fifth-order transfer function model for the TG to be appropriate. The following structure, therefore, models the turbogenerator:

$$G(z) = \frac{K}{(z+P_1)(z+P_2)(z+P_3)(z+P_4)(z+P_5)} \tag{2.12}$$

where:

K: Transfer function gain;
P_1, P_2, P_3, P_4, and P_5 are the transfer function poles.

2.5 SYSTEM IDENTIFICATION

System identification is the technique of dynamic system models from experimental input and output data. The system recognition technique is commonly referred to as evaluating a system or process by mathematical analysis, its input–output relationship.

The need to design better control systems inspired system recognition. In most realistic systems, such as industrial processes, knowledge of a system and its environment is rarely sufficient to create an effective control strategy. We also have to experimentally evaluate significant physical parameters such as heat transfer coefficient, chemical reaction rate, damping factor, etc.

Developing optimal and adaptive control theories increased the need for highly accurate system models. The machine model we're looking for is the mathematical Equation that always connects the input to output. We can test the system with various inputs and observe its responses to obtain such a model. Input–output data are then processed to yield the model based on the degree of a priori system knowledge; we can then classify system recognition problems into two categories:

1. Total identification problem: This means we don't know anything about the system's basic properties, such as linear or non-linear, memoryless or memory-free, etc. It's incredibly hard to solve. Typically, certain assumptions must be made before any meaningful solution can be tried. This form of problem-solving is called the black-box approach.
2. Partial identification problem: certain basic device characteristics, such as linearity, bandwidths, etc., are supposed to be identified in this group. However, we do not know the precise order of the dynamic Equation or the related coefficients. This is often called a gray box problem and is simpler to manage than the black-box approach.

Fortunately, most of the engineering systems and manufacturing processes we experience in practice are of the above sort. In many cases, we know a lot about the system structure such that a precise mathematical model of system dynamics can be guided. Consequently, only several parameters in the Equation model must be calculated.

An essential issue in each model structure is whether the examined system can be adequately described within that structure.

The objective is to create an effective identification model as shown in Figure 2.5, which, when subjected to the same input $u(k)$ as the station, produces an output $y_m(k)$ that approximates the station's actual output $y_p(k)$ and e(k) is the error between actual and estimated output.

The identification model problem consists of setting up a suitably parameterized identification model and adjusting the model's parameters to optimize a performance function based on the error between the station and the identification model outputs.

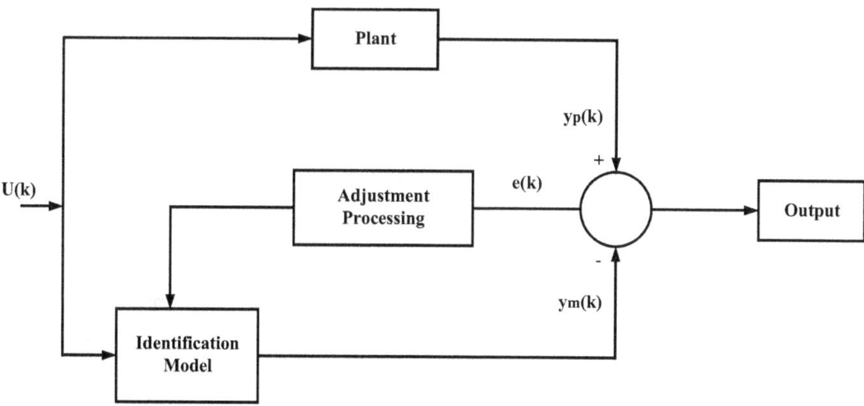

FIGURE 2.5 Identification layout.The objective is to create an effective identification model.

The representation of dynamic system models can be divided into two types:

1. Continuous-time
2. Discrete-time.

The essential difference between the two types is that the system's signals are continuous and discrete in the other. Since continuous systems can be presented in samples taken at appropriate sampling frequencies, continuous systems can be closely approximated by discrete models.

The great majority of system identification techniques are digitally oriented because of the employment of digital computers. Therefore, discrete system models are more convenient to deal with.

2.6 STATION DESCRIPTION

The unit considered in this section has a two-pole turbogenerator driven at (3000 r.p.m., 300 MVA, 20 KV, 50 HZ, and excitation voltage of 625 volts). A conventional fuel-fired boiler produces the steam at pressure of 180 bar, and both main and interceptor valves control steam flow. Figure 2.6 shows a possible physical and computational logic layout for the power station's turbogenerator (TG) identification scheme.

Linear transfer functions indicate station dynamics affecting the sampled data. The turbogenerator is modeled by separating the controller or governor dynamics, A(s), the gates or servo dynamics, B(s), and the turbine and generator, G(s).

The samplers, St, are connected to the input and output of G(s), sampling the gate position, y(t), and the electrical frequency, x(t), to obtain the field data, y(z) and x(z),

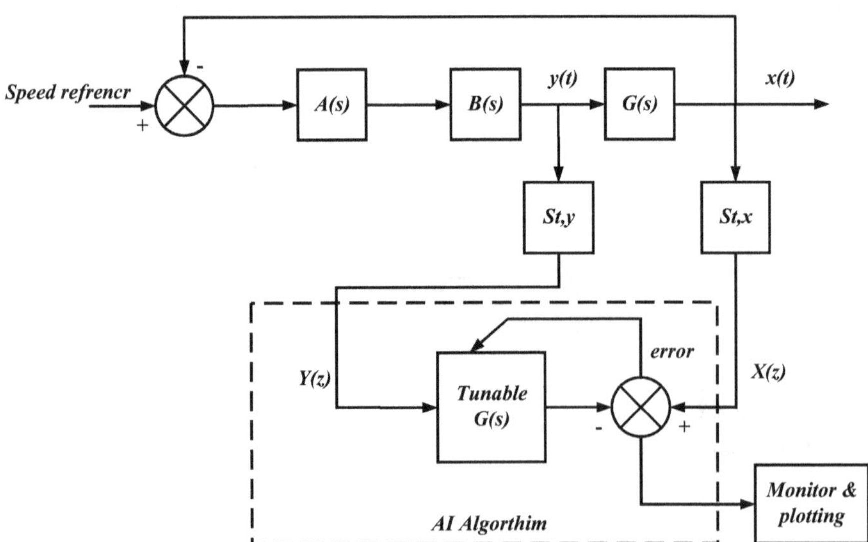

FIGURE 2.6 Turbogenerator Block Diagram with AI technique.

as shown in Figure 2.6. The sample rate was taken to be (100 samples/sec). The Genetic Algorithm (GA) uses the sampled field data for input to develop a discrete-time transfer function, G(z), of the same dynamics as G(s). This is done by inputting the gate position field data, y(z), into the G(z) of the GA, calculating the G(z) output $\bar{X}(z)$, and comparing that output to the frequency field data, x(z). This process is repeated for all G(z) in the GA's population at every generation.

From prior knowledge of the station, turbogenerator dynamics for off-line are known. This knowledge indicates a fifth-order transfer function model for the TG to be appropriate. The following structure, therefore, modeled the turbogenerator:

$$G(z)=\frac{K}{(z+P_1)(z+P_2)(z+P_3)(z+P_4)(z+P_5)} \tag{2.13}$$

A fifth-order turbogenerator transfer function using the power station can be designed as the transfer function G (z) in the following structure:

$$\mathrm{G}(\mathrm{z})=\frac{0.4726}{(\mathrm{Z}-0.1055)(\mathrm{Z}+0.1426)(\mathrm{Z}+0.332)(\mathrm{Z}-0.375)(\mathrm{Z}-0.4785)}$$

The MATLAB script code program following is to draw the system response

```
%G(z) = 0.4726/((Z-0.1055)(Z+0.1426)(Z+0.332)(Z-0.375)(Z-0.4785))

    H = zpk([ ],[0.1055, -0.1426, -0.332, 0.375, 0.4785], 0.4726, 0.1)

pzmap(H)
grid
```

The results:

$$H=\frac{0.4726}{(z-0.1055)(z+0.1426)(z+0.332)(z-0.375)(z-0.4785)}$$

Sample time: 0.1 seconds

Discrete-time zero/pole/gain model.

The results of the program when using the real input of the station as an input to G(z) to produce estimated output $\bar{X}(z)$ as shown in Figure 2.7.

The results of the program when using the real input (the gate position) of the station as an input to G (z) to produce estimated output $\bar{X}(z)$ as shown in Figure 2.8.

For the transfer function G (z) in the following structure:

$$\mathrm{G}(\mathrm{z})=\frac{0.78516}{(\mathrm{Z}-0.082)(\mathrm{Z}-0.1504)(\mathrm{Z}-0.1816)(\mathrm{Z}+0.1036)(\mathrm{Z}+0.04102)}$$

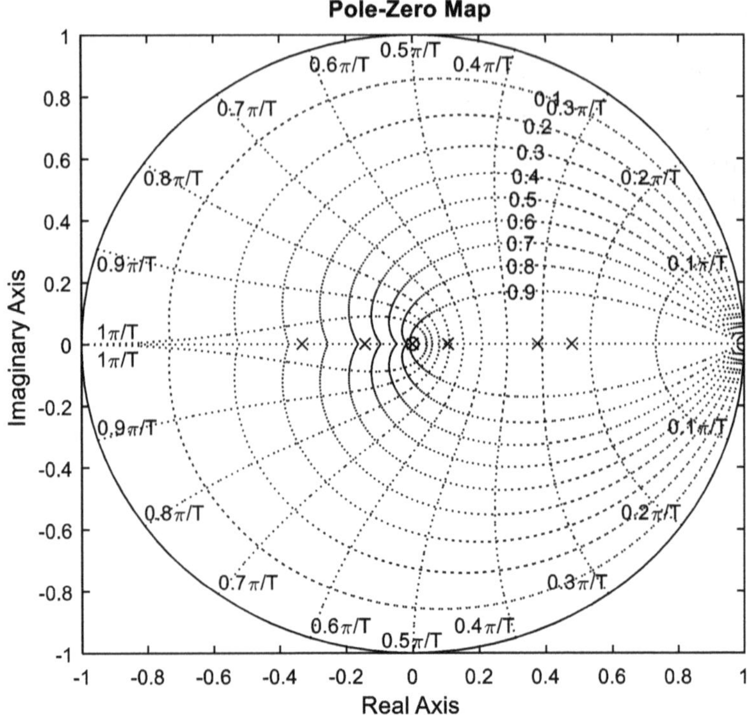

FIGURE 2.7 Pole-zero map sampling rate (30 samplers/sec).

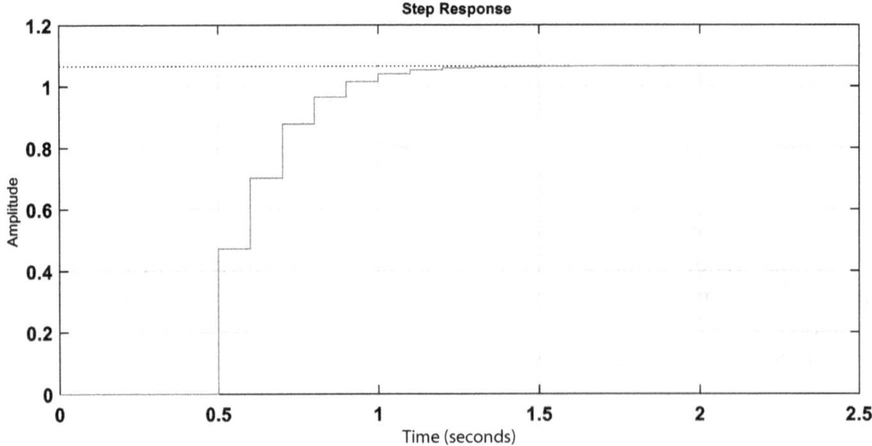

FIGURE 2.8 Result of GA at the sampling rate (30 samplers/sec).

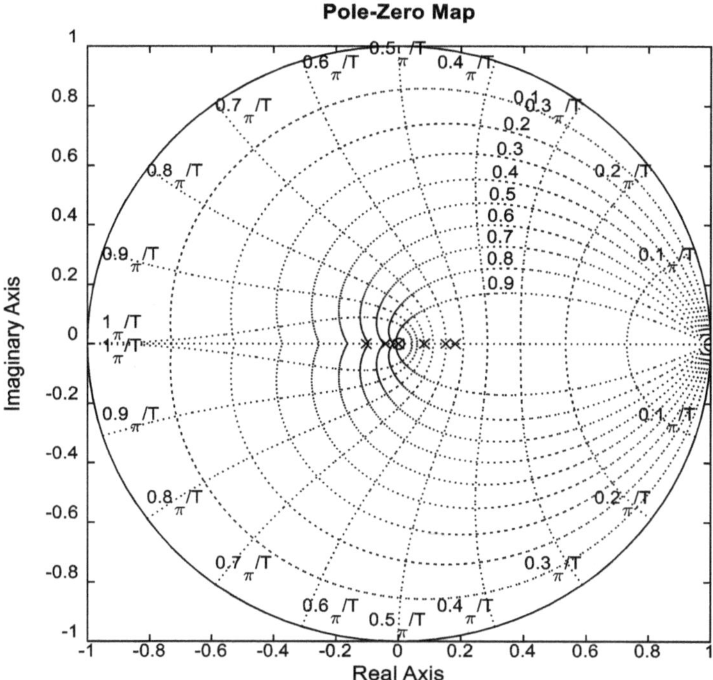

FIGURE 2.9 Pole-zero map sampling rate (50 samplers/sec).

and also the results of the program when using real input (the gate position) of the station as an input to G (z) to produce estimated output $\bar{X}(z)$ as shown in the Figure 2.9:

$$\begin{aligned} &0.78516/((Z-0.082)\\ \%\%G(z) = &\times(Z-0.1504)(Z-.1816)\\ &\times(Z+0.1036)(Z+0.04102)) \end{aligned}$$

$$H = \text{zpk}([\ ],[0.082, 0.1504, 0.1816, -0.1036, -0.04102], 0.78516, 0.1)$$

pzmap(H)
grid
The results:

$$H = \frac{0.78516}{(z-0.082)(z-0.1504)(z-0.1816)(z+0.1036)(z+0.04102)}$$

Sample time: 0.1 seconds
Discrete-time zero/pole/gain model.

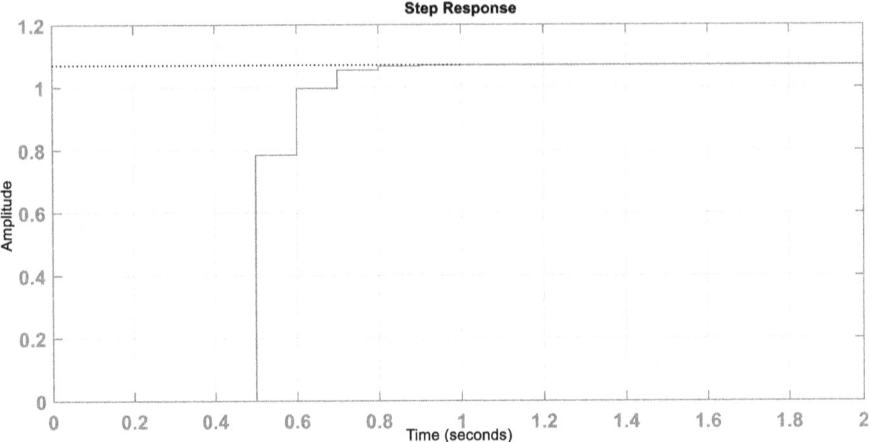

FIGURE 2.10 Result of GA at the sampling rate (50 samplers/sec).

The results of the program when using the real input of the station as an input to G (z) to produce estimated output $\bar{X}(z)$ as shown in Figure 2.9. Figure 2.10 shows a result of GA at the sampling rate (50 samplers/sec).

After the end of the program at several generations (15) and minimum error (3.6671) the transfer function G (z) can be obtained in the following structure:

$$G(z)=\frac{0.8633}{(Z+0.8828)(Z-0.2676)(Z-0.02617)(Z-0.2187)(Z+0.00976)}$$

and also the results of the program when applied the input active power (the gate position) of the station as an input to G (z) to produce estimated output $\bar{X}(z)$ as shown in Figure 2.11.

$$\%\%\%G(z)=\begin{array}{l}0.8633/((Z+0.8828)\\ \times(Z-0.2676)(Z-0.02617)\\ \times(Z-0.2187)(Z+0.00976))\end{array}$$

$$H=\text{zpk}([\ ],[-0.082, 0.2676, 0.02617, 0.2187, -0.00976], 0.8633, 0.1)$$

pzmap(H)
grid

The results:

$$H=\frac{0.8633}{(z+0.082)(z-0.2187)(z-0.2676)(z-0.02617)(z+0.00976)}$$

Sample time: 0.1 seconds
Discrete-time zero/pole/gain model.

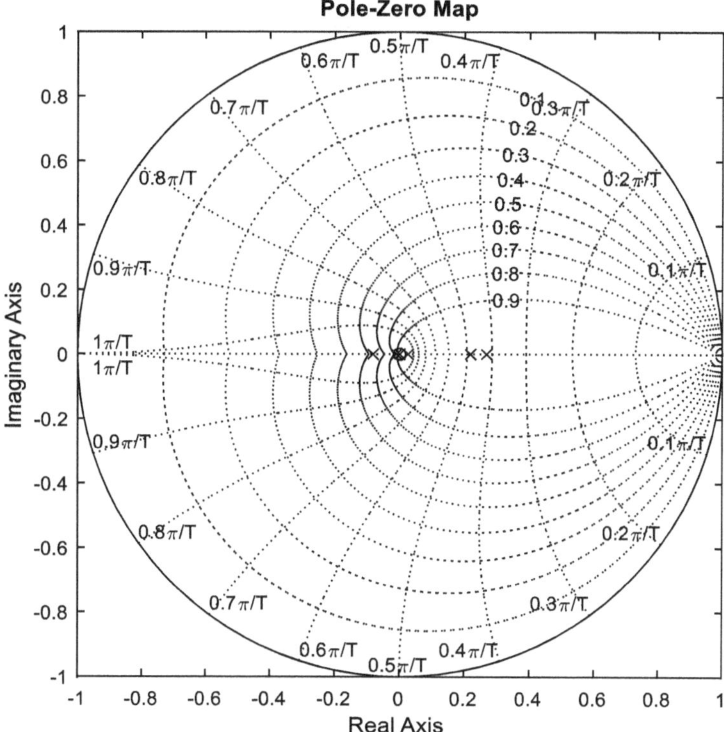

FIGURE 2.11 Pole-zero map sampling rate (100 samplers/sec).

The results of the program when using the real input of the station as an input to G (z) to produce estimated output $\bar{X}(z)$ are shown in Figure 2.11. Figure 2.12 shows a result of GA at the sampling rate (100 samplers/sec).

After the end of the program at the number of generations (8) and minimum error (1.0179) the transfer function G (z) can be obtained in the following structure:

$$G(z) = \frac{1.3047}{(Z-0.1641)(Z-0.5527)(Z+0.5156)(Z+0.8125)(Z+0.2695)}$$

Therefore, we conclude that GA will converge to optimal value using system parameters. So the proposed system response and behavior is satisfactory, and also the results of the program when applied the input active power (the gate position) of the station as an input to *G* (*z*) to produce estimated output $\bar{X}(z)$ as shown in Figure 2.13:

$$\begin{aligned} &1.3047/((Z-0.1641) \\ \%\%G(z) = &\times(Z-0.5527)(Z+0.5156) \\ &\times(Z+0.8125)(Z+0.2695)) \end{aligned}$$

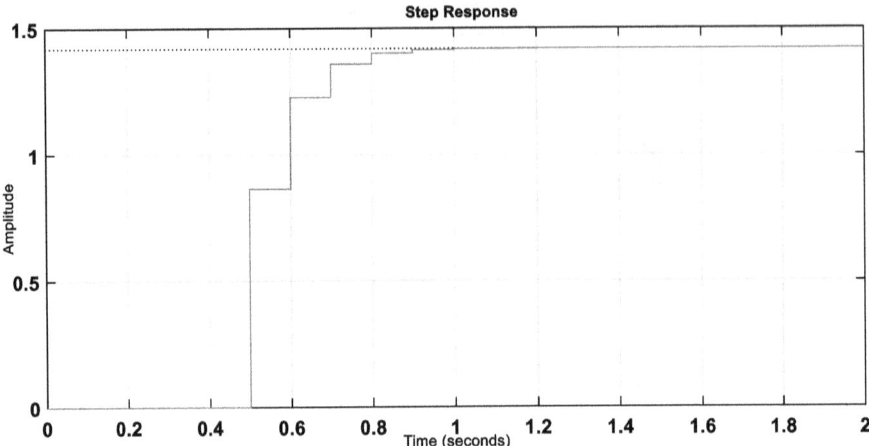

FIGURE 2.12 GA result at the sampling rate (100 samples/sec) and objective function (Root Mean Square Error).

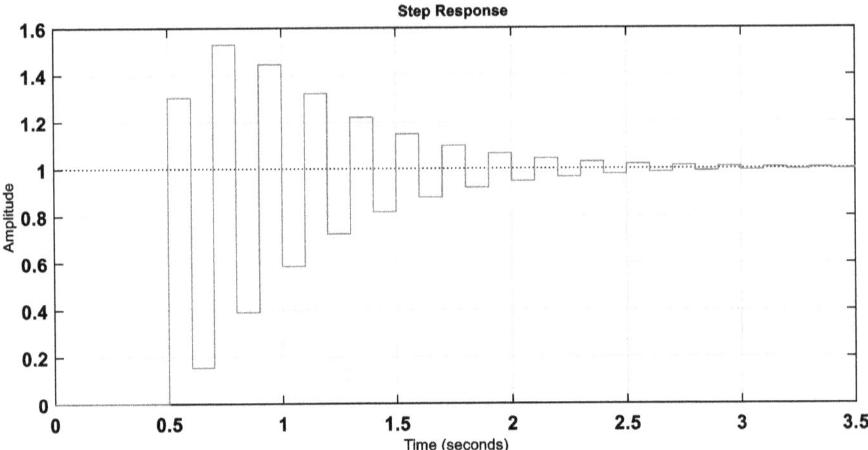

FIGURE 2.13 Result of GA at the sampling rate (100 samples/sec) and objective function (Exponential of Median Error).

$$\mathrm{H} = \mathrm{zpk}\left(\left[\ \right], \left[0.1641, 0.5527, -0.5156, -0.8125, -0.2695\right], 1.3047, 0.1\right)$$

pzmap(H)
grid

The results:

$$H = \frac{1.3047}{(z-0.1641)(z-0.5527)(z+0.5156)(z+0.8125)(z+0.2695)}$$

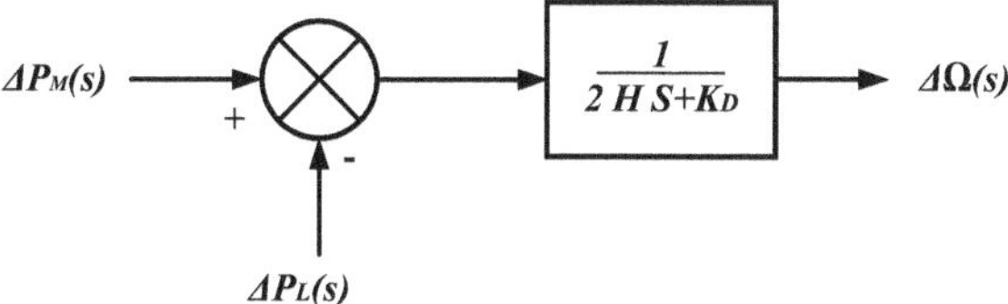

FIGURE 2.14 Block diagram of a Load Model derived from the swing equation.

Sample time: 0.1 seconds
Discrete-time zero/pole/gain model.

The results of the program when d the acitve power of the station as an input to G(z) to produce estimated output $\bar{X}(z)$ is shown in Figure 2.13. Figure 2.14 shows a result of GA at the sampling rate (100 samples/sec) and objective function (Exponential of Median Error).

An important choice for a suitable sampling rate is to translate the system behavior. Therefore, we can select many sampling rates to arrive at a convenient sampling rate value.

2.7 INERTIA CONSTANT AND SWING EQUATION

Synchronous system stability relies on constant inertia and angular momentum. The rotational inertia equations explain the effect of machine imbalance between electromagnetic torque and mechanical torque. With small disturbance and small speed deviation, the swing equation becomes

$$\frac{d\Delta\omega}{dt} = \left(\frac{1}{2H}\right)\left(\Delta P_m - \Delta P_e\right)\frac{\text{p.u}}{\text{sec}} \tag{2.14}$$

Then

$$d\Delta\omega/dt = d^2\delta/dt$$

H= inertia constant
ΔP_m = change in mechanical power
ΔP_e = change in electrical power
$\Delta\omega$ = change in speed (elec. rad/sec)
δ= rotor angle (rad.)

Using Laplace Transformation, Equation (2.14) becomes

$$\frac{d\delta}{dt}\Delta\omega(s) = \left(\frac{1}{2Hs}\right)\left[\Delta Pm(s) - \Delta Pe(s)\right] \tag{2.15}$$

A more appropriate way to explain the swing equation is to provide a damping factor not accounted for in the electrical power *Pe* calculation. Therefore, a speed deviation-proportional term should be used. The speed-load characteristic of a composite load is approximated by

$$\Delta P_e = \Delta P_L + K_D \Delta\omega \tag{2.16}$$

where K_D is the damping factor or coefficient per unit power divided by per unit frequency. $K_D \Delta\omega$ is the frequency-sensitive load change, and ΔP_L is the non-frequency-sensitive load change.

Figure 2.16 presents a block diagram representation of a load change derived from the swing equation with the aid of Equation 2.14 or

$$\Delta\omega(s) = \left[\Delta P_m(s) - \Delta P_L(s)\right]\left[\frac{1}{2Hs + K_D}\right]. \tag{2.17}$$

2.8 SYNCHRONOUS GENERATOR MODELING CONCEPT IN THE POWER SYSTEM

Any type of synchronous generator (Thermal, Hydro, and Gas station) in the electrical grid has power control systems that consist mainly of the two known types of controls

- Load frequency control system through the governor.
- Voltage control system by AVR. As shown in Figure 2.15.

To develop the simulation model, a schematic diagram of the required components for the power system under study and simulation is shown in Figure 2.16.

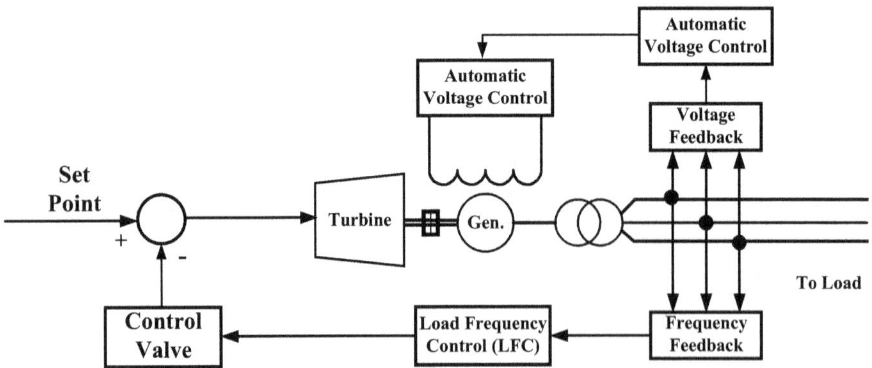

FIGURE 2.15 Schematic diagram of the governor and AVR of the synchronous generator.

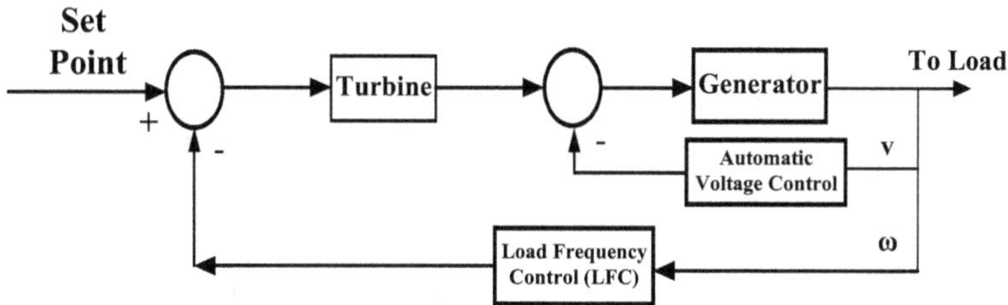

FIGURE 2.16 Block diagram of governor and AVR of the synchronous generator.

Figure 2.15 represents a schematic diagram of the governor and AVR controls of the synchronous generator.

Figure 2.16 represents a simplified corresponding block diagram of the governor's schematic and AVR of the synchronous generator with the two feedback quantities (voltage and frequency).

The schematic diagram and the simplified block diagram give a general view of how the synchronous generator should be modeled. The more important points in the synchronous generator model are obtaining the synchronous generator parameters, time constants, and order. However, to incorporate the functions that can accommodate higher-order time constants, the block diagram in Figure 2.16 will need to be explicitly redefined and illustrated in the next sections.

2.9 EXCITATION SYSTEM CONTROL

Synchronous generator excitation control is one of the most critical steps to increase the power system's reliability and electrical power quality. The excitation system's main control function regulates the terminal voltage generator by changing the field voltage using terminal voltage variance.

Usually, the excitation system is a fast response system with a short time constant. Its basic purpose is to provide a direct winding current. Also, the excitation device performs control and protective functions necessary for system activity by regulating field voltage. Therefore, the field current is within appropriate limits under various operating conditions. The excitation system's defensive functions ensure that the synchronous unit, excitation system, and other controlling equipment are not surpassed. Its control functions include voltage monitoring and reactive flow. These lead to power system stability as an important factor.

The nominal excitation system response is defined in the IEEE standard definitions for synchronous machines' excitation systems. The rate of increase of excitation system output voltage from the excitation system response curve, divided by the rated field voltage, is shown in Figure 2.17. The optimal response to excitation is the path ac rather than ab. Therefore, the excitation's output response is considered linear in most situations, not the case in practice as saturation occurs.

Excitation regulation is one of the most important variables in the transient study of the power system. Usually, a high gain for the excitation controller is needed as an

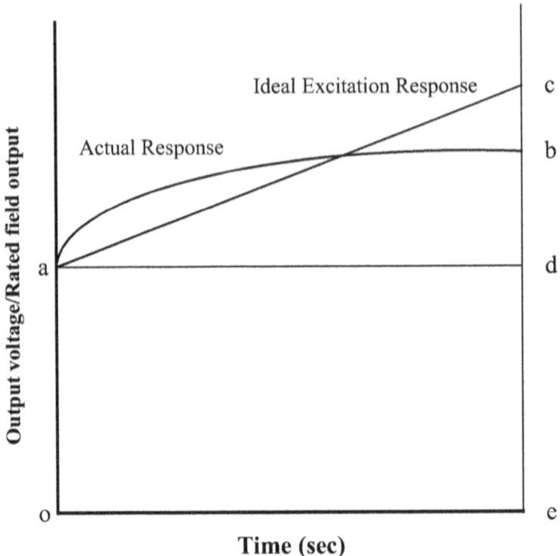

FIGURE 2.17 Nominal excitation system response.

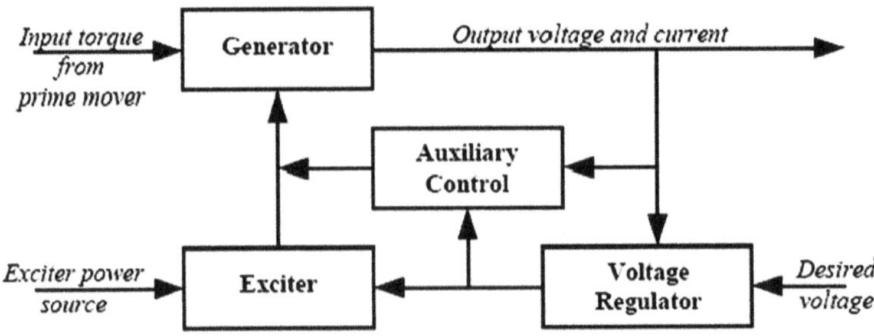

FIGURE 2.18 Typical arrangement of excitation components.

effective means of providing transient stability. Thus, the excitation controller can easily moderate the control signal and provide an excellent system oscillation damping when a disturbance occurs. Figure 2.18 shows a typical relationship between excitation control and the generator.

2.10 TURBINE GOVERNOR CONTROL

In power system studies, knowing the characteristics of the typical turbine is important. The prime mover plays an important role in contributing to the system's stability. The optimum transient response of a closed-loop control system to external disturbance depends on the excitation controller, generator, sensor transfer function, and the speed/load controller. Various types of steam turbines were introduced and categorized according to their functions and characteristics. Several speed regulations were briefly listed, describing different terms in turbine/governor. Instructions for setting regulations for handling specified versions. This standard clearly describes how the steam turbine is illustrated in block diagrams. Regarding speed control, various types of turbines have different ways to measure the regulation. For example, automatic extraction and mixed-pressure turbines will regulate the speed

$$R_s = \left[\frac{(N_o - N_M)}{N_r}\right] \times \left[\frac{P_r}{P_M}\right] \times 100\% \tag{2.18}$$

where:

R_s = steady-state speed regulation
N_o = speed at zero power output
N_r = rated speed
N_M = speed at P_M
P_M = maximum output power at which zero extraction or induction conditions are permitted
P_r = rated power output.

For all other types of turbine, the speed regulation can be expressed as follows

$$R_s = \left[\frac{(N_o - N_r)}{N_r}\right] * 100\% \tag{2.19}$$

In selecting the turbine model, careful consideration is essential; it is obvious that different operating characteristics occur when simulating different turbine models. The stability of the turbine depends on how the speed/load-control system positions the control valves such that a continuous oscillation of the turbine speed (of the power output generated by the speed/load-control system) does not exceed a specified value during operation under steady-state load demand or after a change in new steady-state load requirement. The demand for the steady-state load is expressed in a control band's range of values.

2.10.1 Prime Mover and Governing System Controls

This system provides a means of controlling real power and frequency. A basic characteristic of a governor is shown in Figure 2.19.

Figure 2.19 shows a definite relationship between turbine speed (A) and turbine load (PM) for an environment. An increase in load will reduce speed. This figure shows that if the initial operating point is at A and the load drops to 25%, the speed would increase. To maintain the pace at A, the governor's setting must be changed by adjusting the spring tension in the governor's flywheel. Unlike the excitation mechanism, the governing system is a relatively slow response system due to the turbine machine's slow mechanical operation reaction.

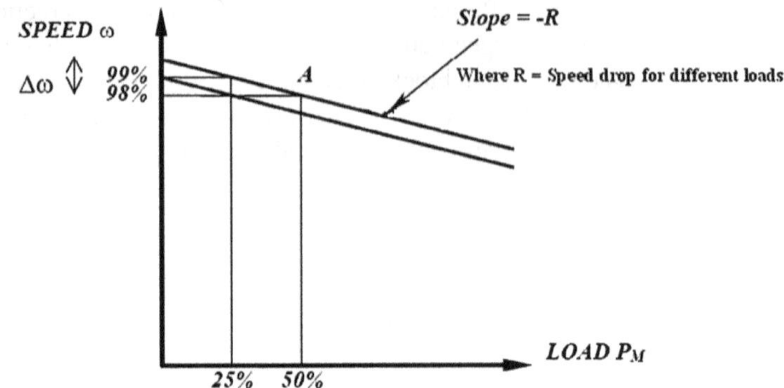

FIGURE 2.19 Governor characteristic.

2.10.2 Governor/Turbine/Generator Relationship

Turbine-governor models are designed to reflect the effects of power stations to control the stability of the generated power. Various turbines are used in various excitation settings. They range from gas turbines hydro turbines to steam turbines. The synchronous machine is vital in understanding how various turbine configurations contribute to power system stability.

2.11 DIVISION OF LOAD BETWEEN GENERATORS

The steam input and electrical output can be changed as needed at a given frequency using the speed changer. Figure 2.20 demonstrates the effect on two machines. Therefore, each machine's performance is not determined by the governor's

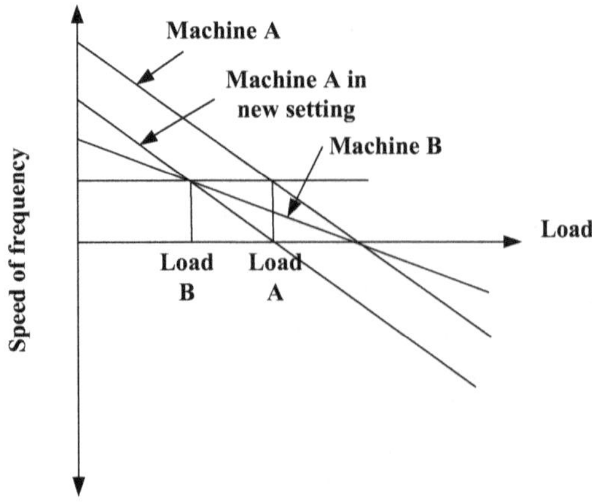

FIGURE 2.20 Characteristics of two machines connected to an infinite busbar.

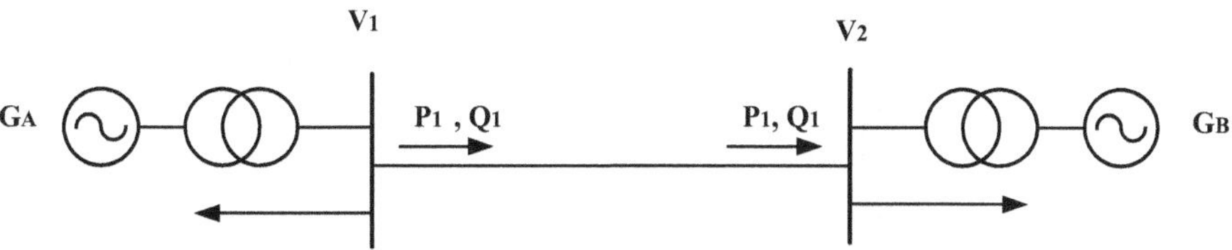

FIGURE 2.21 Two generating stations linked by an interconnector impedance of $(R + jX)$.

characteristics, but the operating staff can differ to meet economic and other considerations. The governor's characteristics determine the machine outputs when a sudden shift in load occurs or when machines must adjust their outputs within a specified range to keep the frequency constant. This latter mode is known as free-governor action. The speed gear of machine A is adjusted so that the machines share the load equally.

Consider two power generators, as shown in Figure 2.21. The angle of generator A advance is due to a greater relative input of turbine A than turbine B. Providing this extra steam (or water) to the turbine of generator A is possible due to the speed gear operation without which the power outputs of generators A and B will be calculated solely by the nominal governor features. Example 2.1 illustrates these concepts.

Example 2.1

Two-Generation Units Maintain 66 kV and 60 kV at the ends of an Inductive Reactance Interconnector per Phase of 40 Ω. A Load of 10 MW is to be Transferred from the 66 kV unit to the other end. Calculate the necessary conditions between the two ends, including the power factor of the current transmitted.

Solution

$$Load\ per\ phase = \frac{10}{3} = 3.33\ MW$$

$$\Delta V_q = \frac{X.P}{V_L} = \frac{40 \times 3.33 \times 10^6}{60 \times 10^3 / \sqrt{3}} = 3840\,V$$

$$\sin\delta = \frac{\Delta V_q}{66 \times 10^3 / \sqrt{3}} = 0.101$$

$$\delta = 5.79°$$

So, the 66 kV busbars are 5.79° in advance of the 60 kV busbars. Hence, Q = 9 *MVAr* for three phases.

$$Power\ factor\ angle\,\theta = \tan^{-1}\frac{Q}{P} = 42°$$

$$Power\ factor = \cos 42° = 0.7431\,lagging.$$

Example 2.2

Two synchronous generators are working in parallel, generating a combined load of 200 MW. The unit capacities are 100 MW and 200 MW, and both have 4% drop governor characteristics from no load to maximum load. Calculate each machine's loading assuming free governor operation.

Solution

Let x megawatts be the power supplied from the 100 MW generator. Referring to Figure 2.22,

$$\frac{4}{100} = \frac{\alpha}{x}$$

$$\frac{4}{200} = \frac{\alpha}{200 - x}$$

Solving to get $x = 66.6\ MW = P_1$ load on the 100 MW machine. The load on the 200 MW machine $=133.3\ MW = P_2$. It will be observed that when the governor drops are the same, the machines share the total load in proportion to their capacities or ratings. Therefore, it is often advantageous for the drops of all turbines to be equal.

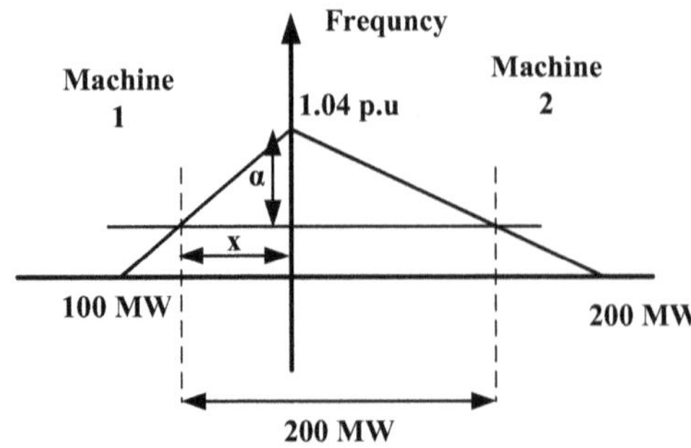

FIGURE 2.22 Power-frequency characteristics of Example 2.2.

Example 2.3

Two synchronous generators work in parallel, delivering a combined load of 1600 MW. Both machines' capacities are 1000 mw, and each generator has 4% and 5% drop governor characteristics from no load to maximum load. Calculate each machine's loading assuming free governor operation.

Solution

Let *x* megawatts be the power supplied from generator one of 1000 MW with 4% speed drop as shown in Figure 2.23

$$\frac{4}{1000} = \frac{4-\alpha}{x}$$

$$\frac{5}{1000} = \frac{5-\alpha}{x}$$

Solving to get $\alpha = 0.889$ and $x = 777.6$ MW = P_1 loads on the 1000 MW machine. The load on the other 1000 MW machine = 7778 MW = P_2. It will be observed that when the governor drops are the same, the machines share the total load in proportion to their capacities or ratings. Therefore, it is often advantageous for the drops of all turbines to be equal.

Example 2.4

Two Turbogenerator units are operating in parallel to supply a total load of 700 MW

Unit 1: Rated output power 600 MW,4% speed drop supplies 400 MW.
Unit 2: Rated output power 500 MW,5% speed drop supplies 300 MW.

If the total load increases to 800 MW, determine

i. The new loading of each unit.
ii. The common frequency change before and supplementary control action occurs neglect losses.

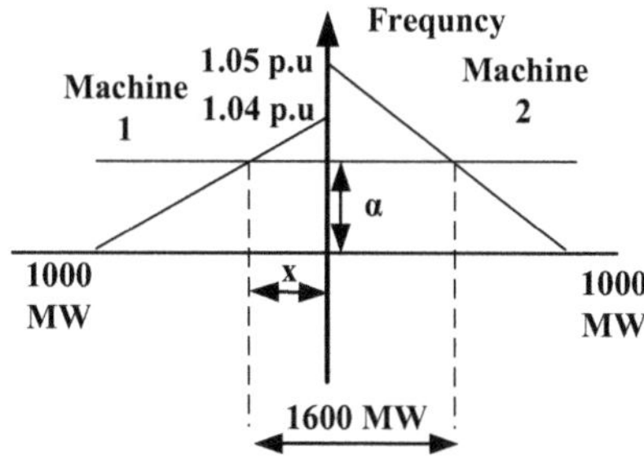

FIGURE 2.23 Power-frequency characteristics (Example 2.3).

$$\frac{\Delta f}{f_R} = \frac{-\Delta P}{\frac{S_{R1}}{R_{u1}} + \frac{S_{R2}}{R_{u2}}}$$

$$\frac{\Delta f}{f_R} = \frac{-(800-700)}{\frac{600}{0.04} + \frac{500}{0.05}} = -0.004\, p.u$$

For

$$f_R = 60\, Hz$$

$$\Delta f = -0.004 \times 60 = -0.24\, Hz$$

The new frequency for both generators

$$f_1 = \Delta f + f_0 = -0.24 + 60 = 59.76\, Hz$$

$$\Delta P_{g1} = \frac{\frac{S_{R1}}{R_{u1}} \Delta P}{\frac{S_{R1}}{R_{u1}} + \frac{S_{R2}}{R_{u2}}} = \frac{\frac{600}{0.04} \times 100}{\frac{600}{0.04} + \frac{500}{0.05}} = 60\, MW$$

So, unit 1 will supply 400 + 60 = 460 MW

Figures 2.24 and 2.25 are the first and second generator's power-frequency characteristics; they illustrate frequency change with power variation

$$\Delta P_{g2} = \frac{\frac{S_{R2}}{R_{u2}} \Delta P}{\frac{S_{R1}}{R_{u1}} + \frac{S_{R2}}{R_{u2}}} = \frac{\frac{500}{0.05} \times 100}{\frac{600}{0.04} + \frac{500}{0.05}} = 40\, MW$$

So, unit 2 will supply 300 + 40 = 340 MW.

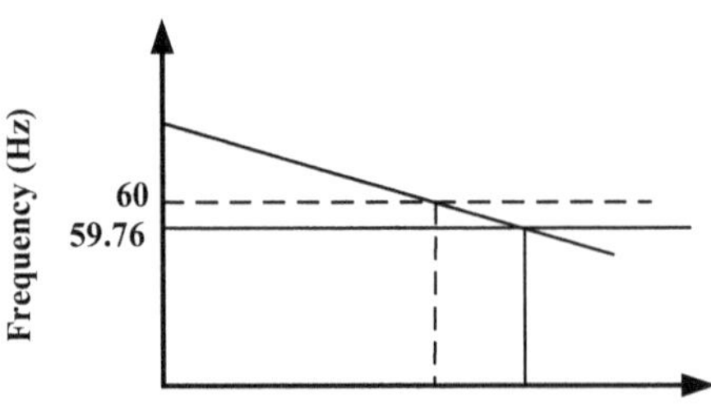

FIGURE 2.24 Power-frequency characteristics of the first generator.

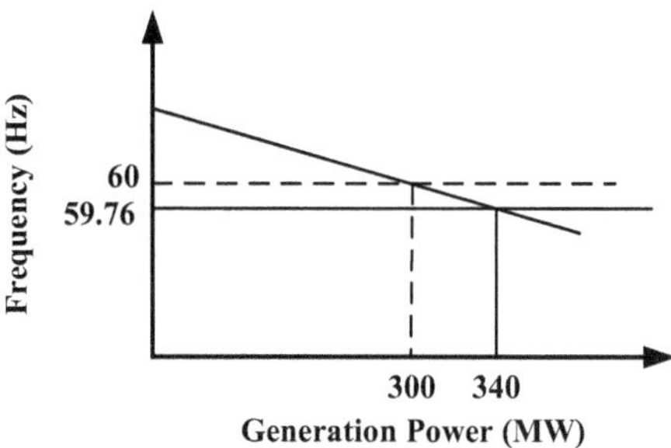

FIGURE 2.25 Power-frequency characteristics of the second generator.

Example 2.5

A 500 MVA, 50 Hz, turbine-generator has a regulation, constant r = 0.05 pu on its rating. If the generator frequency increases by 0.01 Hz in a steady-state, what is the decrease in turbine mechanical power output? Assume a fixed reference power setting.

Solution

$$\text{The per unit change in frequency} = \frac{\Delta f}{f_{base}} = \frac{0.01}{50} = 2\times 10^{-4}\ p.u.Hz$$

$$\Delta P_m = \frac{-1}{R}\Delta f = -\frac{1}{0.05}\times 2\times 10^{-4} = -40\times 10^{-4}\ p.u.MW$$

$$\Delta P_m == -40\times 10^{-4}\times 500 = -2MW.$$

Example 2.6

Prove that the AFRC of an area that has four generating units is:

$$\beta = \frac{1}{R_1}+\frac{1}{R_2}+\frac{1}{R_3}+\frac{1}{R_4}$$

where R_i is the regulation constant for the unit (i). State any assumption you make.

One area of an interconnected 60 Hz power system has three turbogenerator units rated 1000, 800, 700, and 500 MVA. The regulation constant of each unit is R = 0.04 pu based on its rating. Each unit initially operates at one-half of its rating when the system load suddenly increases by 250 MW. Neglect losses and the dependence of load on frequency determine:

i. The PU area frequency response characteristic (β) on a 1000 MVA system base.

ii. The steady-state drop-in area frequency.
iii. The increase in turbine mechanical power output of each unit. Assume that the reference power setting of each unit remains constant. Derive the necessary expressions.

$$\Delta P = -\frac{\Delta f}{R}$$

$$\Delta P_{area} = \Delta P_1 + \Delta P_2 + \Delta P_3 + \Delta P_4$$

$$\Delta P_{area} = -\left(\frac{\Delta f}{R_1} + \frac{\Delta f}{R_2} + \frac{\Delta f}{R_3} + \frac{\Delta f}{R_4}\right)$$

$$= -\left(\frac{1}{R_1} + \frac{1}{R_2} + \frac{1}{R_3} + \frac{1}{R_4}\right)\Delta f$$

$$\Delta P_{area} = -\beta \Delta f$$

$$\therefore \beta = \frac{1}{R_1} + \frac{1}{R_2} + \frac{1}{R_3} + \frac{1}{R_4}$$

$$R_{ui\,new} = R_{ui\,old} \times \frac{S_{base}}{S_{given}}$$

$$R_{u1new} = 0.04 \times \frac{1000}{1000} = 0.04\,p.u$$

$$R_{u2new} = 0.04 \times \frac{1000}{800} = 0.05\,p.u$$

$$R_{u3new} = 0.04 \times \frac{1000}{700} = 0.0571\,p.u$$

$$R_{u4new} = 0.04 \times \frac{1000}{500} = 0.08\,p.u$$

$$\beta = \frac{1}{R_1} + \frac{1}{R_2} + \frac{1}{R_3} + \frac{1}{R_4}$$

$$\beta = \frac{1}{0.04} + \frac{1}{0.05} + \frac{1}{0.0571} + \frac{1}{0.08} = 75$$

$$\Delta P_{area} = -\beta \Delta f$$

$$\frac{250}{1000} = -75\Delta f$$

The change in frequency

$$\Delta f = -3.33 \times 10^{-3}$$

The change of generation power for each unit

$$\Delta P_1 = -\frac{\Delta f}{R_1} = -\frac{-3.33 \times 10^{-3}}{0.04} = 0.0833 p.u = 83.33 MW$$

$$\Delta P_2 = -\frac{\Delta f}{R_2} = -\frac{-3.33 \times 10^{-3}}{0.05} = 0.0666 p.u = 66.67 MW$$

$$\Delta P_3 = -\frac{\Delta f}{R_3} = -\frac{-3.33 \times 10^{-3}}{0.0571} = 0.05837 p.u = 52.377 MW$$

$$\Delta P_4 = -\frac{\Delta f}{R_4} = -\frac{-3.33 \times 10^{-3}}{0.08} = 0.04166 p.u = 41.67 MW.$$

2.12 AMPLITUDE AND FREQUENCY ESTIMATION OF POWER SYSTEM

Harmonics pose a big challenge to power quality systems. To mitigate the effect of these harmonics, it is important to estimate the harmonic parameters. To estimate the harmonic parameters an Adaptive Hopf Oscillator (AHO) is used. The method exploits AHO's frequency and amplitude learning ability. Because the power system signal is typically a multi-frequency component signal, AHO's pool will be used in a negative feedback loop to teach each AHO a frequency component. The AHOs in the pool are not related, and each AHO learns a different harmonic independent of each other.

Non-linear loads and time-varying devices like arc furnaces, grinders, uninterrupted power supply, thyristors, diodes, etc., have become ubiquitous features of an electrical power system. Consequently, periodic disturbances, including harmonics, inter-harmonics, and subharmonics, are inducted into the electrical power system's current and voltage waveform. These harmonics degrade power-quality structures. Harmonics will also affect generator performance, transformer saturation, copper loss, relay misoperation, and voltage dips. Therefore, these harmonics need to be calculated to enhance the power system signal's efficiency. A suitable filter or compensator can be built to curtail the harmonic impact.

2.12.1 Adaptive Hopf Oscillator

Adaptive Hopf Oscillator (AHO) has a correlation-based learning form found in neural networks. Therefore, learning is called dynamic Hebbian learning. The AHO

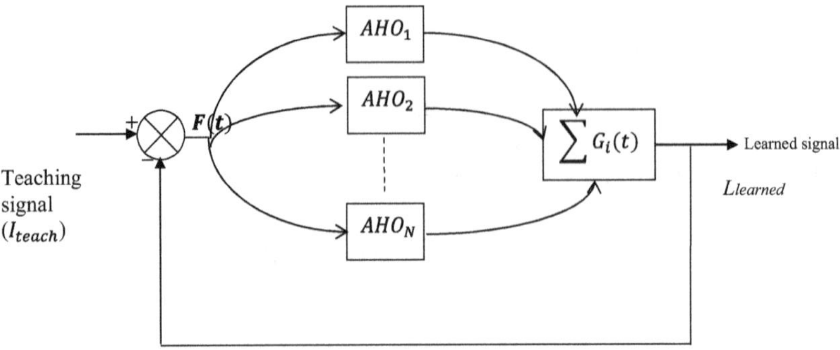

FIGURE 2.26 A pool of N AHOs with no coupling term. Each AHO is independent in a feedback structure and can track its frequency component.

differs from the conventional Hopf oscillator since its intrinsic frequency becomes a dynamic device condition. This means the AHO's intrinsic frequency can respond to any teaching signal frequency. Figure 2.26 shows a pool of non-coupling N AHOs. Each AHO is independent of its feedback structure and can monitor its frequency portion.

$$\dot{r}(t) = \gamma\left(\mu - r^2(t)\right)r(t) + \epsilon F(t)\cos\varphi(t) \tag{2.20}$$

$$\dot{\varphi}(t) = \omega(t) - \frac{\epsilon}{r(t)}F(t)\sin\varphi(t) \tag{2.21}$$

$$\dot{\omega}(t) = -\epsilon F(t)\sin\varphi(t) \tag{2.22}$$

$$\dot{\alpha}(t) = \rho F(t)\cos\left(\varphi(t)\right)r(t) \tag{2.23}$$

$$G = \alpha(t)r(t)\cos\varphi(t) \tag{2.24}$$

where $r(t)$ is the limit cycle radius, $\mu > 0$ regulates the limit cycle radius, $F(t)$ is the driving signal or disturbance, π defines how quickly the oscillator returns to the limit cycle after disturbance, and $\pi(t)$ is the oscillator step. $\alpha(t)$ and α(t) represent the oscillator frequency and amplitude, while the oscillator output is $G(t)$. Learning rates of frequency and amplitude are expressed by π, respectively. These values determine how quickly AHO converges to teaching signal frequency and amplitude.

The AHO can be used as the building block of a dynamic frequency analysis method. The system is generated using an N AHO pool through negative feedback. AHO's F(t) pool disturbance is the teaching signal minus the experienced signal (which summarizes each AHO output in the pool).

The idea behind the adaptive oscillator pool is when an AHO learns an I_{teach} frequency component, it becomes part of the feedback signal. The negative feedback excludes it from the disturbance signal $F(t)$.

To learn a signal with a completely unknown frequency and amplitude components, many AHOs (with either uniform or random frequency initialization) must be used in the pool until the error between I_{teach} and $L_{learned}$ is minimal (ideally zero). This may not be optimal as the pool may need many AHOs due to many oscillators learning the same I_{teach} frequency component.

2.12.2 Power System Signal Modeling

The voltage or current of a power system can be described as

$$y(k) = \sum_{i=1}^{n} A_i \sin\left(\omega_i t_s + \varphi_i\right) + k_g e(k) \tag{2.25}$$

where A_i, φ_i, and ω_i are the amplitude, phase, and angular frequency of the ith harmonic, $e(k)$ is the Gaussian white noise with zero mean and unit frequency, while $\boldsymbol{k_g}$ is the noise gain factor, and n is the number of harmonics in the signal. The angular frequency is given by

$$\omega_i = 2\pi \boldsymbol{f_0} \tag{2.26}$$

where $\boldsymbol{f_0}$ is the fundamental frequency.

In assessing this estimation method, three separate signals will be considered. The first signal is a static, harmonic-only signal. The second signal is a time-varying signal with just harmonics. The third signal is a static, inter-harmonic, and sub-harmonic signal. Because the approximation concerns frequency and amplitude, the step does not model the power system signals.

2.13 POWER SYSTEM STABILIZER

The basic function of a power system stabilizer (PSS) is to add damping to the generator rotor oscillations by controlling its excitation using auxiliary stabilizing signal(s). The stabilizer must produce a component of electrical torque in phase with the rotor speed deviations to provide damping. Continuous low-frequency oscillations result from insufficient damping of the systems' automatic mode. Since the purpose of a PSS is to introduce a damping torque component, a logical signal to use for controlling generator excitation is the speed deviation $\Delta\omega_r$.Usually, PSS is designed separately for each machine, and the rest of the system is considered an infinite bus from the point of view of that machine.

Figure 2.27 shows a power system modeling without PSS and the voltage response as in Figure 2.28. The power system stabilizer is connected to the power system, as Figure 2.29, to improve the response. The effect of PSS connection to the power system is shown in Figure 2.30, where the voltage response will modify.

If the exciter transfer function $G_{ex}(s)$ and the field circuit transfer functions were pure gains, direct feedback would result in a damping torque component. However, the generator and the exciter (depending on its type) exhibit frequency-dependent gain and phase characteristics in practice. Therefore, the PSS transfer function, $G_{PSS}(s)$, should have appropriate phase compensation circuits to compensate for the lag between the exciter input and the electrical torque. In the ideal case, with the phase characteristics of $G_{PSS}(s)$ being an exact inverse of the exciter and field circuit phase characteristic to be compensated, the PSS would result in a pure damping torque at all oscillating frequencies. PSS aims to increase the damping of the system

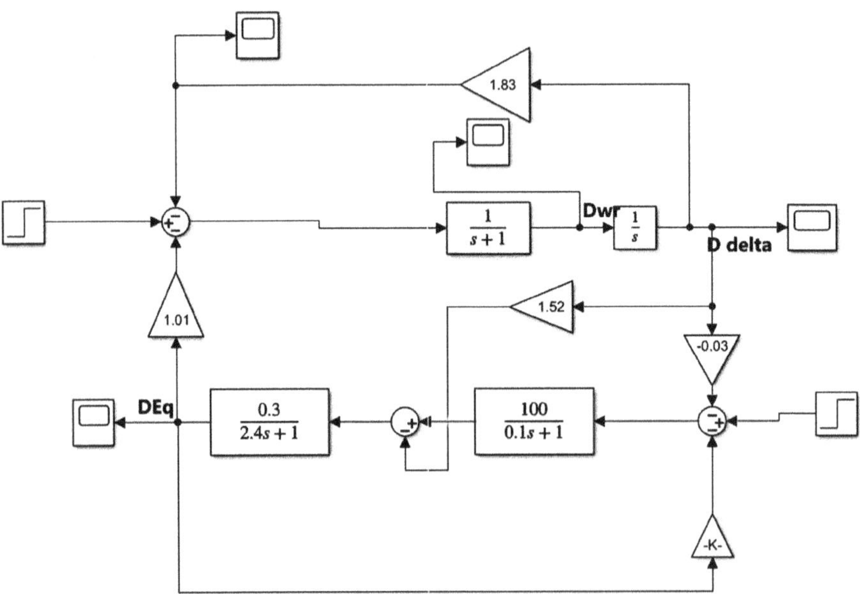

FIGURE 2.27 Power system modeling without PSS.

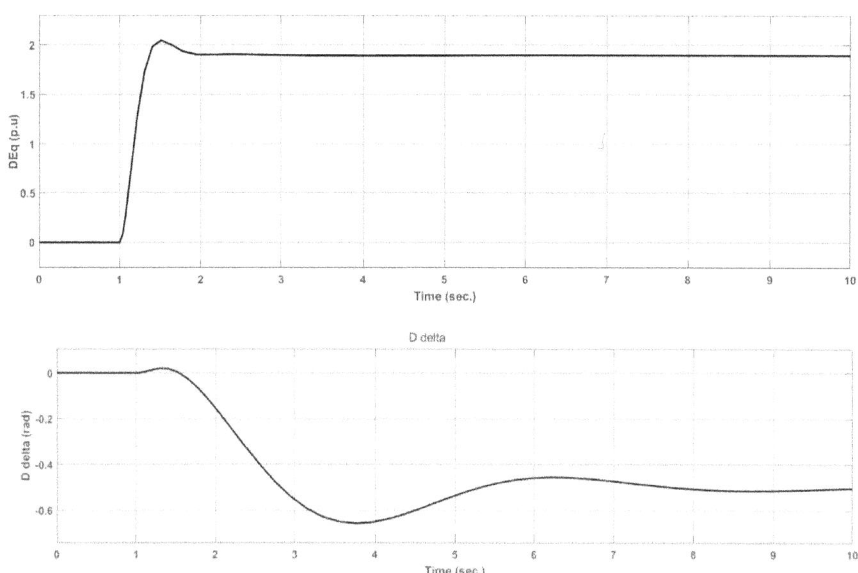

FIGURE 2.28 The voltage response without PSS.

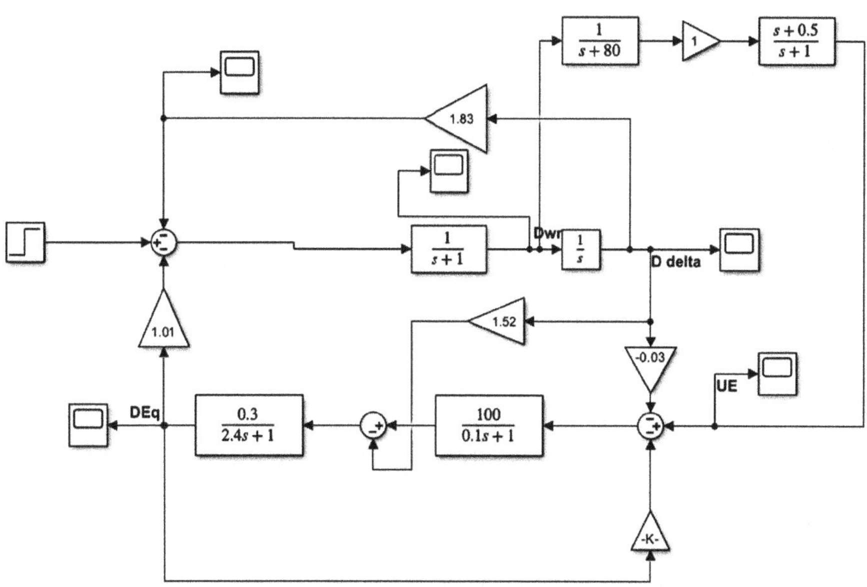

FIGURE 2.29 Power system modeling with PSS.

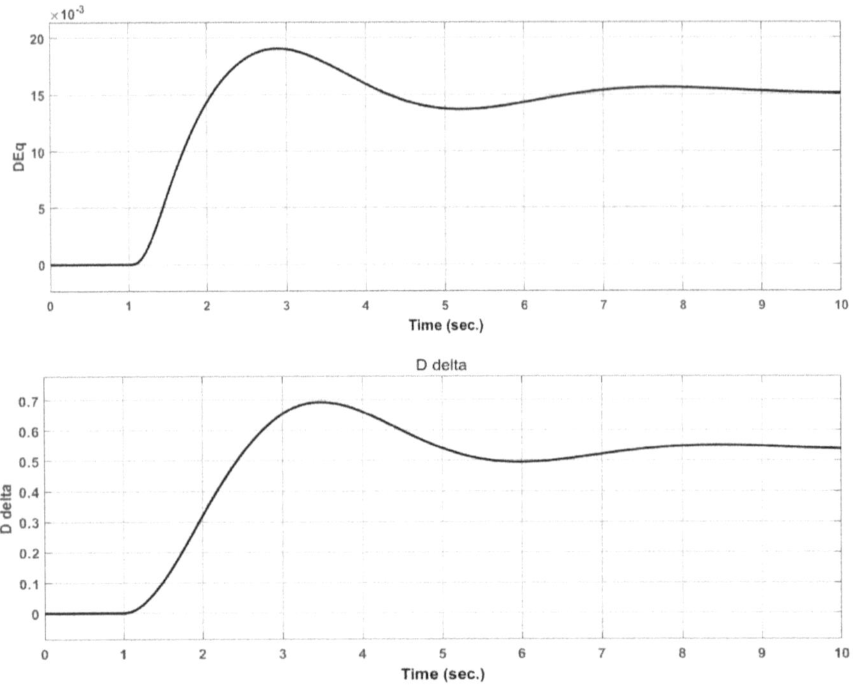

FIGURE 2.30 The voltage response with PSS.

PSS can be designed based on the undamped natural frequency or the complex frequency of the mechanical mode.

PROBLEMS

2.1 State the state-variable model's advantages.
2.2 What is AGC?
2.3 What are the conditions for load sharing between the two synchronous machines?
2.4 Defining area control error.
2.5 What is machine load classification?
2.6 What is meant by load frequency control?
2.7 Why the frequency and voltage in a power system?
2.8 Compare the "Speed Governor" and "Speed Changer" functions of a turbine generator.
2.9 What is meant by a coherent generator group?
2.10 A speed control system cannot completely eradicate the frequency error induced by a shift in the power system. Justify the point.
2.11 How is the real power structure controlled?
2.12 What is Free Governor Operation?

2.13 What is the power system load frequency control function?
2.14 Define speed droop.
2.15 Draw complex frequency shift responses for single-area load change.
2.16 What is the use of a secondary loop in the ALFC system?
2.17 AFRC's sense.
2.18 State if changes will be reflected in the ALFC loop.
2.19 List multi-area operational advantages.
2.20 Explain the principle of tie-line bias control.
2.21 The integrated 60 Hz power system has three turbogenerators. Units 1200, 900, 750 MVA. Each unit's constant regulation is R=0.04 pu based on its ranking. Initially, each unit works at half its rating when the device load rises by 300 MW. Neglect losses and frequency dependence determine
   i. Characteristic *p. u* area frequency response β on a 1200 MVA system base.
   ii. Steady-state drop-in pace.
   iii. Increase each unit's mechanical turbine output. Assume each unit's reference power setting remains unchanged. Draw the required expressions.

2.22 A three power control area is shown in Figure 2.31, with the AGC system comprising the interconnected 60 Hz system. The speed drop characteristics for each area are shown in the table

Area	R_u(p.u)	S_u(MW)	B (MW/0.1 Hz)
A	0.025	14000	1100
B	0.012	11000	1300
C	0.015	8500	900

Each area has a load level equal to 90 % of its rating on the capacity. For economic consideration, area C imports 450 MW of its load requirements from area B, and 150 MW of this interchange passes over the tie-line of area A.

Determine
   i. The system frequency deviation.
   ii. Generation changes in each area.

If the fully loaded 450 MW generator is forced out of service in area B.

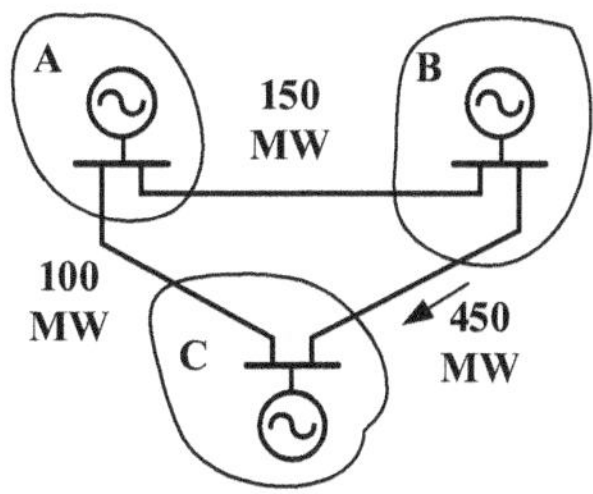

FIGURE 2.31 Power system configuration for Problem 2.22.

3 Load Frequency Control

The power and frequency regulation in a power system is related only to reactive power and voltage control. For certain reasons, it is possible to assume the governors' working regulating the power of the prime movers of the generating units independently of the AVRs controlling the excitation and the generators' reactive power and voltage. A greater appreciation of power systems' operation can be achieved by dealing with power and frequency independently from voltage and reactive power management.

The current is a major factor in power and frequency control and system behavior in a single area and multiple areas. It also presents the power-frequency characteristic of an interconnected system and the effect of governor characteristics.

3.1 STRUCTURES OF INTERCONNECTION SYSTEM

Various synchronous generators, large and small, are directly connected to a large integrated grid and have the same frequency. In certain power systems (e.g., in Great Britain), power management is carried out by the control engineers' decisions and actions, as opposed to systems in which the control and distribution of the load to the machinery occur entirely automatically. Completely automated control systems are often based on computers' continuous load-flow estimation.

Power systems operate in interconnection to:

i. Increase supply security and reliability.
ii. Reduce the amount of reserve capacities.
iii. Make use of load diversity.
iv. Operate economically.

Before a load is connected, the allocation of necessary power available from generators must be determined. It is also important to estimate the load in advance. The loads encountered in previous years during the same period are analyzed; the load's value immediately before the period under review and the weather forecast are both taken into consideration. Having been determined, the likely load to be predicted is then assigned to the separate turbine generators.

It is necessary to have the machines' capacity to rapidly raise their output from zero to maximum load and reduce their output power.

It is highly unlikely that the machines' performance would precisely equal the device's load at any moment. If the output exceeds the demand, the machines' speed will begin to rise, and the frequency will increase, and vice versa. Thus, the frequency is not a fixed number, but it varies continuously; these fluctuations are typically small and significantly impact most customers. The frequency is constantly tracked against normal time sources, and the control engineers take suitable

DOI: 10.1201/9781003293965-3

action by controlling generator outputs when long-term trends to rise or fall are observed.

The frequency will decrease if the overall generation available is inadequate to satisfy the demand. The decreased speed of power station pumps and fans restrict power stations' performance if the frequency drops by more than approximately 1 Hz and a serious situation occurs.

When the load demand increases, the frequency, and speed of all interconnected generators decrease, as the additional energy input is satisfied by the load kinetic energy, due to the governors' working, which induces a rise in steam or water admitted to the turbines, a new load balance is achieved. Initially, the boilers have a thermal reserve of steam in their boiler drums, from which sudden adjustments can be made before a new firing rate has been developed. Modern gas turbines provide an overload capacity useful in emergency scenarios for a few minutes.

Both generators and rotating loads in the system are supplied with the stored kinetic energy by

$$KE = \frac{1}{2} I \omega^2 \tag{3.1}$$

where:

I: the moment of inertia of all generators (kgm^2)
ω: rotational speed of all generators (rad/s).

When $P_m = P_e$ (Figure 3.1), the rotational speed of the generators is maintained, and the frequency is constant at 50Hz or 60Hz. When $P_m < P_e$ the rotational speed of the generators, and hence system frequency reduces. When $P_m > P_e$, the rotational speed of the generators, and hence system frequency, increases.

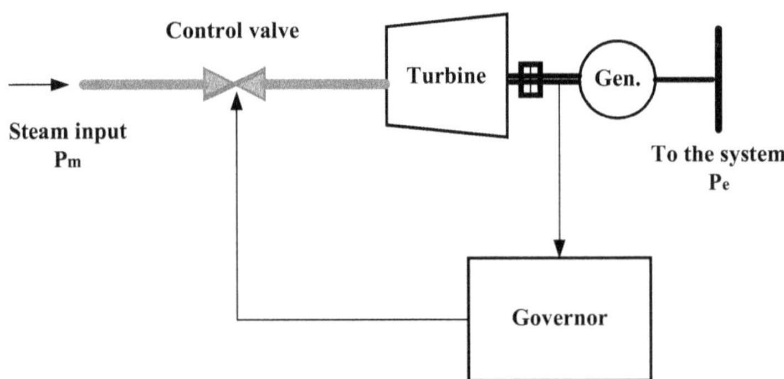

FIGURE 3.1 Control of frequency. Angular speed is measured and controls an inlet valve for steam input to turbine.

The torque balance of any spinning mass determines the rotational speed as

$$T_m - T_e = I\frac{d\omega}{dt} \tag{3.2}$$

In power systems, it is conventional to express the inertia per unit as an H constant

$$H = \frac{\frac{1}{2}I\,\omega_s^2}{S_{rated}} \tag{3.3}$$

S_{rated} is the MVA rating of either an individual generator or the entire power of the system, vs. is the angular velocity (rad/s) at synchronous speed. Thus,

$$\frac{d\omega}{dt} = \frac{\omega_s^2}{2\,H\,S_{rated}}\left(T_m - T_e\right) \tag{3.4}$$

which is per unit may be written as

$$\frac{d\omega}{dt} = \frac{1}{2\,H}\left(T_m - T_e\right) \tag{3.5}$$

Thus, the rate of change of rotational speed and frequency depends on the power imbalance and the inertia constant.

3.2 THE TURBINE GOVERNOR

A simplified schematic of a typical governor structure is seen in Figure 3.2. The sensing unit, responsive to speed adjustment, is the time-honored Watt Centrifugal Governor. In this, two weights shift radially outward as their rotating speed increases, pushing a sleeve on the central spindle. This sleeve movement is conveyed to the pilot-valve piston by a lever mechanism. Thus, the servo-motor is controlled. A dead band is present in this system, i.e., speed must adjust by a certain amount before the valve begins working due to friction and mechanical backlash. Due to delays in hydraulic pilot-valve and servo-motor systems, the time required to move the main steam valve is appreciable, 0.2 – 0.3 sec.

Figure 3.3 shows the governor characteristic of a large steam turbo-alternator, and there is a percent reduction in speed between no load and maximum load. Because of the need for fast reaction time, low dead band, and precision in speed and load control, the mechanical governor was replaced by electro-hydraulic control in large modern turbo-generators. The system usually used to determine speed is based on a toothed wheel on the generator shaft and pick-up magnetic-probe. Using traditional servo-valves, electronic controls include an electro-hydraulic conversion step.

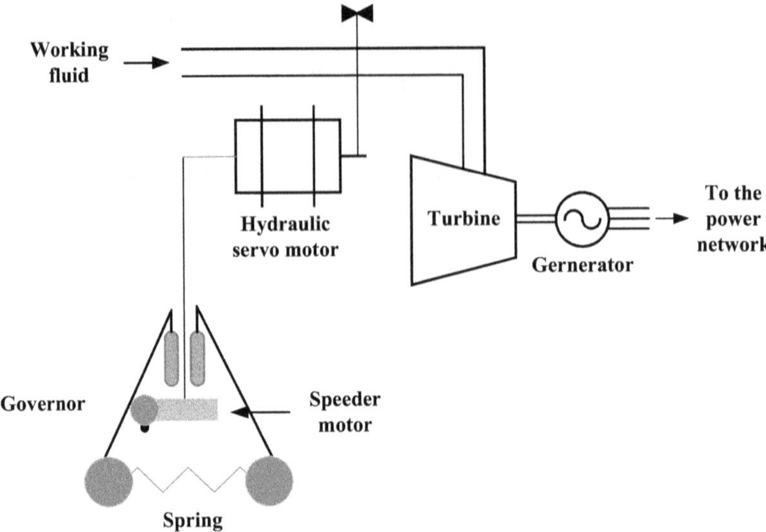

FIGURE 3.2 Governor control system employing the Watt governor as a sensing device and a hydraulic servo-system to operate the main supply valve.

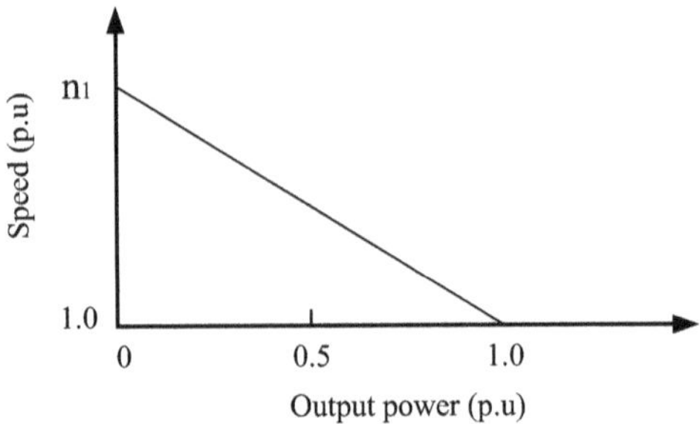

FIGURE 3.3 Idealized governor characteristic of a turbo-alternator with n% droop from zero to full load.

A significant feature of the governor system is how the governor sleeves, and therefore the main-valve positions, can be altered and modified apart from when the velocity varies. The speed-changer does this, or 'speeder engine,' as often called. This modification results in a family of parallel features, as seen in Figure 3.4. Therefore the generator power output at a given speed can be modified independently of device frequency, which is highly necessary when running at the maximum economy.

The turbine torque can equal the main inlet valve displacement d. The expression for velocity torque shift can be expressed roughly by the equation

$$T = T_0\left(1 - k \cdot N\right) \tag{3.6}$$

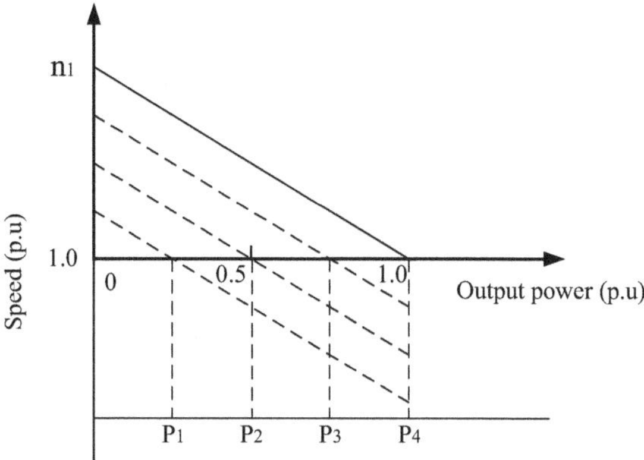

FIGURE 3.4 Effect of speeder gear on governor characteristics. P_1, P_2, and P_3 are power outputs at various settings but the same speed.

where T_0 is the torque at speed N_0 and T the torque at speed N; k is a constant for the governor system. As the torque depends on both the main-valve position and the speed,

$$T = f(d, N) \tag{3.7}$$

There is a time difference between load change and new operating conditions. This is due to the governor mechanism and the new steam or water flow rate that must accelerate or decelerate the rotor to achieve the new speed.

A significant aspect of turbines is the risk of overspeeding with potential mechanical failure when the shaft load is completely lost. Special valves are mounted to automatically cut the turbine's energy supply to stop this.

3.3 CONTROL LOOPS

The machine and its associated controller and voltage-regulator systems may reflect the block diagram in Figure 3.5. Two factors affect the prime mover's complex response:

i. Educated steam between the inlet valves and the turbine's first stage (in large machines, this may be sufficient to cause synchronization failure after the valves are closed);
ii. The energy stored in the reheater causes the low-pressure turbine output to lag behind the high-pressure side.

The transfer function

$$G(s) = \frac{\textit{Prime mover torque}}{\textit{Valve opening}} = \frac{k_{1s} \cdot k_{2r}}{(1+\tau_1 s)(1+\tau_2 s)} \tag{3.8}$$

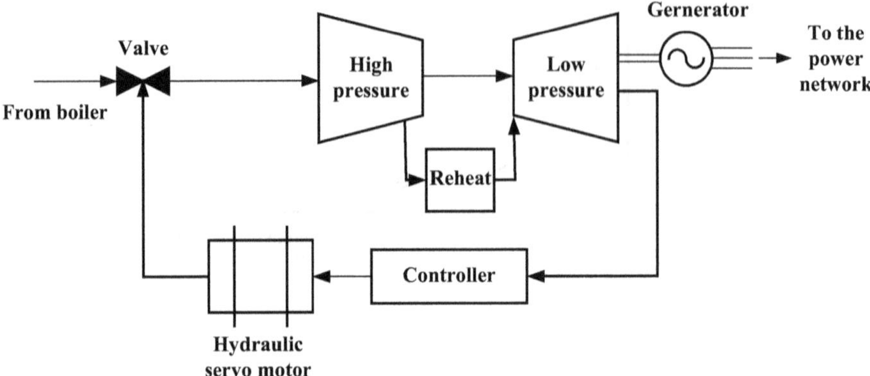

FIGURE 3.5 Block diagram of complete turboalternator control systems.

where:

k_{1s}: Entrained steam constant;
k_{2r}: Reheater gain constant;
τ_1: Entrained steam time constant;
τ_2: Reheater time constant.

The transfer function relating steam-valve opening $\Delta\delta$ to changes in speed $\Delta\omega$ due to the governor feedback loop is

$$\frac{\Delta\delta}{\Delta\omega} = \frac{k_3 \cdot k_4 \cdot k_5}{(1+\tau_s s)(1+\tau_1 s)(1+\tau_2 s)} \tag{3.9}$$

where τ_s: Governor-relay time constant.

3.4 SYSTEM BEHAVIOR/SINGLE AREA

A single area system is not connected to any other system, and hence the demand for the system is fully met by its generation. Sudden load changes accommodated in such systems in the following ways:

i. Change in stored energy.
ii. Change in effective load.
iii. Change in a generation.

Energy stored in the spinning capacity of the system is a function of the square of the frequency and can be expressed as:

$$W^o = kf_o^2 \tag{3.10}$$

$$W^1 = kf_1^2 \tag{3.11}$$

Assuming

$$f_1 = f_o + \Delta f \tag{3.12}$$

So,

$$\frac{W^1}{W^o} = \frac{k(f_o + \Delta f)^2}{k f_o^2}$$

$$\therefore W^1 = \left(\frac{f_o + \Delta f}{f_o}\right)^2 W^o \tag{3.13}$$

In Equation 3.13, W^o and W^1 are stored energies at frequencies f_o(nominal) and f_1, respectively, and k is a constant.

Hence, the change in stored energy

$$\Delta W = W^1 - W^o = \frac{2\Delta f}{f_o} W^o \tag{3.14}$$

Dividing by P_r (rated power of the system),

$$\frac{\Delta W}{P_r} = \frac{W^1 - W^o}{P_r} = \frac{2\Delta f}{f_o} \cdot \frac{W^o}{P_r} \tag{3.15}$$

$$\frac{\Delta W}{P_r} = \frac{2\Delta f}{f_o} \cdot H \tag{3.16}$$

$$\Delta P_G - \Delta P_D = \frac{d}{dt} \Delta W + D \cdot \Delta f \tag{3.17}$$

where ΔP_G is the change in generation due to an increase in demand ΔP_D. Substituting the value of ΔW and dividing all through by P_r (base power) yield

$$\frac{\Delta P_G - \Delta P_D}{P_r} = \frac{d}{dt} \cdot \frac{\Delta W}{P_r} + \frac{D \cdot \Delta f}{P_r} \tag{3.18}$$

$$\frac{\Delta P_G - \Delta P_D}{P_r} = \frac{d}{dt} \cdot \left(\frac{2\Delta f}{f_o} \cdot \frac{W^o}{P_r}\right) + \frac{D \cdot \Delta f}{P_r} \tag{3.19}$$

In the above relation, ΔP_G, ΔP_D, and D are expressed in per unit. Hence, in per unit

$$\Delta P_G - \Delta P_D = \left(\frac{2}{f_o} \cdot H\right)\frac{d\Delta f}{dt} \cdot + D \cdot \Delta f \tag{3.20}$$

This is a linear differential equation.
The static frequency drop can be obtained by putting $\frac{d\Delta f}{dt} = 0,$

$$\Delta P_G - \Delta P_D = D \cdot \Delta f \tag{3.21}$$

but

$$\Delta P_G = -\frac{\Delta f}{R} \tag{3.22}$$

$$-\frac{\Delta f}{R} - \Delta P_D = D \cdot \Delta f \tag{3.23}$$

$$-\Delta P_D = D \cdot \Delta f + \frac{\Delta f}{R} \tag{3.24}$$

$$\Delta f\left(D + \frac{1}{R}\right) = -\Delta P_D \tag{3.25}$$

$$\Delta f = \frac{-\Delta P_D}{\left(D + \frac{1}{R}\right)} \tag{3.26}$$

$$\Delta f = \frac{-\Delta P_D}{\beta} \tag{3.27}$$

where

$$\beta = D + \frac{1}{R} \tag{3.28}$$

Thus, the step-load change is positive, which signifies a sudden increase in demand; there will be a state frequency drop. Alternatively, if a step load change is negative, which means a sudden decrease in demand, there will be a static frequency rise.

β represents what is known as the area frequency response characteristics (AFRC). It is unit in p.uMW/Hz and it combines the effects of the frequency characteristic of the area's load and the inherent governing system of the arts. It is thus dependent on the operating conditions, i.e., on the level of the area load and magnitude and the nature of spinning capacity. AFRC is generally expressed in percentage of the area spinning capacity, and its typical values fall in the range of 1–5% per 0.1 Hz.

3.5 THE POWER-FREQUENCY CHARACTERISTIC OF AN INTERCONNECTED SYSTEM

Changing power for a given frequency shift in an integrated system is known as system stiffness. The lower the frequency shift for a given load, the stiffer the mechanism. A straight line and ΔP/Δf = K, where K is a constant (MW per Hz) depending on the governor and load characteristics, will approximate the power-frequency characteristic.

Let ΔPG be the generation transition of governors running free-acting,' arising from a sudden rise in DPL load. The resulting out-of-balance strength

$$\Delta P = \Delta P_L - \Delta P_G \tag{3.29}$$

Example 3.1

i. Start from $W = k\,f^2$, *W = Energy, k = constant, f = frequncy,* prove that

ii. $$f_1 = f_o\sqrt{\frac{W^o + \Delta W}{W^o}}$$

where:
f_1= an arbitrary system with a new frequency;
f_o=nominal system frequency;
W^o= Stored energy at f_o;
$W^1 = W^o + \Delta W$ =stored energy at f_1.

iii. A 150 MVA,50 Hz turbo-generator operates at no load at 3600 rpm. A load of 40 MW is suddenly applied, and the steam valves commence to open after 0.5 sec due to the time-lag in the governor system. If the turbo-generator inertia is constant, H=5 MW.sec/MVA, calculate the frequency to which the generated voltage drops before steam flow begins to increase to meet new load requirements.

Solution

i. From

$$W^1 = k\,f_1^2, f_1 = \sqrt{\frac{W^1}{k}}$$

Also

$$W^o = k\,f_o^2, k = \frac{W^o}{f_o^2}$$

So

$$f_1 = f_o \sqrt{\frac{W^o + \Delta W}{W^o}}$$

The energy stored at no load $W^o = 5 \times 150 = 750\ MJ$.
The energy lost in the time lag of 0.5 sec.

$$\Delta W = 40 \times 0.5 = 20\,MJ$$

$$f_1 = f_o \sqrt{\frac{W^o + \Delta W}{W^o}}$$

$$= 50\sqrt{\frac{750 - 20}{750}} = 49.329\,Hz.$$

Example 3.2

An independent 75 MVA synchronous generator feeds its load and runs at 3000 rpm, 60 Hz. A 30 MW load is unexpectedly applied, and the turbine steam valves start to open after 0.4 sec due to the time lag in the governor mechanism. Calculate the frequency at which the produced voltage drops until it reaches the new load. The machine's stored energy is 4 kW/kVA of generator power.

Solution

$$W^o = H.S = 4 \times 75 = 300\,MJ$$

$$\Delta W = \Delta P.t = 30 \times 0.4 = 12\,MJ$$

$$f_1 = f_o \sqrt{\frac{W^o + \Delta W}{W^o}}$$

$$= 60 \times \sqrt{\frac{300 - 12}{300}} = 58.787\,Hz.$$

Example 3.3

A 250 MVA, 60 Hz turbo-generator operates at no load at 3000 rpm. A load of 40 MW is suddenly applied, and the steam valves commence to open after 0.5 sec due to the time-lag in the governor system. If the turbo-generator inertia constant, H = 5 MW.sec/MVA, calculate the frequency to which the generated voltage drops before steam flow commence to increase to meet the new load.

$$W^o = H.S = 5 \times 250 = 1250\,MJ$$

$$\Delta W = \Delta P.t = 40 \times 0.5 = 20\,\text{MJ}$$

$$f_1 = f_o \sqrt{\frac{W^o + \Delta W}{W^o}}$$

$$= 60 \times \sqrt{\frac{1250 - 20}{1250}} = 59.518\,\text{Hz}$$

Therefore,

$$k = \frac{\Delta P_L}{\Delta f} - \frac{\Delta P_G}{\Delta f} \tag{3.30}$$

$\frac{\Delta P_L}{\Delta f}$ measures the effect of the frequency characteristics of the load and

$$\Delta P_G \propto \left(P_T - P_G\right)$$

where P_T is the turbine capacity connected to the network and P_G the output of the associated generators. When steady conditions are again reached, the load PL is equal to the generated power P_G (neglecting losses).

3.6 SYSTEM CONNECTED BY LINES OF RELATIVELY SMALL CAPACITY

Let K_A and K_B be the respective power-frequency constants of two separate power systems A and B, as shown in Figure 3.6. Let the change in the power be transferred from A to B, if a transition leading to an out-of-balance power ΔP occurs in system B, be ΔP_t, where ΔP_t is positive when power is moved from A to B. The frequency change in system B is $(-(\Delta P - \Delta P_t))/K_B$ due to an extra load and an extra ΔP_t input from A (the negative sign indicates a fall in frequency).

The decrease in frequency in A due to the extra load ΔP_t is $-\Delta P_t = K_A$. But the frequency changes in each device must be equal as they are electrically related. Therefore,

$$\frac{-\left(\Delta P - \Delta P_t\right)}{K_B} = \frac{-\Delta P_t}{K_A}$$

$$\frac{-\Delta P}{K_B} = \frac{-\Delta P_t}{K_A} - \frac{\Delta P_t}{K_B}$$

$$\frac{-\Delta P}{K_B} = -\Delta P_t\left(\frac{1}{K_A} + \frac{1}{K_B}\right)$$

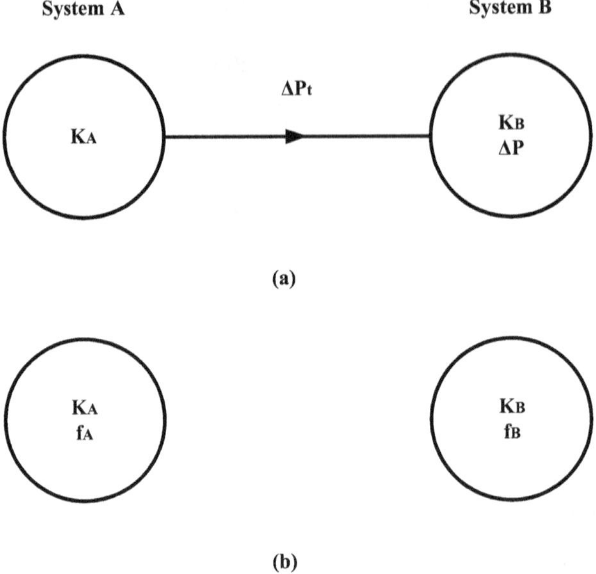

FIGURE 3.6 (a) Two interconnected power systems connected by a tie-line; (b) The two systems with the tie-line open.

$$\frac{-\Delta P}{K_B} = -\Delta P_t\left(\frac{K_A + K_B}{K_A K_B}\right)$$

So

$$\Delta P_t = \frac{K_A}{K_A + K_B}\Delta P \tag{3.31}$$

Next, consider the two systems operating at a common frequency f with A exporting ΔP_t to B. The connecting link is now opened, and A is relieved of ΔP_t and assumes a frequency f_A and B has ΔP_t more load and assumes f_B. Hence,

$$f_A = f + \frac{\Delta P_t}{K_A} \tag{3.32}$$

and

$$f_B = f - \frac{\Delta P_t}{K_B} \tag{3.33}$$

From which

$$\frac{\Delta P_t}{f_A - f_B} = \frac{K_A K_B}{K_A + K_B} \tag{3.34}$$

Hence, by opening the link and measuring the resultant change in f_A and f_B, K_A and K_B's values can be obtained.

3.6.1 Effect of Governor Characteristics

Fuller's treatment of the output of two interconnected systems in a steady-state requires further consideration of the generators' control aspects. Figure 3.8 provides a complete block diagram for steam turbine generators linked to power systems.

$\Delta P'$: The change in speed-changer setting.
ΔP: The change in power output of prime movers.
ΔP_L: The change in load power;
Δf: The change in frequency;
R: governor droop, that is drop in speed or frequency when machines of an area range from no load to full load.

Example 3.4

Two power systems A and B are connected by a tie-line $\boldsymbol{K_A} = 750$, $\boldsymbol{K_B} = 500$ MW/Hz. The tie-line trip is out of service when the power transfer exceeds 250 MW. At $\boldsymbol{f} = 50$ Hz. A is exporting 220 MW to B, a load surge of 80MW took place in B. Calculate the consequent frequencies attained by the two systems.

The tie line will trip out due to the 300 MW power flow. So the frequencies for system A:

$$\Delta P_A = -K_A\left(f_A - f_0\right)$$

$$-220 = -750\left(f_A - 50\right)$$

$$f_A = 50.29\text{Hz}$$

For system B:

$$\Delta P_A + \Delta P_B = -K_B\left(f_B - f_0\right)$$

$$300 = -500\left(f_B - 50\right)$$

$$f_B = 49.4\text{Hz}$$

Example 3.5

Two power systems, A and B, are connected by a tie-line. $K_A = 750$, $K_B = 500$ MW/Hz. The tie-line trip is out of service when the power transfers exceed 450 MW. At $f = 60$ Hz A is exporting 413 MW to B, a load surge of 87 MW took place in B. Calculate the consequent frequencies attained by the two systems (Figure 3.7).

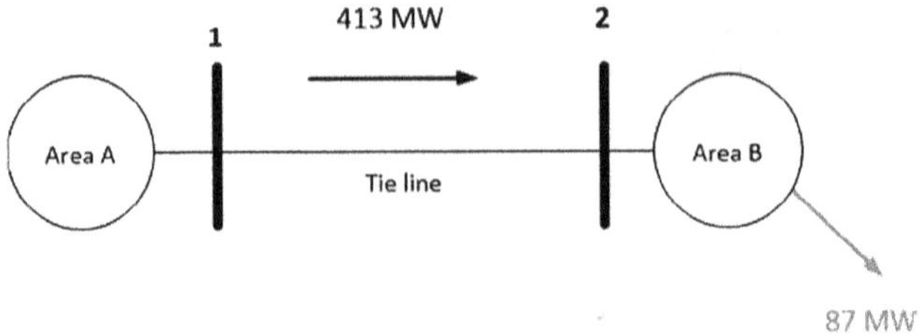

FIGURE 3.7 Two interconnected power systems connected by a tie-line configuration of Example 9.5.

Solution

The tie line will trip out due to the 500 MW power flow. So the frequencies for system A:

$$\Delta P_A = -K_A\left(f_A - f_0\right)$$

$$-413 = -750\left(f_A - 60\right)$$

$$f_A = 60.55\text{Hz}$$

For system B:

$$\Delta P_A + \Delta P_B = -K_B\left(f_B - f_0\right)$$

$$500 = -500\left(f_B - 60\right)$$

$$f_B = 59\text{Hz}.$$

Example 3.6

Two 60 Hz power systems are interconnected by a tie-line, carrying 1200 MW from system A to system B. After the line's outage, the frequency in system A increases to 60.5 Hz, while system B's frequency decreases to 59 Hz.

i. Calculate the stiffness of each system.
ii. If the systems operate interconnected with 1000 MW being transferred from A to B, calculate the line's flow after an outage of a 600 MW generator in system B.

Solution
For system A

$$\Delta f_A = \frac{\Delta P_t}{k_A}$$

$$(f_A - f_o) = \frac{\Delta P_t}{k_A}$$

$$k_A = \frac{1200}{(60.5 - 60)} = 2400\,\text{MW/Hz}$$

$$\Delta f_B = \frac{\Delta P_t}{k_B}$$

$$(f_B - f_o) = \frac{-\Delta P_t}{k_B}$$

$$k_B = \frac{-1200}{(59 - 60)} = 1200\,\text{MW/Hz}$$

Let the change in load in system B = ΔP.
Hence, the change in frequency in system B is

$$\Delta f_B = -\frac{(\Delta P - \Delta P_t)}{k_B}$$

The negative sign indicates a fall in frequency.
The drop in frequency in system A due to extra export to system B,ΔP_t is

$$\Delta f_A = -\frac{\Delta P_t}{k_A}$$

As two systems are still interconnected,

$$\Delta f_A = \Delta f_B$$

or

$$-\frac{\Delta P_t}{k_B} = -\frac{(\Delta P - \Delta P_t)}{k_B}$$

solving to get

$$\Delta P_t = \frac{k_A}{k_A + k_B}\Delta P = \frac{2400}{2400 + 1200} \times 1200 = 800\,\text{MW}$$

So

$$P_t = 1200 + 800 = 2000\,\text{MW}.$$

Example 3.7

i. The two area power systems are connected through a relatively weak tie-line. The power /frequency characteristics of the two systems when acting independently may be represented

$$\frac{\Delta P_A}{\Delta f_A} = -k_A$$

and

$$\frac{\Delta P_B}{\Delta f_B} = -k_B$$

When they were operating at common frequency and leading of P_{LA} and P_{LB}, suddenly demand ΔP_{LB} occurred on system B. If ΔP_{tA} represents system -A contribution to the demand via the tie-line, derive the expression of the frequency decrease of the two areas (Δf_A *and* Δf_B).Also, prove that

$$\Delta P_{tA} = \frac{k_A}{k_A + k_B} \Delta P_{LB}$$

ii. If $k_A = 3500\frac{MW}{Hz}$ and $k_B = 4500\frac{MW}{Hz}$, f = 60Hz, determine the change in frequency in each area and the new frequency after the change in the power flow in the tie -line due to the increase of the load in area B by 120 MW.

Solution

i. $\text{The change in freqyency in area } A = \Delta f_A = -\frac{\Delta P_{tA}}{k_A}.$

$\text{The change in freqyency in area } B = \Delta f_B = -\frac{\Delta P_{LB} - \Delta P_{tA}}{k_B}$

Since both systems are interconnected,

$$\Delta f_A = \Delta f_B$$

or

$$-\frac{\Delta P_{tA}}{k_A} = -\frac{\Delta P_{LB} - \Delta P_{tA}}{k_B}$$

Solving to get

$$\Delta P_{tA} = \frac{k_A}{k_A + k_B} \Delta P_{LB}.$$

ii. For $k_A = 3500\dfrac{\text{MW}}{\text{Hz}}$ and $k_B = 4500\dfrac{\text{MW}}{\text{Hz}}$, $f = 60\text{Hz}$.

$$\Delta P_{tA} = \frac{k_A}{k_A + k_B}\Delta P_{LB}$$

$$= \frac{3500}{3500 + 4500} \times 120$$

$$= 52.5\,\text{MW}$$

$$\Delta f_A = -\frac{\Delta P_{tA}}{k_A}$$

$$= -\frac{52.5}{3500} = -0.015\text{Hz} = \Delta f_B = \Delta f$$

$$f_{new} = \Delta f + f_0 = -0.015 + 60 = 59.985\text{Hz}.$$

Figure 3.13 can be used to represent several coherent generators with similar characteristics.

For this system, the following equation holds:

$$Ms \cdot \Delta f + D \cdot \Delta f = \Delta P - \Delta P_L \tag{3.35}$$

where:

M is the angular momentum of the spinning generators and loads.

D is the damping coefficient of the load, the change of power drawn by the load with frequency. Therefore, the change from normal speed or frequency,

$$\Delta f = \frac{1}{Ms + D}\left(\Delta P - \Delta P_L\right) \tag{3.36}$$

The high water inertia must be accounted for hydro-turbines, and analysis is more difficult.

The representation of two connected structures is seen in Figure 3.9.

The basic analysis is as before but due to the tie-line for additional power terms. Machines in each independent power system are closely coupled with one identical rotor.

From System A

$$M_A\, s\Delta f_A + D_A \Delta f_A + T_{AB}\left(\delta_A - \delta_B\right) = \Delta P_A - \Delta P_{LA} \tag{3.37}$$

From System B

$$M_B\, s\Delta f_B + D_B \Delta f_B + T_{AB}\left(\delta_B - \delta_A\right) = \Delta P_B - \Delta P_{LB} \tag{3.38}$$

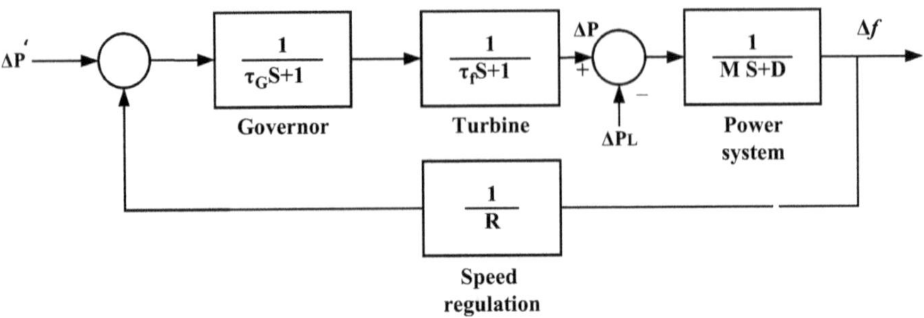

FIGURE 3.8 Block diagram for a turbine generator connected to a power system.

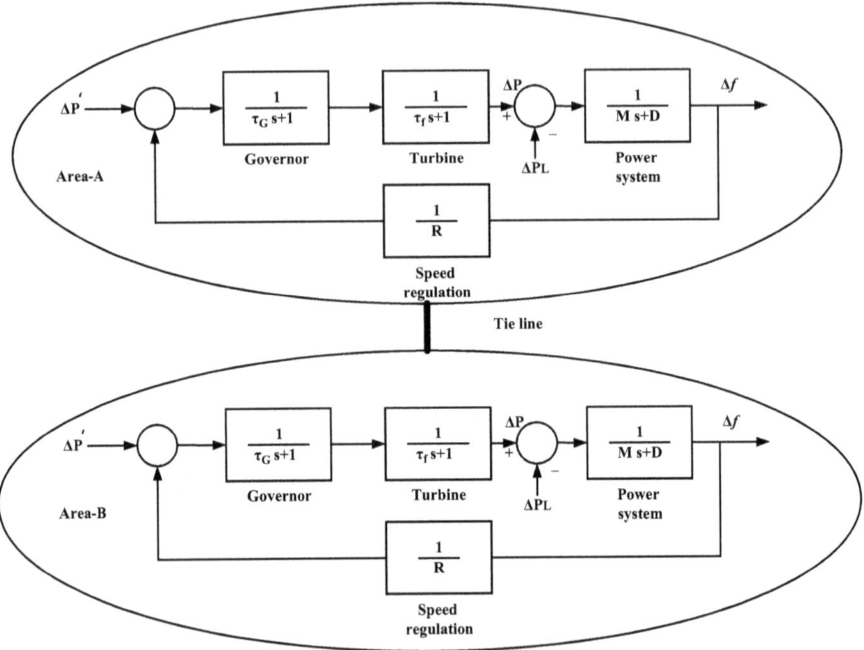

FIGURE 3.9 Block control diagram of two power systems connected by a tie-line.

The response of the governor is given by

$$\Delta P = \frac{-1}{\tau_g\, s+1}\left(\frac{1}{R}\Delta f + \Delta P'\right)$$

$$\Delta P_A = \frac{1}{R_A}\Delta f$$

$$\Delta P_B = \frac{1}{R_B}\Delta f$$

$$\left(D_A + \frac{1}{R_A}\right)\Delta f_A + T_{AB}\left(\delta_A - \delta_B\right) = -\Delta P_{LA}$$

$$\left(D_B + \frac{1}{R_B}\right)\Delta f_B + T_{AB}\left(\delta_B - \delta_A\right) = -\Delta P_{LB}$$

$$\Delta f_A = \Delta f_B = \Delta f$$

$$\left(D_A + \frac{1}{R_A}\right)\Delta f_A + \left(D_B + \frac{1}{R_B}\right)\Delta f_B = -\Delta P_{LA} - \Delta P_{LB}$$

$$\Delta f = \frac{-\Delta P_{LA} - \Delta P_{LB}}{D_A + D_B + \frac{1}{R_A} + \frac{1}{R_B}} \tag{3.39}$$

$$T_{AB}\left(\delta_A - \delta_B\right) = \frac{-\Delta P_{LA}\left(D_A + \frac{1}{R_A}\right) - \Delta P_{LB}\left(D_B + \frac{1}{R_B}\right)}{D_A + D_B + \frac{1}{R_A} + \frac{1}{R_B}}. \tag{3.40}$$

Example 3.8

Find the static frequency drop if the load is suddenly increased by 10 % on a system having the following data:

Rated system capacity = 500 MW.
Operating load = 350 MW.
Inertia constant = 4 sec.
Governor regulation = 4 Hz/PU MW.
Frequency = 60 Hz.
Also, find the additional generation.

Solution
The increase in power $\Delta P_D = 10\% \times 350 = 35$ MW

$$P_D = \frac{350}{500} = 0.7\ p.u.\text{MW}$$

$$D = \frac{P_D}{f} = \frac{0.7}{60} = 0.011667\ p.u.\text{MW/Hz}$$

$$\Delta P_D = \frac{35}{500} = 0.07\,p.u.\text{MW}$$

$$\Delta f = \frac{-\Delta P_D}{D + \frac{1}{R}} = \frac{-0.07}{0.011667 + \frac{1}{4}} = -0.302157\,\text{Hz}$$

$$\Delta f = f_1 - f_0 = -0.302157 = f_1 - 60$$

$$f_1 = 59.6978421\,\text{Hz}$$

$$\Delta P_G = -\frac{\Delta f}{R} = -\frac{-0.302157}{4} = 0.075539\,p.u.\text{MW}$$

$$\Delta P_G = 0.075539 \times 500 = 37.77\,\text{MW}.$$

Example 3.9

Find the static frequency drop if the load is suddenly increased by 25 MW on a system having the following data:

Rated system capacity = 600 MW.
Operating load =350 MW.
Inertia constant = 5 sec.
Governor regulation = 3 Hz/p.u MW.
Frequency = 60 Hz.
Also, find the additional generation.

Solution

$$P_D = \frac{350}{600} = 0.5833\,p.u.\text{MW}$$

$$D = \frac{P_D}{f} = \frac{0.5833}{60} = 9.722 \times 10^{-3}\,p.u.\text{MW/Hz}$$

$$\Delta P_D = \frac{25}{600} = 0.041667\,p.u.\text{MW}$$

$$\Delta f = \frac{-\Delta P_D}{D + \frac{1}{R}} = \frac{-0.041667}{9.722 \times 10^{-3} + \frac{1}{3}} = -0.121457\,\text{Hz}$$

$$\Delta f = f_1 - f_0 = -0.121457 = f_1 - 60$$

$$f_1 = 59.87854\,\text{Hz}$$

$$\Delta P_G = -\frac{\Delta f}{R} = -\frac{-0.121457}{3} = 0.04048\,p.u.\text{MW}$$

$$\Delta P_G = 0.04048 \times 600 = 29.29\,\text{MW}.$$

Example 3.10

Two power systems, A and B, both have 0.1 p.u regulation. A damping factor D of 1 each (on their respective capacity bases). System capacity of system A is 1800 MW and for B is 1600 MW. Both systems are interconnected via a tie-line at 60 Hz. If system A raises the load by 200 MW, calculate the change in the stable frequency and power transfer values.

$$D_A = 1800\frac{\text{MW}}{\text{Hz}}$$

$$D_B = 1600\frac{\text{MW}}{\text{Hz}}$$

$$R_A = 0.1\times\frac{60}{1800} = \frac{1}{300}\text{Hz / MW}$$

$$R_B = 0.1\times\frac{60}{1600} = \frac{3}{800}\text{Hz / MW}$$

$$\Delta f = \frac{-\Delta P}{\left(D_A + D_B + \frac{1}{R_A} + \frac{1}{R_B}\right)}$$

$$\Delta f = \frac{-200}{\left(1800 + 1600 + \frac{1}{\frac{1}{300}} + \frac{1}{\frac{3}{800}}\right)} = -0.0504\text{Hz}$$

$$f_{new} = \Delta f + f_0 = -0.0504 + 60 = 59.949\text{Hz}$$

$$P_{AB} = \frac{-\Delta P\left(D_B + \frac{1}{R_B}\right)}{\left(D_A + D_B + \frac{1}{R_A} + \frac{1}{R_B}\right)}$$

$$P_{AB} = \frac{-200\times\left(1600 + \frac{1}{\frac{3}{800}}\right)}{\left(1800 + 1600 + \frac{1}{\frac{1}{300}} + \frac{1}{\frac{3}{800}}\right)} = -94.11\text{MW.}$$

Example 3.11

Two thermal generating units are operating in parallel, as shown in Figure 3.10, at 60 Hz to supply a total load of 700 MW. Unit 1,600 MW and 4% speed-droop characteristics, supply 400 MW, and Unit 2,500 MW and 4% speed-droop characteristics, supply 300 MW of the load if the total load increases to 800 MW, determine the new loading of each unit and the common frequency change before any supplementary control action occurs.

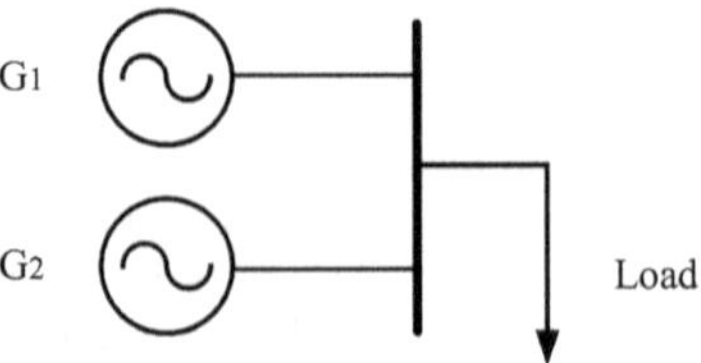

FIGURE 3.10 Power system configuration of Example 3.11.

Unit 1: P_r = 600 MW, R_{u1} = 4% speed drop characteristics, P_{g1} = 400 MW.
Unit 2: P_r = 500 MW, R_{u1} = 4% speed drop characteristics, P_{g2} = 300 MW.

$$P_{load} = 700\,\text{MW}$$

Solution

$$\frac{\Delta f}{f_R} = -\frac{\Delta P}{\left(\frac{P_{R1}}{R_{1u}} + \frac{P_{R2}}{R_{2u}}\right)}$$

$$\frac{\Delta f}{f_R} = -\frac{100}{\left(\frac{600}{0.04} + \frac{500}{0.05}\right)} = -0.004\,p.u$$

For f_R = 60 Hz base frequency, then Δf = – 0.24Hz

$$\Delta f = f_2 - f_1$$

$$-0.24 = f_2 - 60$$

$$f_2 = 59.76\text{ Hz.}$$

3.6.2 Frequency-Bias-Tie-Line Control

Consider three interconnected power systems A, B, and C, as seen in Figure 3.11, of comparable scale. Suppose A and B initially export to C their previously negotiated power transfers. If C increases load, the average frequency decreases, increasing the generation in A, B, and C. This enhances power transitions from A and B to C. However, these transfers are restricted to previously accepted values by the tie-line power controller. Thus instructions are given for A and B to minimize generation, and C is not supported.

This is a serious downside to what is known as straight tie-line control, which can be solved if the devices are managed using both load transfer and frequency since the following equation holds:

$$\Sigma\Delta P + k.\Delta f = 0 \tag{3.41}$$

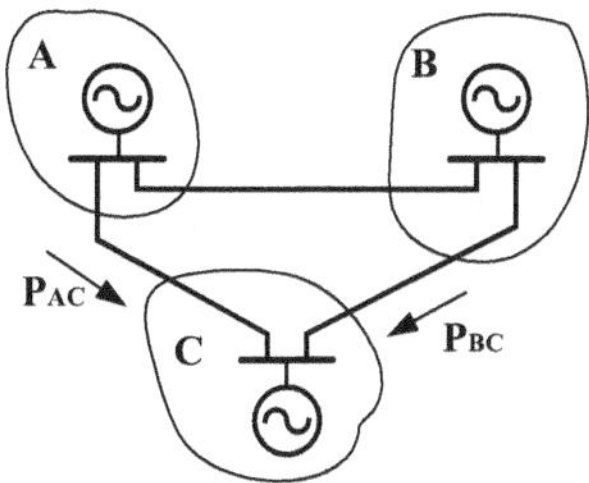

FIGURE 3.11 Three area power tie-lines connected.

where:

- ΔP is the net transfer error and depends on the size of the system and the governor characteristic,
- Δf is the frequency error and is positive for high frequency, and
- k is known as the frequency bias factor.

In the case above, after the load change in C, the frequency error is negative (i.e., Low frequency) for A and B, and the sum of DP for the lines AC and BC is positive.

For correct control,

$$\sum \Delta P_A + k_A \cdot \Delta f = \sum \Delta P_B + k_B \cdot \Delta f = 0 \tag{3.42}$$

Systems A and B take no regulatory action despite dropping frequency. In C, though BC is negative as it imports from A and B, speeder motors in C work to increase output and restore frequency. This method is called frequency-bias-tie-line control and is mostly automatically applied in interconnected systems.

For a network of power in a system

$$\Delta P = \Delta P_{gen} - \Delta P_{load} \tag{3.43}$$

The P–f relation, which is assumed to vary linearly related to frequency, is given by:

$$\frac{\Delta P}{\Delta f} = -k \tag{3.44}$$

K is a constant dependent upon the system's load and governing characteristics in MW/Hz.

$\Delta P + k\Delta f = 0$ for an isolated system.

For the two coupled systems,

$$\Delta P_{total} = \Delta P_1 + \Delta P_2 = -k_1 \Delta f - k_2 \Delta f$$

$$= -(k_1 + k_2) \cdot \Delta f \tag{3.45}$$

System stiffness is the change in power for a given change in the frequency in an interconnected system. The smaller the change in frequency for a given load change, the stiffer the system.

Following the load frequency MATLAB®/Simulation-based on 100 MVA. Figure 3.13 shows the deviation in load power time response.

Example 3.12

The three power systems A, B, and C are connected by a tie-line.K_A = 650, K_B = 500 and K_C = 450 MW/Hz. The tie-line AB and BC trip out of service when the power transfers exceed 300 MW and 120, respectively. At f = 60 Hz.A fault takes place, and line AC is tripped out of services. Calculate the consequent frequencies attained by the two systems.

Solution

After the trip of line AC, 100 MW is now transferred to C through AB and BC.
So, AB carries 350 MW and trips out of service.
For system A: −350 = − 650(f_A − 60)

$$f_A = 60.538\,\text{Hz}$$

For system B+C: 350 = − (500 + 450) (f_{BC} − 60)

$$f_{BC} = 59.63\,\text{Hz}$$

The 350 MW increase in demand is supplied by systems B and C

$$\Delta PB = -500(59.63 - 60) = 185\,\text{MW}$$

$$\Delta PC = -450(59.63 - 60) = 166.5\,\text{MW}$$

As a consequence, the power transfer for C to B increases by

$$166.5 - 100 = 66.5\,\text{MW}$$

And line BC carries 75 + 66.5 = 141.5 MW and will trip.
So, the frequencies for B and C.
For system B: 141.5 = − 500(f_B − 59.63)

$$f_B = 59.347\,\text{Hz}$$

For system C: −141.5 = − 450(f_C − 59.63)
f_C = 59.944 Hz.

Example 3.13

i. Two areas are connected by a tie line operating at a normal state. Area 1 with R_1 and D_1, area-2 with R_2 and D_2, all given in per unit, start from $\Delta f = \frac{-\Delta P_D}{\left(D + \frac{1}{R}\right)}$

Prove that for a change in load in area-1 of ΔP_{L1}, the frequency deviation

$$\Delta f = -\frac{\Delta P_{L1}}{\left(D_1 + D_2 + \frac{1}{R_1} + \frac{1}{R_2}\right)}$$

Also, prove that the tie-power flow deviation is

$$\Delta P_{tie} = \frac{\Delta P_{L1}\left(D_2 + \frac{1}{R_2}\right)}{\left(D_1 + D_2 + \frac{1}{R_1} + \frac{1}{R_2}\right)}$$

If

$$D_1 = 0.8\, p.u, D_2 = 1.0\, p.u, R_1 = 0.01\, p.u, \text{and}\, R_2 = 0.02\, p.u$$

A load change of 0.2 p.u occurs in area-1. What is the new steady-state frequency, and what is the change in tie-line flow? Assume base values of 60 Hz and 100 MVA.

Solution

i. Start from

$$-\Delta P_D = \Delta f\left(D + \frac{1}{R}\right)$$

For area-1

$$+\Delta P_{tie} - \Delta P_{L1} = \Delta f\left(D_1 + \frac{1}{R_1}\right)$$

For area-2

$$-\Delta P_{tie} = \Delta f\left(D_2 + \frac{1}{R_2}\right)$$

Solving to get

$$\Delta f = -\frac{\Delta P_{L1}}{\left(D_1 + D_2 + \frac{1}{R_1} + \frac{1}{R_2}\right)}$$

And

$$\Delta P_{tie} = \frac{\Delta P_{L1}\left(D_2 + \frac{1}{R_2}\right)}{\left(D_1 + D_2 + \frac{1}{R_1} + \frac{1}{R_2}\right)}$$

ii.

$$\Delta f = -\frac{\Delta P_{L1}}{\left(D_1 + D_2 + \frac{1}{R_1} + \frac{1}{R_2}\right)}$$

$$\Delta f = -\frac{0.25}{\left(0.85 + 1.0 + \frac{1}{0.01} + \frac{1}{0.025}\right)} = 1.7624 \times 10^{-3}\, p.u$$

$$\Delta f = -1.7624 \times 10^{-3} \times 60 = -0.1057\,\text{Hz}$$

$$f_{new} = \Delta f + f_0 = -0.1057 + 60 = 59.89425\,\text{Hz}$$

$$\Delta P_{tie} = \frac{\Delta P_{L1}\left(D_2 + \frac{1}{R_2}\right)}{\left(D_1 + D_2 + \frac{1}{R_1} + \frac{1}{R_2}\right)}$$

$$= \frac{0.25 \times \left(1.0 + \frac{1}{0.025}\right)}{\left(0.85 + 1.0 + \frac{1}{0.01} + \frac{1}{0.025}\right)} = 0.02722\, p.u$$

$$\Delta P_{tie} = 0.02722 \times 500 = 36.13\,\text{MW}$$

$$P_{tie-new} = \Delta P_{tie} + P_{tie-old} = 36.13 + 0.25 \times 500 = 161.13\,\text{MW}$$

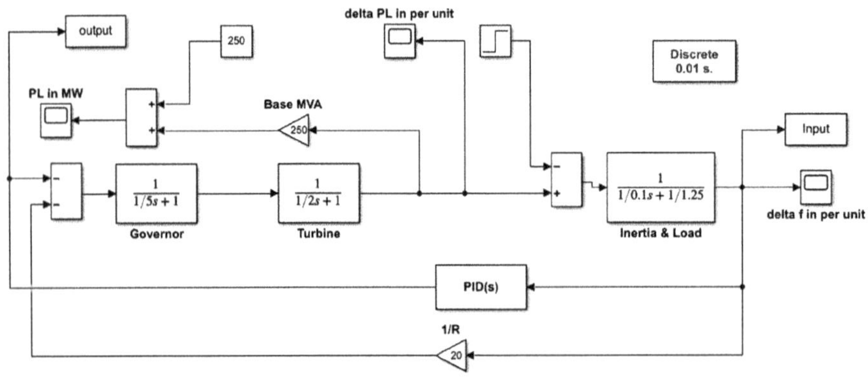

FIGURE 3.12 Load frequency simulation.

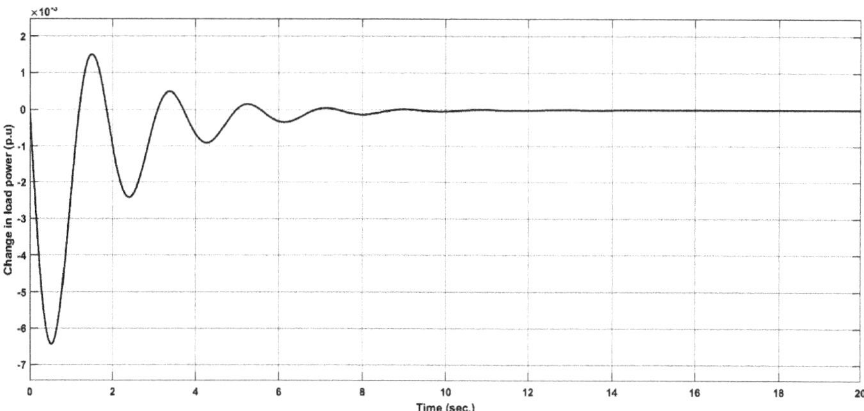

FIGURE 3.13 Load power deviation.

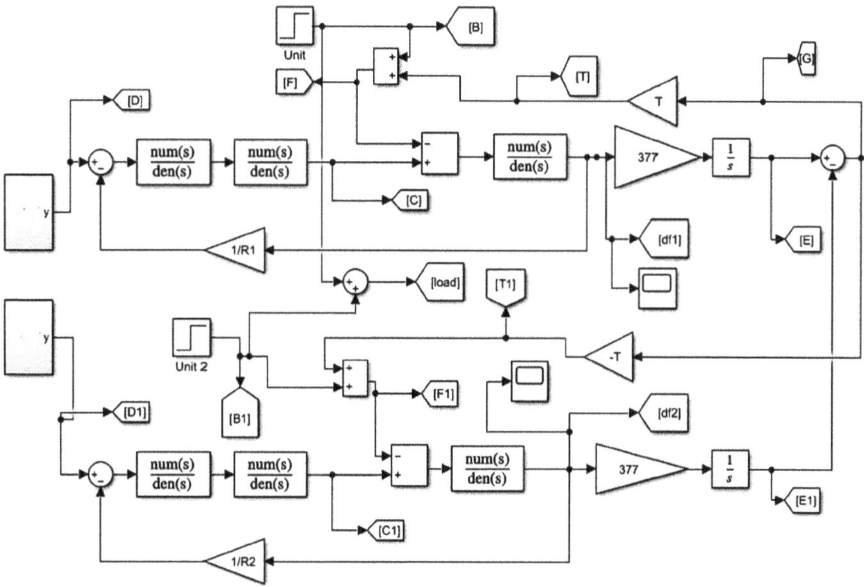

FIGURE 3.14 Load frequency simulation of two areas.

3.7 LOAD SHEDDING

If a complete blackout in the electrical network occurred due to the frequency dropping below 50 Hz or 60 Hz and the load shedding relay devices not working, which lead to the exit of the generating stations in succession.

Devices to remove the load from the protection devices are very important in the stability of the electrical network, and the purpose of using it: "is to get out of some loads from the network when there is a decrease in the frequency so that we do not reach the stage of complete darkness."

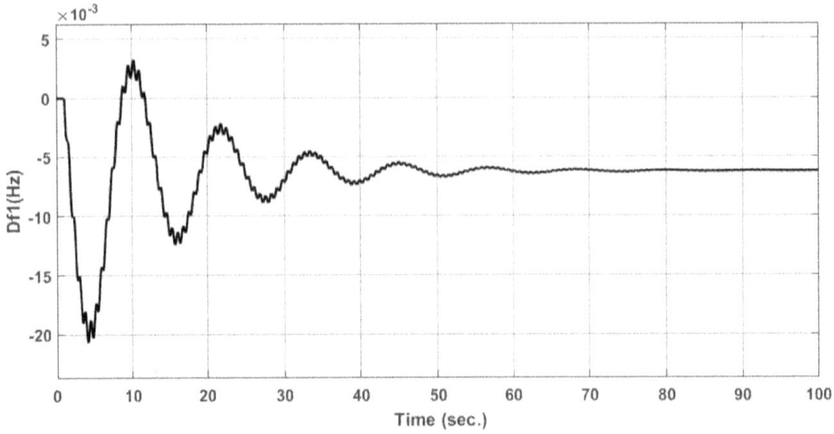

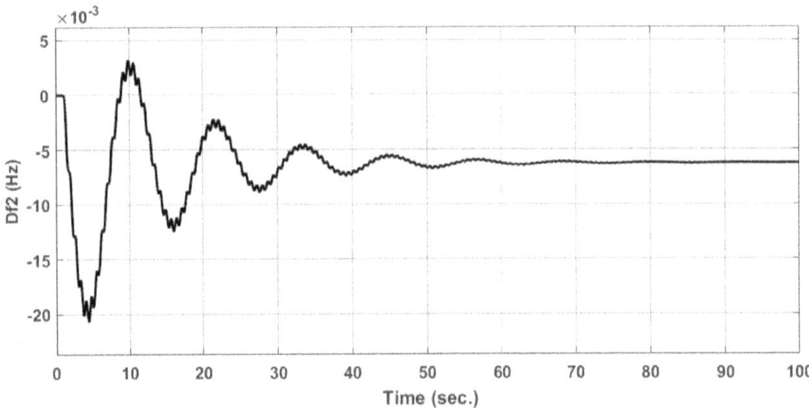

FIGURE 3.15 Frequency deviation response of load frequency simulation of two areas.

The stages of pregnancy shedding load for nominal frequency 50 Hz are as follows:

1. The first stage: This stage prefers 6% of the value of the network loads at a frequency of 49.2 cycle/s, and loads of this stage are the rural area loads.
2. The second stage: This stage prefers 3% of the network loads' value at a frequency of 49.1 cycle/s, and these loads are rural areas close to the city.
3. The third stage: This stage prefers 4% of the network loads' value at a frequency of 49 cycles/s and may reach loads of the city's edges.
4. The fourth stage: This stage prefers 7% of the network loads' value at a frequency of 48.9 cycles/s, and loads of this stage are completed by the city's outskirts.

5. The fifth stage: This stage favors 20% of the network loads' value at a frequency of 48.8 cycles/s, and small city loads enter this stage.
6. The sixth stage: This stage prefers 20% of the network loads' value at a frequency of 48.7 cycles/s. Large city loads enter this stage.

Some loads in the electrical network are outside load disposal devices such as hospitals and airports. The loads return gradually according to the control instructions when the frequency is stabilized.

PROBLEMS

3.1 A turbo-alternator, rated of 600 MVA, two poles deliver 500 MW to a 60 Hz system. If the generator circuit breaker is suddenly opened, and the main steam valves take 350 ms to operate, what will the generator's over-speed? The stored energy of the machine (generator and turbine) at synchronous speed is 8 kW.s/kVA

3.2 Two identical 60 MW synchronous generators operate in parallel. The governor settings on the machines are such that they have 4% and 6% droops (no-load to full-load percentage speed drop).

Determine

i. each machine takes the load for a total of 100 MW;
ii. the percentage adjustment in the no-load speed is to be made by the speeder motor if the machines share the load equally.

3.3

i. Explain how the output power operating in a constant frequency system is controlled by adjusting the governor's setting. Show the effect on the generator power-frequency curve.
ii. Generator A of 250MW and generator B of 400 MW have governor droops of 5% and 8%, respectively, from no-load to full-load. They are the only supply to an isolated system whose nominal frequency is 60 Hz.
iii. The corresponding generator speed is 3000 r.p.m. Initially, generator A is at 0.55 p.u. Load and generator B is at 0.75 p.u load, and both are running at 60 Hz. Find the no-load speed of each generator if it is disconnected from the system. Also, determine the total output when the first generator reaches its rating.

3.4 Two power systems A and B are interconnected by a tie-line and have power frequency constants KA and KB MW/Hz. An increase in the load of 600 MW on system A causes a power transfer of 360 MW from B to A. System A's frequency is 59.4 Hz when the tie-line is open, and system B 60 Hz. Determine the values of KA and KB, deriving any formulae used.

3.5 Two power systems, A and B, are connected as shown in Figure 3.12, having 3200 MW and 2200 MW, respectively, are interconnected through a tie-line. Both operate with frequency bias- tie-line control. The frequency bias for each area is 1% of the system capacity per 0.1 Hz frequency deviation. Calculate the steady-state change in frequency in Hz if the tie-line interchange for A is set at 200 MW and for B is set (incorrectly) at 300 MW calculate the steady-state change in frequency.

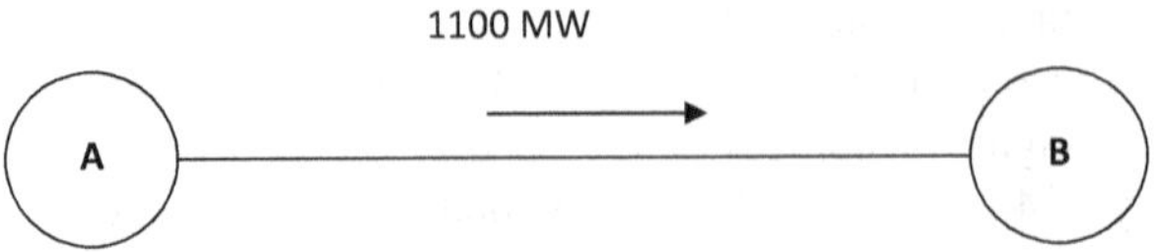

FIGURE 3.16 Interconnected power system of Problem 3.5.

3.6a.

i. Why do power systems operate in an interconnected arrangement?
ii. How is the frequency controlled in a power system?
iii. What is meant by the stiffness of a power system?

b. Two 50 Hz power systems are interconnected by a tie-line, which carries 1100 MW from system A to system B, as shown in Figure 3.16. After the outage of the line shown in the figure, system A's frequency increases to 50.67 Hz, while system B's frequency decreases to 49.4 Hz.

i. Calculate the stiffness of each system.
ii. If the systems operate interconnected with 1150MW being transferred from A to B, calculate the line's flow after an outage of a 650MW generator in system B.

3.7 a.

i. Why do power systems operate in an interconnected arrangement?
ii. What is meant by the stiffness of a power system?

b. Two 50 Hz power systems are interconnected by a tie-line, which carries 1000 MW from system A to system B, as shown in figure. After the outage of the line shown in the figure, the frequency in system A increases to 50.6 Hz, while the frequency in system B decreases to 49.2 Hz.

i. Calculate the stiffness of each system.
ii. If the systems operate interconnected with 1000 MW being transferred from A to B, calculate the line's flow after the outage of a 620 MW generator in system ***B***.

3.8 a. Why do power systems operate in an interconnected arrangement?

b. What is meant by the stiffness of a power system?

c. Prove that: $f_1 = f_0\sqrt{\left[\left(W_0 + W_1\right)/W_0\right]}$ and $\Delta f = -\Delta P/\beta$

where β: area frequency response characteristic.

3.9 A 50 Hz power system consists of two interconnected areas. Area-1 has 1500 MW of total generation and an area frequency response characteristic β_1 = 600 MW/Hz, area-2 has 3800 MW of total generation, and β_2 = 1200 MW/Hz. Each area initially generates one-half of its total generation at $\Delta P_{tie} = 0$ and at 60 Hz when the load at area-1 suddenly increases by 150 MW. Determine the steady-state frequency change and the steady-state tie-line power change ΔP_{tie} of each line, neglecting losses and dependence of load on frequency. Assume load frequency controlled status OFF.

3.10 A 500 MVA, two poles, turbo-alternator delivers 400 MW to a 50 Hz system. If the generator circuit breaker is suddenly opened and the main steam valves take 400 ms to operate, what will be the generator's over-speed? The

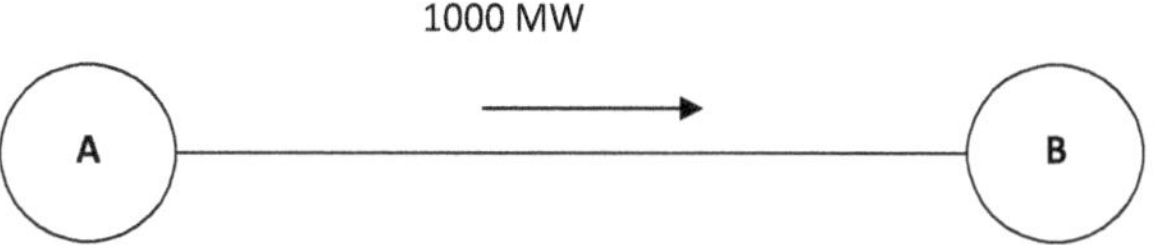

FIGURE 3.17 System configuration of Problem 9.7.

machine's stored energy (generator and turbine) at synchronous speed is 3.8 kW.s/kVA.

3.11 Two identical 60 MW synchronous generators operate in parallel. The governor settings on the machines are such that they have 5% and 4% droops (no-load to full-load percentage speed drop). Determine (a) the load taken by each machine for a total of 150 MW; (b) the percentage adjustment in the no-load speed to be made by the speeder motor if the machines share the load equally.

3.12 Two synchronous generators are operating in parallel. Their capacities are 300 MW and 400 MW. The droop characteristics of their governors are 3% and 5% from no load to full load. Assuming that the generators are operating at 50 HZ at no load, how would be a load of 580MW shared between them. What will be the system frequency at this load? Assume free governor action.

3.13 A two power area systems, A and B, are connected. Each has power/frequency characteristics of

$$\frac{\Delta P_A}{\Delta f_A} = -K_A = 4500\,\text{MW/Hz}$$

$$\frac{\Delta P_B}{\Delta f_B} = -K_B = 5500\,\text{MW/Hz}$$

The two systems are connected via a tie line; the load on system B is increased by 150 MW. Determine the resultant reduction in frequency and change in power transfer through the tie line.

3.14 Two synchronous generators operating parallel are rated 120 MW and 250 MW, respectively. Each generator has 4% and 3% governor droop characteristics, respectively, from no load to full load. Assuming free governor action. Calculate

i. The load is taken by each machine for a total load of 200 MW.
ii. The system frequency at two loads.
iii. Repeat i and ii above if both generators have a drop of 4% in pf frequency.

4 Voltage and Reactive Power Control

The generating station's electrical energy is distributed to the industrial or residential customer through a transmission and distribution network in a modern power grid. The customer's main concern is reliable power within the given voltage and current specifications. The last part of a power transmission system the customer sees is converting voltage from a high voltage level to a level that can be used within the home or industrial facility. Figure 4.1 is the last leg of a power system with a power transformer and attached overcurrent protective system.

To ensure a satisfactory operation of generators, lamps, and other loads, customers should be supplied with a significantly constant voltage within specified higher or lower voltage limitations. Significant voltage fluctuations can cause irregular activity or even malfunctioning of consumer appliances. The regulatory voltage deviation maximum is ± 5% of the declared voltage at the user terminal. Electrical voltage management is critical for the proper functioning of electrical power equipment to avoid damage, such as overheating generators and motors, minimize transmission losses, and preserve the system's ability to withstand and prevent voltage collapse.

This chapter presents the concept of voltage and reactive power control. The voltage control methods and the relation between voltage, power, and reactive power at a node are presented in this chapter. Also, the reactive power compensation and harmonic effect on load balancing are presented in this chapter.

4.1 TYPES OF VOLTAGE VARIATION

The principal cause of voltage variation in the user's premises is a difference in the load on the delivery grid. As the device load rises, the voltage at the user terminal decreases due to a high voltage drop. The voltage drop in power systems is due to the following:

i. The synchronous impedance of the alternator.
ii. The impedance of the transmission line.
iii. The impedance of the transformer.
iv. Distributions and current limiting reactors.

To ensure effective and stable operation of the power grid, voltage and reactive power management should fulfill the following objectives:

i. Voltages at all terminals of all devices in the device are within reasonable limits.
ii. The reliability of the system is improved to optimize the use of the transmission system.
iii. The reactive power flow is minimized to reduce I^2R and I^2X losses.

DOI: 10.1201/9781003293965-4

FIGURE 4.1 Power distribution. The last leg of a power system with a power transformer and attached overcurrent protective system.

Almost all electricity transported in alternating current (AC) networks supplies or absorbs two powers: real power and reactive power. Real power does useful work while reactive power supports the voltage regulated for device stability. Reactive control is necessary to deliver active power to the consumer via the transmission and distribution grid. The voltage and current pulsate at the device's frequency for AC systems. Simultaneously, AC voltage and current pulsate are the algebraic product of voltage and current peaks at different times. Real power is the total power determined by watts. A part of the power with zero average value, reactive power, is calculated in volt-ampers reactive, or VARs.

The estimated relationship between the value of the voltage difference between two nodes in the network and the movement of power is

$$\Delta V \approx \Delta V_P = \frac{RP + XQ}{V} \tag{4.1}$$

Also, it was shown that the transmission angle δ is proportional to

$$\delta \propto V_q = \frac{XP - RQ}{V} \tag{4.2}$$

Consider the transmission line that links two A and B generating stations, as shown in Figure 4.2(a).

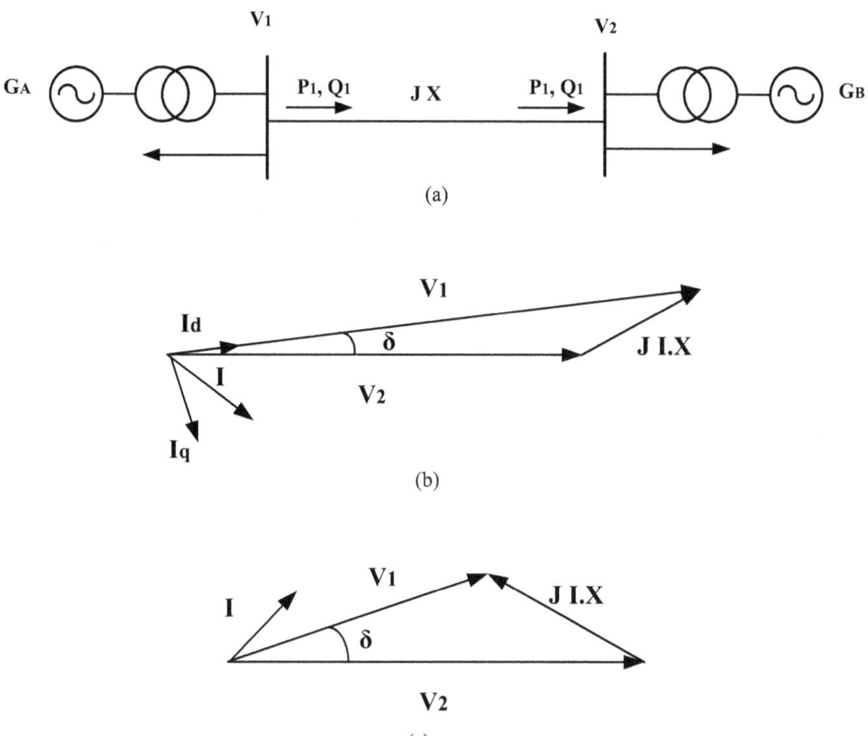

FIGURE 4.2 (a) System of two generators interconnected; (b) Phasor diagram when $V_1 > V_2$. I_d and I_q are components of I. (c) Phasor diagram when $V_2 > V_1$.

If the system is initially only considered reactive, and R is neglected. The amount of voltage angle at machine A is greater of that at machine B; therefore, the real power flow from A to B. The reactive power given for receiving bus is given by

$$Q_2 = \frac{V_2 \Delta V}{X} \tag{4.3}$$

This can be shown from the schematic of phases seen in Figure 4.2(b) when the resistance of the transmission line is ignored as compared in value with the reactance of the transmission line. It is stated that I_d and the voltage at sending end are greater than receiving end voltage (i.e., $V_1 > V_2$) due to voltage drop across the transmission line's reactance connected between the two busses. Therefore, Q will be at a loss at this reactance. For $V_1 > V_{2,}$ the reactive power is shifted from A to B in this situation. As seen in Figure 4.2(c), the reactive power direction is reversed by modifying the generator excitations such that $V_2 > V_1$. Thus, actual power can be sent from A to B or B to A by altering the amount of steam or water) admitted to the turbine accordingly, and reactive power can be sent by changing the voltage magnitudes in either direction.

This is a convenient way to explain the effect of the transferred current's power factor. While it might seem new initially, the ability to think in terms of VAR flows would make it much simpler to analyze power networks rather than only with power factors and phasor diagrams if Q_2 in the scheme in Figure 4.2(a) can be arranged to be zero, there will be no voltage decrease between A and B, a very satisfactory state of affairs.

Suppose a resistance to the interconnecting mechanism is seen in Figure 4.2(a) and that V_1 is constant. Take into account the effect of keeping V_2, and hence the ΔV voltage decrease, steady.

$$Q_2 = \frac{V_2 \Delta V - RP_2}{X} = K - \frac{R}{X} P_2 \tag{4.4}$$

where K is a constant and R is the resistance of the system. If this value of Q_2 does not exist naturally in the circuit, it will have to be obtained by artificial means, such as the connection at B of capacitors or inductors.

$$Q_2' - Q_2 = \frac{R}{X}\left(P_2' - P_2\right) \tag{4.5}$$

An increase in actual power allows V_2 to be maintained by the reactive power available. However, the shift is proportional to (R/X), which is normally minimal.

4.2 REACTIVE POWER GENERATION AND ABSORPTION

Following are the types of reactive power generation and absorption in power systems.

4.2.1 Synchronous Reactance of Synchronous Generators

To produce or consume reactive electricity, synchronous generators may be used. An under-excited machine consumes it; an over-excited machine, that is, one with greater than negligible excitation, produces reactive power. The key cause of the supply of both positive and negative VARs to the power grid is synchronous generators. An output of a synchronous generator shows the ability to produce or consume reactive electricity. Until the rotor currents lead to overheating, reactive power generation (lagging power factor operation) is restricted by the maximum excitation voltage permitted.

4.2.2 Transformers and Overhead Lines

Overhead lines consume reactive power when fully loaded. The absorbed VARs are I^2X per step where I is the current flowing in a line of reactance per stage X(V). The shunt capacitance of longer lines will become dominant on light loads.

Overhead lines with high voltage then become generators of VAR.

Transformers often absorb reactive power. A helpful expression for the amount of reactance of the transformer X_T in per unit can be obtained for the total load of $3V_f.$ I_{rated}. The Ohmic reactance

$$X = X_T \frac{V_\phi}{I_{rated}} \tag{4.6}$$

Therefore,

$$Q_{absorbed} = 3 \times I^2 \times X_T \frac{V_\phi}{I_{rated}} \tag{4.7}$$

$$= 3 \times I^2 \times X_T \frac{V_\phi}{I_{rated}} \times \frac{3.V_\phi}{3.V_\phi} \tag{4.8}$$

$$= \frac{\left(3 \times I.V_\phi\right)^2}{3 \times I_{rated}.V_\phi} X_T \tag{4.9}$$

$$= \frac{\left(Load\,VA\right)^2}{Transformer\ rated} X_T. \tag{4.10}$$

4.2.3 Cables

Cables are generators of reactive power due to their high shunt capacitance, and absorbs of reactive power due to inductance specially in case long cables.

4.2.4 Loads

In planning a network, it is desirable to assess the reactive power requirements to ascertain whether the generators can operate at the required power factors for the load's extremes to be expected. For example, a load of 0.95 pf absorbs a reactive power demand of 0.33 kVAR per kW of power.

Example 4.1

In the radial transmission system shown in Figure 4.3, all per unit values are referred to as the voltage bases shown and 100 MVA. Determine the power factor at which the generator must operate.

Data: Eech transformer reactane is 0.05 p.u
Each transmission line reactance is 0.1 p.u
Load 1:60 MW, unity power factor.
Load 2:200 MW,0.8 lagging power factor.

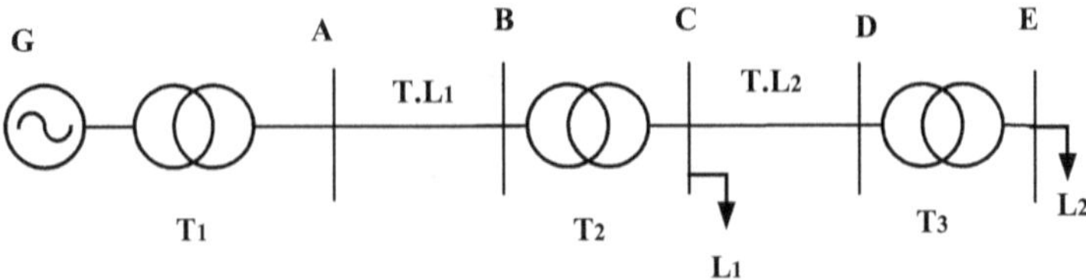

FIGURE 4.3 Radial transmission system with intermediate loads of Example 4.1.

Solution

The voltage decreases will be ignored, and the nominal voltages assumed. The VARs for each part of the circuit are inserted to obtain the complete busbar A. Starting with the load,

$$VAR\,losses\ between\,C\ to\,E = \frac{P^2+Q^2}{V^2} = \frac{0.6^2}{1.0^2}\times(0.1+0.05) = 0.054\,p.u$$

$$Active\ Power\ in\,AC = 2.0+0.6 = 2.6\,p.u$$

$$Reactive\ power\ in\,AC = 2.0\times0.6+0.054 = 1.254\,p.u$$

$$VAR\,losses\ between\,A\,to\,C = \frac{P^2+Q^2}{V^2} = \frac{2.6^2+1.254^2}{1.0^2}\times(0.1+0.05)$$
$$= 1.25\,p.u$$

The losses in the large generator-transformer may be neglected
So

$$The\ power\ generated\ from\ the\ generator = 2.6+j2.504\,p.u$$

$$p.f = \frac{P}{\sqrt{P^2+Q^2}} = \frac{2.6}{\sqrt{2.6^2+2.504^2}} = 0.72\,lagging.$$

4.3 RELATION BETWEEN VOLTAGE, POWER, AND REACTIVE POWER AT A NODE

The voltage V at a node is a function of P and Q at that node as in Figure 4.4.

$$V = f(P,Q) \tag{4.11}$$

The voltage also depends on adjacent nodes, and the present treatment assumes that these are infinite busbars.

The total differential of the load voltage V,

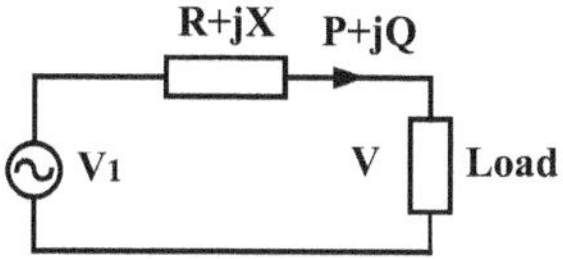

FIGURE 4.4 The voltage V at a node is a function of P and Q at that node.

$$dV = \frac{\partial V}{\partial P}.dP + \frac{\partial V}{\partial Q}.dQ \quad \frac{\partial P}{\partial V}\frac{\partial V}{\partial P} = 1 \tag{4.12}$$

$$\text{and} \quad \frac{\partial Q}{\partial V}\frac{\partial V}{\partial Q} = 1$$

$$dV = \frac{dP}{\frac{\partial P}{\partial V}} + \frac{dQ}{\frac{\partial Q}{\partial V}} \tag{4.13}$$

From Figure 4.4

$$\left(V_1 - V\right)V - PR - XQ = 0 \tag{4.14}$$

$$\frac{\partial P}{\partial V} = \frac{\left(V_1 - 2V\right)}{R} \tag{4.15}$$

$$\frac{\partial Q}{\partial V} = \frac{\left(V_1 - 2V\right)}{X} \tag{4.16}$$

$$dV = \frac{dP}{\frac{\partial P}{\partial V}} + \frac{dQ}{\frac{\partial Q}{\partial V}} = \frac{R.dP + X.dQ}{\left(V_1 - 2V\right)} \tag{4.17}$$

$$\frac{\partial Q}{\partial V} = \frac{\left(V_1 - 2V\right)}{X} \tag{4.18}$$

assuming

$$R \ll X$$

$$I = \frac{V_1}{X}. \tag{4.19}$$

Example 4.2

As shown in Figure 4.5 are three supply points A, B, and C that are connected to a common busbar D. If, at a particular system load, the line voltage of D falls below its nominal value by 5 kV, calculate the magnitude of the reactive volt-ampere injection required at D to re-establish the original voltage.

The p.u values are expressed on a 500MVA and 220 kV base at common busbar D, and resistance may be neglected throughout.

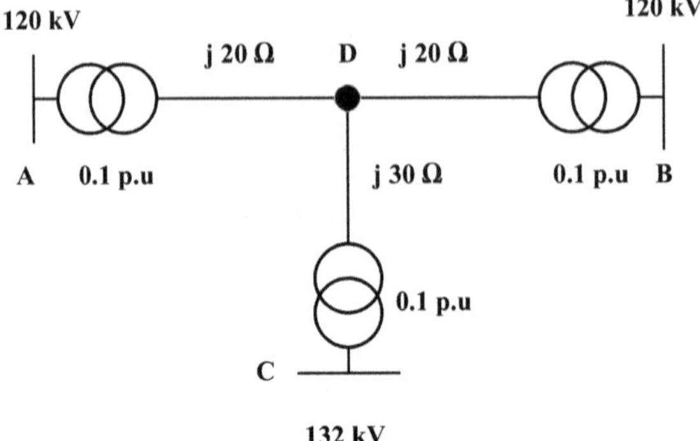

FIGURE 4.5 Power system configuration of Example 4.2.

Solution

The reactance diagram for the power system is given in Figure 4.6(a), and the equivalent Thevenin single-phase circuit is shown in Figure 4.6(b). It is necessary to determine the value of dQ/dV at the node or busbar D; hence the current flowing into a three-phase short-circuit at D is required.

$$Z_{base} = \frac{V_b^2}{S_b} = \frac{220^2}{500} = 98.8\Omega$$

$$X_{TL\,p.u} = \frac{X_{given}}{Z_{base}} = \frac{20}{98.8} = j0.2066\,p.u$$

$$X_{TL\,p.u} = \frac{X_{given}}{Z_{base}} = \frac{30}{98.8} = j0.3036\,p.u.$$

(a) Reactance diagram. (b) Equivalent Thevenin circuit.

$$X_{eq.} = j0.1018\,p.u$$

$$S_{s.c} = \frac{S_b}{X_{eq.}} = \frac{500}{0.1018} = 4911.59\,\text{MVA}$$

$$I_f = \frac{S_{s.c}}{\sqrt{3}.V_f} = \frac{4911.59}{\sqrt{3}\times 220} = 12.89\,\text{kA}$$

$$\frac{\partial Q}{\sqrt{3}\,\partial V} = I_f$$

$$\frac{\partial Q}{\partial V} = \sqrt{3}\,I_f = \sqrt{3}\times 12.89 = 22.32\,\text{MVAr/kV}$$

The potential drop at bus D is 5 kV, so

$$\partial Q = 22.32\times 5 = 111.62\,\text{MVAr}.$$

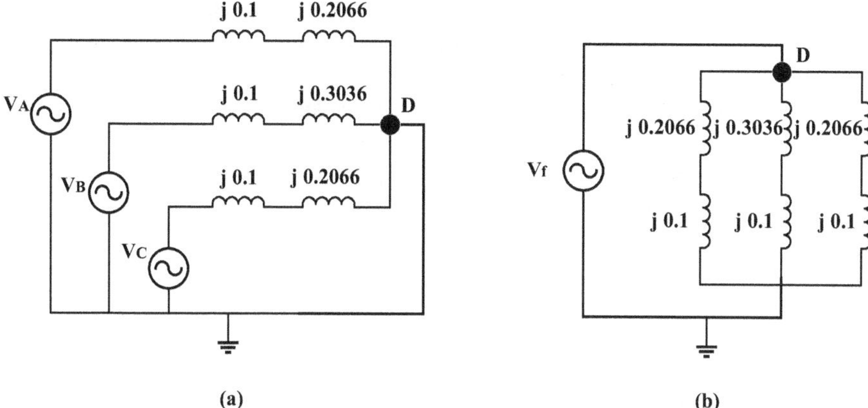

FIGURE 4.6 Reactance diagram of Example 4.2.

4.4 METHODS OF VOLTAGE CONTROL

The methods of voltage control in the power systems are as follows.

4.4.1 Injection of Reactive Power

Busbar voltages may be regulated by injection or absorption of reactive power in $X \gg R$ transmission systems. However, in distribution networks where higher circuit resistances contribute to reactive power flows having less impact on voltage and causing a rise in actual power losses, regulating network voltage through reactive power flow is less effective.

Although reactive power does not achieve real work, it increases the magnitude of the network current and the real power losses. Via imposing charges based on kVAh (or even kVARh) in addition to kWh or even basing part of the fee on peak kVA drawn, energy providers also penalize loads with a low power factor. It has long been established that static capacitors are provided to enhance the power factors of factory loads. The capacitance necessary for the load enhancement of an optimal economy's power factor is calculated.

The consumer's tariff is focused on both kVA and kWh as:

$$charge = \$A \times kVA + \$B \times kWh \tag{4.20}$$

A load of P kilowatts has a lagging power factor ($\cos\varphi_1$). As this power factor is increased to ($\cos\varphi_2$), the saving due to increasing of power factor is

$$saving = \$.PA\left(\frac{1}{\cos\varphi_1} - \frac{1}{\cos\varphi_2}\right) \tag{4.21}$$

It is interesting to note that the optimum power factor is independent of the original one. In such a way, the increase of load power factors would help mitigate The whole VAR flow dilemma in the distribution chain.

In general, four reactive power injection methods are available, including the utilization of the following:

1. Static shunt capacitors.
2. Static series capacitors.
3. Synchronous compensators.
4. SVCs and STATCOMs.

4.4.1.1 Reactors and Shunt Capacitors

Shunt capacitors are used to compensate for a lagging power factor, when reactors are used on VARs generating circuits such as lightly filled cables. The effect of these shunt devices are to supply or absorb the required reactive power to sustain the voltage's magnitude. Capacitors are directly connected to a busbar or a primary transformer's tertiary winding. Often installed along the routes of distribution circuits to minimize the losses and voltage drops.

But as the voltage reduces, the reactive power produced by a shunt capacitor or absorbed by a reactor falls as the square of the voltage. With light network load when the voltage is high, the capacitor output is large, and the voltage tends to rise to excessive levels, requiring some capacitors or cable circuits to be switched out by local overvoltage relays.

4.4.1.2 Series Capacitors

Capacitors can be linked with overhead lines in series and are then used to reduce inductive reactance between the supply point, and the load. One major downside of series capacitors is the high overvoltage caused by a short circuit through the capacitor.

The current passes through the circuit, and special safety equipment (e.g., spark gaps) and non-linear resistors must be integrated. For a line with a series capacitor, the phasor diagram is shown in Figure 4.7.

It is important to summarize the relative merits between the shunt and series capacitors as follows:

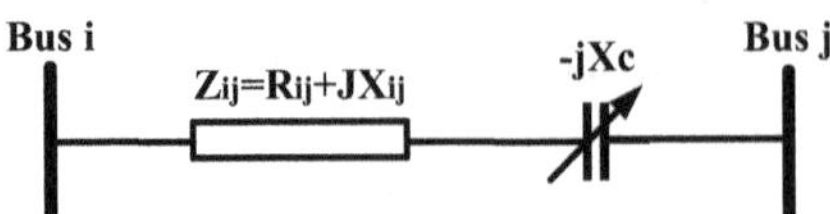

FIGURE 4.7 Model of transmission line with TCSC.

1. Series capacitors are of little benefit for the minimal load VAR requirement.
2. The reduction in line current is limited with series capacitors; thus, if thermal considerations limit the current, little benefit is gained, and shunt compensation achieved should be used.
3. If the limiting factor is a voltage drop, so using a series capacitors is effective. Fluctuations are equalized attributable to arc furnaces, as an example.
4. When the overall reactance of the line is high, series condensers are very efficient in reducing

Voltage decreases and system stability increases.

It is important to add both shunt and series capacitors with caution since they can both lead to resonance with the power system's inductive reactance. Figure 4.8 shows a capacitor connection in series with the transmission line.

From Figure 4.8(b) and the phasor diagram in Figure 4.8(c),

$$-E + I_a\left(R_a + jX_s\right) + V = 0$$

$$V = E - I_a\left(R_a + jX_s\right) \tag{4.22a}$$

The terminal voltage at the capacitor

$$V_1 = V - I_a\left(R + jX\right) \tag{4.22b}$$

and the receiving end voltage:

$$V_R = V_1 - I_a\left(-jX_c\right). \tag{4.22c}$$

4.4.1.3 Synchronous Compensators

The synchronous compensator is a synchronous motor operating without a mechanical load that produces or consumes reactive power based on excitation value. The power factor is not zero compared to static capacitors because the losses are considerable. When used with a voltage regulator, the compensator will automatically operate over-excited at heavy load times and under-excited at light load.

A standard installation of a synchronous compensator linked to the tertiary (delta) winding of the main transformer is seen in Figure 4.8(a). The Automatic Voltage Regulator (AVR) on the compensator is regulated by a combination of the voltage and current output system, resulting in a decrease in the voltage-VAR output curve that can differ, as seen in Figure 4.7, for the appropriate relation of the synchronous compensator and the related Volt-VAR output characteristic in Figure 4.10. In 2.5 minutes and then synchronized, the compensator is run-up as an induction motor.

The simplicity of operation with all load conditions is a great benefit. While the expense of such installations is high, it is justified under some situations, for example, at the receiving-end busbar of a long high-voltage line where it is difficult to

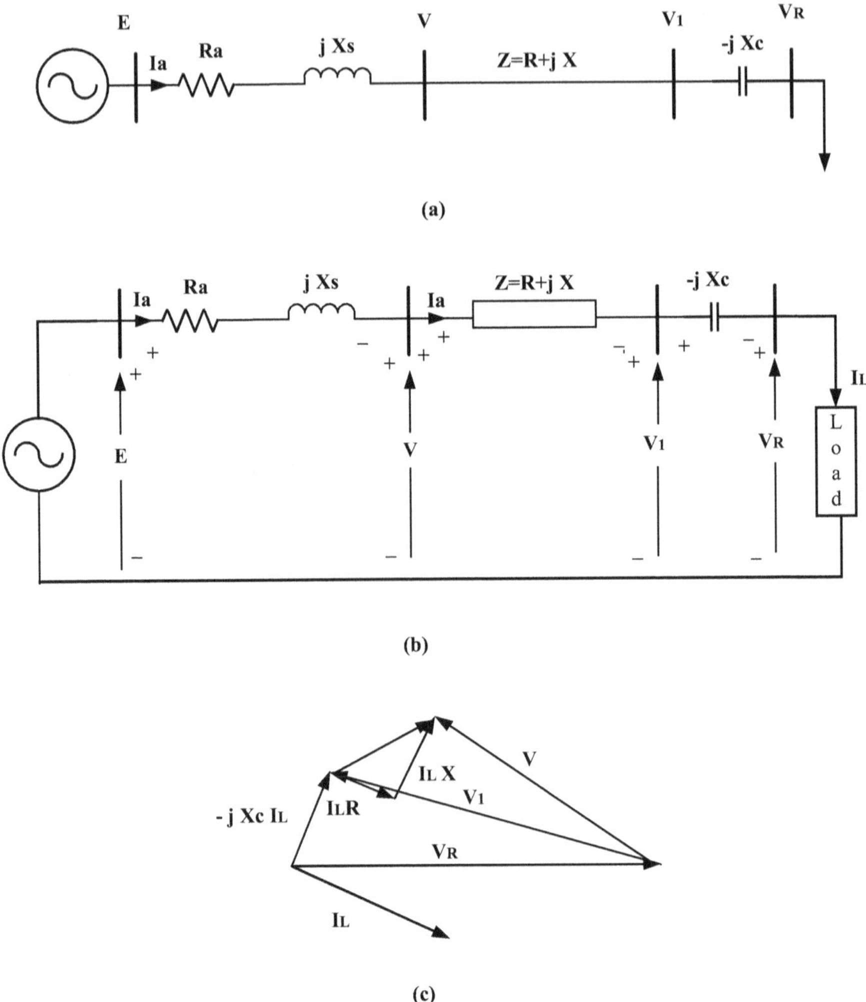

FIGURE 4.8 Capacitor connection in series with the transmission line. Power system schematic diagram. (b) Reactance diagram; (c) Phasor diagram.

accept transmission at power factors smaller than the unit. As a revolving engine, the stored energy is valuable for increasing the power system's inertia and driving through intermittent disruptions, including voltage drops.

4.4.1.4 Static Reactive Compensators and Static Synchronous Compensators

Synchronous compensators are revolving devices, costly, and are prone to a mechanical failure. Thus, electronic compensators are gradually replacing them: SVCs and STATCOMs. SVCs use thyristor-controlled shunt reactors and

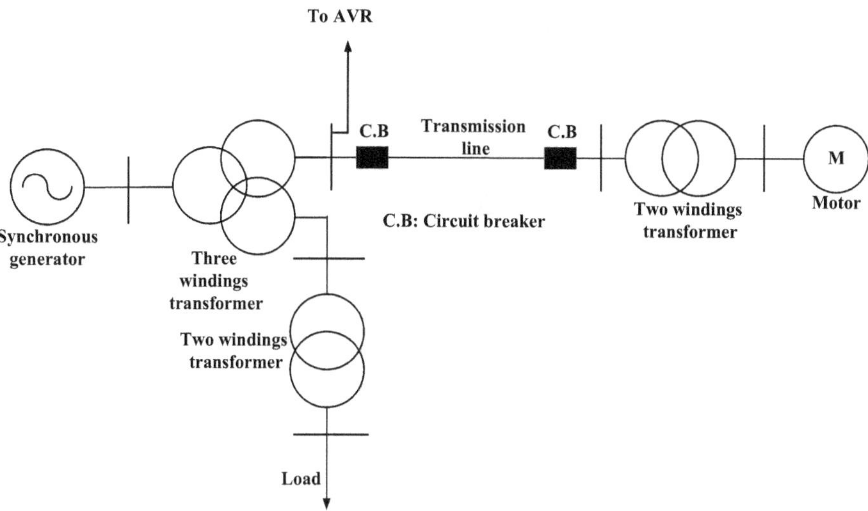

FIGURE 4.9 Synchronous compensator connection to the power system.

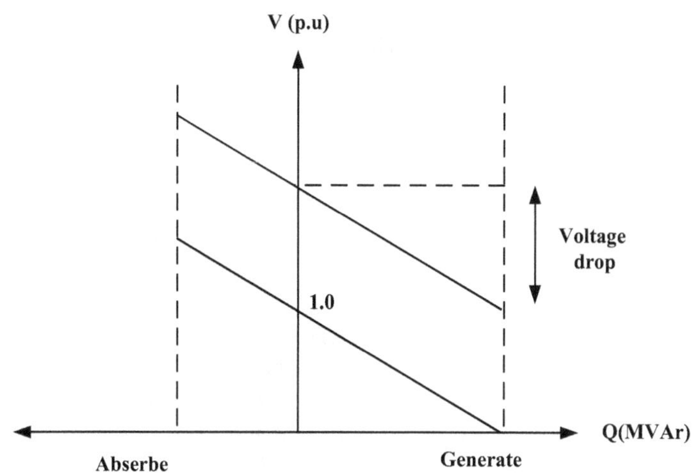

FIGURE 4.10 Voltage-reactive power output characteristics.

condensers. The reactive power is given by shunt elements (capacitors and inductors). Figure 4.10 presents voltage-reactive output of a synchronous 40MVAR compensator.

The Thyristor-Controlled Reactor (TCR) is regulated throughout the 50/60 Hz interval by delaying the thyristor turn ON time. When the current decreases to zero, the thyristor turns off. The thyristor's firing angle can be varied within each loop, and the TCR can regulate the VAR absorption. TCRs can be used with mechanically or

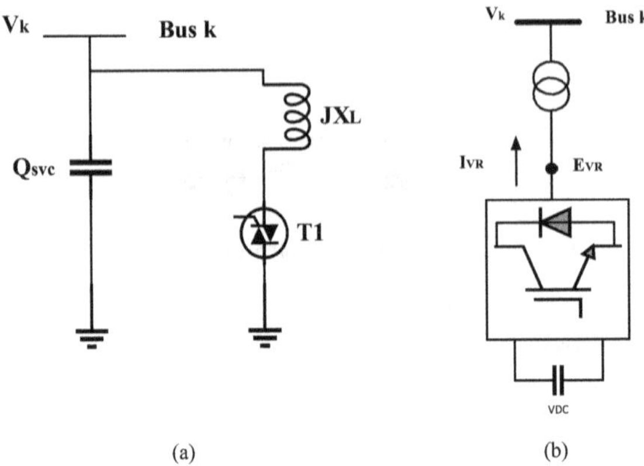

FIGURE 4.11 SVC firing angle model and STATCOM model. (a) Possible combinations of controlled reactors and capacitors, forming an SVC; (b) STATCOM model.

thyristor-switched capacitors to create an SVC to export and import VARs, as seen in Figure 4.11(a).

Rather high currents will flow when a capacitor is linked to a high voltage source. Thyristor Switched Capacitors (TSCs) are also worked only in integral intervals, and the thyristor operation is timed to switch when there is no instantaneous voltage around the condenser.

A STATCOM (Static Compensator) is also a reactive control electronic system, but it works on another principle (Figure 4.11(b)) STATCOM A.

It consists of a voltage source converter (VSC) attached to the power grid by a coupling reactance (L). The VSC utilizes very large transistors to synthesize a voltage sine wave of any magnitude and phase that can be switched on and off.

$V_{STATCOM}$ is a sinusoidal waveform of 50/60 Hz retained in phase with terminal voltage of $V_{terminal}$. If the magnitude of STATCOM voltage $V_{STATCOM}$ is greater than that of the $V_{terminal}$, STATCOM produces reactive power. In comparison, if the magnitude of $V_{STATCOM}$ is less than that of $V_{terminal}$, STATCOM consumes reactive power. Between $V_{STATCOM}$ and $V_{terminal}$, a very limited step angle is added. A small volume of actual power flows through the STATCOM to charge the DC capacitor to provide for the converter losses. However, the operation theory is that two voltage magnitudes around the reactor provide reactive power. The DC capacitor is only used to regulate the ripple current and run the power electronics.

4.4.2 Tap-Modifying Transformers

The transition ratio adjusts the voltage in the secondary circuit of a transformer. Hence, regulation of voltage and reactive power is obtained. Tap-changing transformers in distribution circuits are the main instruments in controlling voltage. The tap-changer in a delivery transformer compensates for the tap-changer fall in voltage through the transformer reactance, and the transformer reactance changes.

Reactive power is dispatched in transmission circuits by changing the transformer taps and regulating the network's voltages.

i. For the tap-changing at the receiving end

The following example states the effect of the tap-changing transformer when it is installed at the receiving end.

Example 4.3

For the power system shown in Figure 4.12.

Grid: 66 *kV*, 80 *MVA*, $X = 10\%$.

T.L: $j0.2\frac{\Omega}{mile}, 100\frac{mile}{phase}$.

Transformer: 66/33 *kV*, 75 *MVA*, $X = 15\%$.
Load: 33*kV*, 100 MVA, 0.8 *lagging power factor*.
Find the tap-changer transformer ratio to compensate for voltage drops.

Solution
Select the base values of 100 MVA and 66 kV on the transformer's high voltage side and 33kV on the low voltage side. Referring to the impedance diagram shown in Figure 4.13.

$$X_{Grid} = X_{given} \times \frac{S_{base}}{S_{given}} = j0.1 \times \frac{100}{80} = j0.125\,p.u$$

$$X_{T.L} = j0.2\frac{\Omega}{mile} \times 100\,mile = j20\,\Omega$$

$$Z_{base} = \frac{(V_{base})^2}{S_{base}} = \frac{(66)^2}{100} = 43.56\,\Omega$$

$$X_{T.L(p.u)} = \frac{X_{T.L}}{Z_{base}} = \frac{j20}{43.56} = j0.4591\,p.u$$

$$X_{tr} = j0.15 \times \frac{100}{75} = j0.2\,p.u.$$

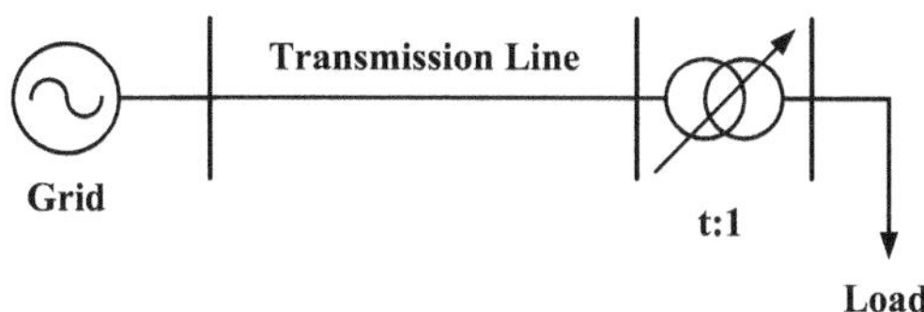

FIGURE 4.12 Power system configuration of Example 4.3.

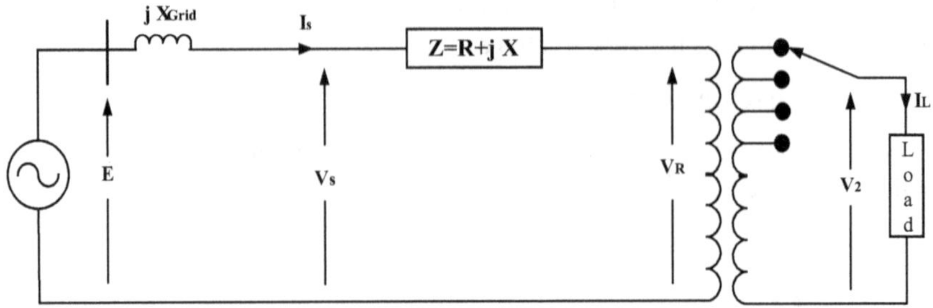

FIGURE 4.13 Impedance diagram of Example 4.3.

Load:

$$33kV, 100\,MVA, 0.8\,lagging\ power\ factor.$$

$$S_L = \frac{100}{100}\angle -\cos^{-1}0.8 = 1.0\angle -36.87°p.u$$

$$S_L = V_L \times I_L^*$$

$$1.0\angle -36.87° = 1.0\angle 0° \times I_L^*$$

$$I_L = 1.0\angle 36.87°p.u$$

$$V_R = 1.0\angle 36.87° \times 0.2\angle 90° + 1.0\angle 0° = 0.8944\angle 10.3°$$

To compensate for the voltage drop set t = 1: 0.8944 ∠ 10.3°p.u.

ii. For the tap-changing at the sending and receiving end,t_sV_1, and t_rV_2 are the real voltages at the ends of the circuit given in Figure 4.14(a). The tap-changing ratios needed to compensate for the voltage drop in the line fully must be calculated. The product of t_s. t_r will be made unity; this means that the average voltage level is in the same order and that all transformers use the minimum range of taps. (Note that all values are per unit; t is the off-nominal tap ratio.) pass the load circuit to all amounts.

$$The\ arithmetic\ voltage\ drop = V_S - V_R = V_1 \times \frac{t_s}{t_r} - V_2 \approx \frac{RP + XQ}{t_r^2 V_2} \tag{4.23}$$

Solving to get

$$V_2 = \frac{1}{2} \times \left[t_s^2.V_1 \pm t_s.\left(t_s^2.V_1^2 - 4 \times \left(RP + XQ\right)\right)^{\frac{1}{2}} \right] \tag{4.24}$$

where

$$t_s.t_r = 1 \tag{4.25}$$

When t_s is specified, so t_r is also specified. For a given V_1, there are then two values of V_2.

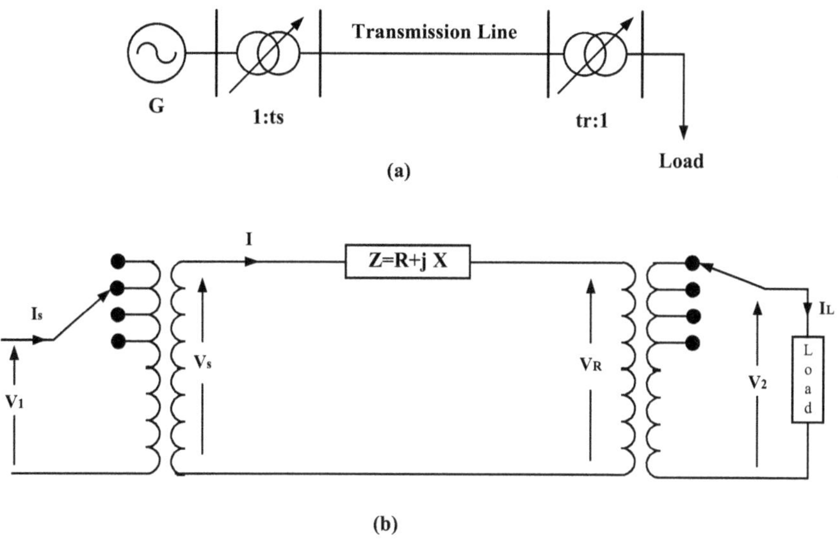

FIGURE 4.14 Connection of tap changing transformers at receiving and sending end. (a) Coordination of two tap-changing transformers in a radial transmission link. (b) Equivalent circuits for dealing with off-nominal tap ratio of two transformers.

Example 4.4

A 230-kV line is fed from a constant 66-kV supply through a 66/230-kV transformer, as given in Figure 4.14(a). At the load end of the line, another 230/66 kV nominal transformer reduces voltage. Complete line and transformer impedance at 230 kV is (100 + *j*200) Ω. Both transformers are fitted with tap-changing facilities arranged so that the product of the two off-nominal settings is unity. Suppose the load is 100 MW at 0.8 p.f. Lagging, measure tap-changer settings needed to maintain load busbar voltage at 66 kV. Using a 100 MVA as the base value.

Solution

$$Z_{base} = \frac{V_{base}^2}{S_{base}} = \frac{230^2}{100} = 529\,\Omega$$

$$Z_{T.L} = \frac{100 + j200}{529} = 0.189 + j0.378\,p.u$$

From the equivalent circuit shown in Figure 4.14(b), the load voltage is given as:

$$V_2 = \frac{1}{2} \times \left[t_s^2.V_1 \pm t_s.\left(t_s^2.V_1^2 - 4 \times (RP + XQ)\right)^{1/2} \right]$$

$$V_1 = V_2 = 1.0\,p.u$$

$$\varphi = \cos^{-1} 0.8 = 36.87°$$

$$Q = P\tan\varphi = 100 \times \tan 36.87° = 75\,\text{MVAr}$$

$$S_{Load} = \frac{100 + j75}{100} = 1.0 + j0.75\,p.u$$

$$1.0 = \frac{1}{2} \times \left[t_s^2.1.0 \pm t_s.\left(t_s^2.1.0^2 - 4 \times \left(0.189 \times 1 + 0.378 \times 0.75\right)\right)^{1/2}\right]$$

$$2 = t_s^2 \pm t_s.\left(t_s^2 - 1.89\right)^{\frac{1}{2}}$$

$$2 - t_s^2 = \pm t_s.\left(t_s^2 - 1.89\right)^{1/2}$$

$$\left(2 - t_s^2\right)^2 = t_s^2.\left(t_s^2 - 1.89\right)$$

$$t_s = 1.377$$

$$t_r = \frac{1}{t_s} = \frac{1}{1.377} = 0.726.$$

4.5 VAR COMPENSATOR

For dynamic power factor correction and terminal voltage stabilization, phase–phase unbalanced reactive power compensation is required in distribution systems. Load compensation is related to phase–phase unbalanced reactive power compensation issues observed in realistic implementation. The phase–phase unbalanced reactive power demand of large single-phase, traction, and fluctuating industrial loads such as electric arc furnaces and rolling mills is typically minimized or canceled by load compensation requirements. These loads drain unbalanced reactive power from power supply devices, resulting in unbalanced device voltages and voltage spikes, leading to large supply voltage fluctuations and causing effects such as incandescent light flicker, blurred television image, electrical control circuit interruption, and computer equipment. In general, these heavy industrial loads are clustered in one plant and supplied from one network bus, so a local compensator attached to the same bus can accommodate them better.

To retain the terminal voltage within the appropriate variations, the AC system is too weak, and it is also not feasible or realistic to supply the system's reactive power demand. Load compensation eliminates the adverse effects of an unbalanced single set of loads on the device without adjusting the bus voltage's external control.

Generally, many traditional approaches are used for reactive power compensation in control systems. There are shunt capacitors, SVC, and synchronous compensator, sequence capacitors. One of the strategies commonly used to minimize the cost of delivering a low power factor load is the shunt reactive power compensation, typically called power factor correction. It is typically necessary to inject much of the needed reactive power as near as possible into the load required in practice. This usually decreases the power loss and voltage decrease in transmission and distribution systems induced by reactive power flow.

Some important practical applications for the compensation of reactive power in the power system are as follows:

i. Maintaining voltage under slowly changing conditions at or above a constant level due to changes in loads.
ii. The changes in voltage due to unforeseen occurrences, such as load rejections, transformer outages, and line outages.
iii. Reducing flicker caused by loads like arc furnace loads that fluctuate rapidly.
iv. Improving the power grid's reliability by supporting the voltage at key points.
v. Improving voltage and power factor.
vi. Locate near HVDC converter/inverter terminals to regulate the voltage.
vii. Provide unbalanced reactive capacity phase–phase.

The basic types of SVCs are TSC with a fixed reactor, TCR with a fixed capacitor, and TSC with a mixture of TCR and TSC. Two types of SVCs are used in the current work, capacitor tap, and TCR. The TCR consists of a reactor in series with a thyristor switch and is designed to absorb reactive power according to conduction angles. Depending on the firing angle, which is determined from a zero voltage crossing, the thyristors operate on alternating half-cycles of the supply voltage. Essentially, the current is reactive and sinusoidal.

The action of thyristor-controlled compensators can advantageously meet the phase–phase unbalanced and varying-cyclic load reactive power requirements in the device at reasonable conduction angles. However, such an operation pollutes the power supply differently by injecting harmonic currents into the AC grid. In such cases, either the provision of external harmonic filters or the internal minimization of harmonic generation is required; the latter term is less burdensome for the expenditure of harmonic filters.

4.6 OBJECTIVES OF LOAD COMPENSATION

The control of reactive power to maximize supply efficiency in AC power systems is load compensation. The compensating machinery is normally mounted next to the load on the consumer's premises. There are three main targets for load compensation:

1. Correction to the power factor.
2. Improvement in the control of voltage.
3. Balancing the load.

A short reference to the three compensation goals will be given in the following three subsections.

4.6.1 Correcting Power Factor

There are lagging power components in most industrial loads, which is, they consume reactive power. Therefore, the load current continues to be greater than is needed to provide the actual power alone, which increases the loss of active power and decreases the productive ability to generate the system's units and components. For these purposes, power providers in many countries set a lower limit (cos) at which the user must pay for its excess reactive energy usage in the form of fines.

The power factor correction usually includes producing (some or all) reactive power as near as possible to the load needed, rather than delivering it from a distant power plant. Therefore, design engineers have applied shunt capacitors at medium voltage (11–33) kV. The individual compensation system is one of the techniques used to maximize the force factor.

There are many advantages to be obtained from incentive implementations to increase the power factor, some of which are the following:

a. Increasing the apparent potential for power.
b. Line failure elimination.
c. Avoidance of losses from transformers.

4.6.2 Controlling Voltage

The steadiness of the supply voltage at the point of use is one indicator of service efficiency. The designer of the delivery system's standard aim is to retain the consumer's average supply voltage in the nominal (percent) range.

All loads differ in their reactive power demand; although, their range and variation rate vary greatly. In both situations, the shift in demand for reactive power allows the voltage at the point of supply to vary, which may conflict with the effective operation of all the plants connected to the point of supply, giving rise to the likelihood of interference between loads to different customers.

Strengthening the power grid, increasing the scale and number of generating units and making the network more tightly integrated would be one way to boost voltage control. In general, this method would be uneconomic and would add voltage levels and switchgear rating-related issues. Controlling the voltage and VARs on the load side is more practical and economical.

The variation of the transmission voltage can be minimized using the following:

a. Swapped banks of shunt capacitors,
b. Switched banks of shunt reactors,
c. Transformer taps alter,
d. Synchronous compensators are used,
e. Condensers in sequence.

When a certain amount of reactive power is pumped into a feeder, the reactive current flow will decrease to the point of injection, decreasing the voltage drop and increasing the feeder's voltage profile.

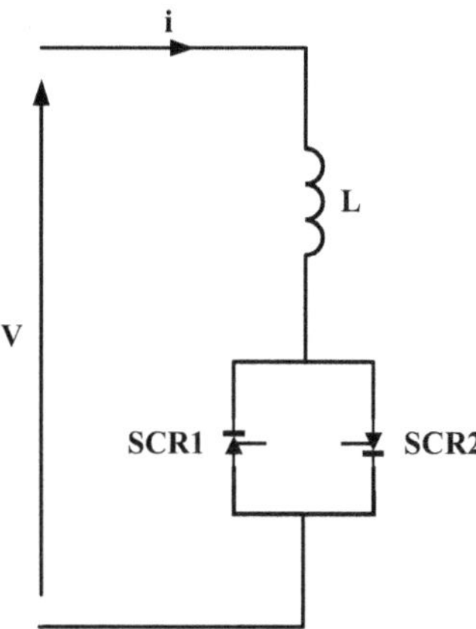

FIGURE 4.15 Thyristor-controlled reactor.

4.6.3 Balancing the Load

Most AC power systems are three-phase and are designed for balanced operation. The unbalanced operation gives rise to current components in the negative and zero sequence components. Such components have undesirable effects, including an additional loss in motors and generating units, oscillating torque in the AC machines, increased ripple in rectifiers, malfunction of several types of equipment, a saturation of transformers, and excessive neutral currents.

The load-balancing goal is achieved by phase control of static VAR compensators. There are many methods used for phase balancing, such as:

i. Balancing an unsymmetrical resistive load.
ii. The compensation of admittance network (the ideal load compensator is conceived as any passive three-phase admittance network that will present a real and symmetrical load to the supply).

4.7 REACTIVE POWER COMPENSATION TYPES

There are different categories of compensators which are as follows.

4.7.1 Series Capacitor

The series capacitors are applied to compensate for the inductive reactance of the transmission line. They may be installed at the ends or an intermediate point on a long transmission line. Their benefits include the following:

a. Increase Steady-State Stability limits.
b. Reduced voltage drops in load areas during severe disturbances.
c. Reduced transmission losses.
d. Better adjustment of line loadings.

There are many problems associated with this method of compensation which are as follows:

1. Excessive fault currents.
2. The electrical resonance of the synchronous generator and the capacitor compensated transmission line causes sub-synchronous torsional oscillations. This phenomenon is generally known as sub-synchronous resonance (SSR).
3. The series capacitor must carry the line current and the anticipated fault current; thus, it must be protected by a spark gap to prevent excessive currents from flowing through it.

4.7.2 Synchronous Capacitors

Synchronous condensers or synchronous compensators are designed to compensate for reactive shunt power. They fall into the class of shunt compensators that are working. The three phases caused by the synchronous spinning machine's e.m.f (e_1, e_2, e_3) are in phase with the device voltages for solely reactive power flow (V_1, V_2, V_3). The reactive power can be controlled by manipulating the machine's excitation.

The synchronous machine can be used as a dynamic reactive power compensator because it guarantees steady-state and dynamic voltage regulation, lower line losses, and a smoother voltage profile at the transmission line's receiving end.

4.7.3 Shunt Capacitors

Shunt Capacitors are used for steady-state voltage control and reactive power flow control. They may include the following:

- ***Fixed Capacitors (FC):*** They are installed across the supply at the receiving end near load centers. Their use has increased due to advanced manufacturing techniques, which result in a reduction of their price, improved reliability, smaller size, and weight.
- ***Thyristor Switched Capacitors (TSCs):*** The VARs released to the host AC system are controlled by switching capacitors in and out of the circuit using thyristors. Capacitors are switched in when the phase voltage is minimum and switched out at zero current crossings.
- ***Automatic (Variable) Capacitors:*** Power factor relays can regulate this type of capacitors to keep the power factor fixed at the required value by keeping the same proportion between active and reactive power during the complete load cycle.

4.7.4 Shunt Reactors

Shunt reactors are used for steady-state voltage control, reactive power flow control, and for reduction of switching surge overvoltages.

Shunt reactors are mainly classified as the following:

a. Linear reactor.
b. The saturated reactor.
c. Thyristor-Controlled Shunt Reactor (TCR).

The basics of TCR are shown in Figure 4.15. The thyristor controller is the controlling element as two oppositely poled thyristors that conduct on alternating half-cycles of the supply voltage waveform.

4.8 CONTROLS OF SWITCHED SHUNT CAPACITORS

Switched capacitors supply the variable amount of reactive power needed as the load increases. Switching of capacitor banks may be done either manually or automatically. Several types of automatic control can be used to initiate the switching of capacitor banks. These include the following:

1. Time switch control.
2. Voltage control.
3. Current control.
4. Voltage-time control.
5. Voltage-current control.
6. Temperature control.

Time switch control is the least costly type of control but can only be used where power factor and demand vary on a firm time basis. Voltage control is used where objectionable voltage changes occur with varying voltages. Current control is used when loads change, but voltage is well regulated, or load power factors remain substantially constant. Current control is also effective when the power factor varies predictably with the load.

4.9 IN POWER SYSTEM HARMONIC DISTORTION

Harmonic studies have become an important component of power system analysis and design. They are used to quantify the distortion in voltage and current waveforms at various points in a power system. Such studies are important because the presence of harmonic producing equipment is increasing. As harmonics propagate through the system, they increase losses and possible equipment loss of life. Equipment can be damaged by overcurrents or overvoltage resulting from resonances.

4.10 SOURCES OF HARMONICS

Many sources generate harmonics in power systems. Some important sources are as follows:

1. Non-linear loads, such as rectifiers, inverters, welders, arc furnaces, voltage controllers…etc.

2. Transformer magnetizing currents.
3. Static VARs compensators.
4. HVDC power conversion.
5. Cycloconverters are used for low-speed, high-torque machines.

4.11 HARMONIC MEASUREMENT

The effects of harmonics in the system are generally measured by calculating defined performance indices, Distortion Factor (D), and Telephone Interference Factor (TIF).

4.11.1 Distortion Factor

The distortion factor is a measure of the deviation of the voltage and current wave from a sinusoidal shape and is used as a measure of the effect of harmonics on the power system and is given by:

Voltage distortion factor

$$D_v = \sqrt{\left(\sum_{h=2}^{n} V_h^2\right) \Big/ V_f^2} \tag{4.26a}$$

and current distortion factor

$$D_I = \sqrt{\left(\sum_{h=2}^{n} I_h^2\right) \Big/ I_f^2} \tag{4.26b}$$

In the current work, only the effect of harmonics on power systems is considered. Therefore, the harmonic measurement is done by calculating the current distortion factor.

4.11.2 Telephone Interference Factor

The harmonic current or voltage in the power system generates noise, coupled with wire communication circuits through magnetic and electrostatic fields. Weighting factors have been developed to compare the relative effectiveness of different frequencies in interfering with telephone conversations. One of the widely used weighting factors is the TIF, which is given by the following formula:

$$TIF = \left[\frac{1}{I_f}\sum_{h=2}^{n}\left(W_h I_h\right)^2\right]^{\frac{1}{2}} \tag{4.27}$$

W_h is a weight factor accounting for audio effects induction coupling at the hth harmonic frequency.

4.12 HARMONIC REDUCTION METHODS

The elimination of harmonic or minimization of its generation is necessary for the proper operation of the power system. Several techniques of harmonics filtering can control the harmonics.

Current harmonics are often treated as a local problem, at least for one feeder in the distribution network. The impedance of the distribution network dampens the harmonic propagation. Therefore, harmonic filtering should be performed near the source of the current harmonics for the best result. Other equipment will be unaffected by the harmonic producing load if this is done. Harmonic filtering or compensation is accomplished by using shunt filters or series filters.

4.12.1 Shunt Filters

For current harmonics, shunt filters have low impedance paths. Present harmonics flow into shunt filters instead of returning to supply. Figure 4.16 demonstrates a standard passive shunt filter. The passive filter consists of a series of LC filters tuned for particular harmonics in conjunction with a general high-pass filter used to remove the higher-order current harmonics that remain.

The performance of a passive filter is strongly dependent on the system impedance at the harmonic frequencies. The system impedance depends on the distribution network configuration and loads. Therefore, passive filters' design involves a thorough system analysis to obtain adequate filtering performance. The disadvantages of this filter are as follows:

i. It only filters a single (tuned) harmonic frequency.
ii. Other non-linear loads will import harmonics.

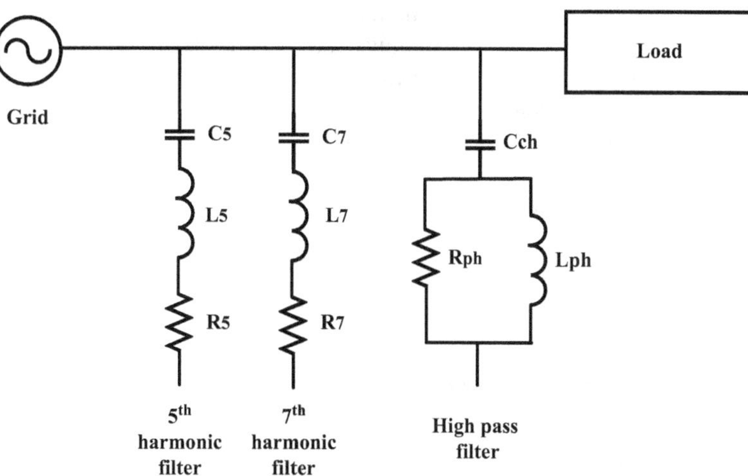

FIGURE 4.16 Typical shunt filters for the reduction of current harmonics.

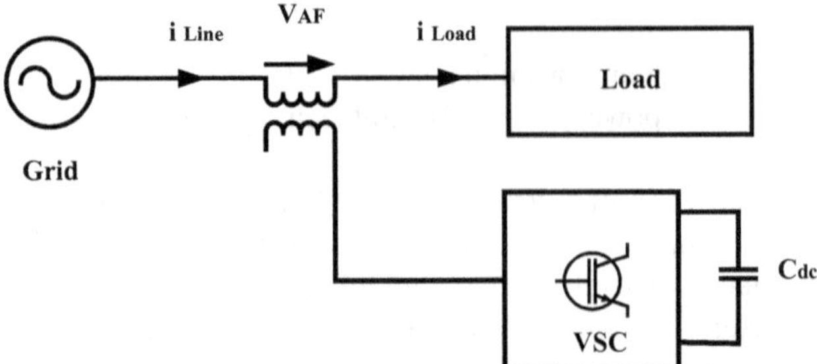

FIGURE 4.17 Principal configuration of a series AF system.A matching transformer connects the active filters in series to a transmission line. VSCs are suitable for series AFs as the regulated source. It consists of Cdc as a DC-link capacitor, electronic switches, and a series transformer.

iii. Multiple filters are needed to reach typical harmonic limits.
iv. Harmonic analysis is important before adding this filter to the power system.

4.12.2 Filter Sequence

A matching transformer connects the active filters in series by a transmission line. VSCs are suitable for series AFs as the regulated source (see Figure 4.17). It consists of C_{dc} as a DC-link capacitor, electronic switches, and a series transformer.

The operation principle of series AFs is based on the harmonics' isolation between the load and the supply. This is obtained by injection harmonic voltages across the transformer line side. The injected harmonic voltages affect the equivalent line impedance seen from the load. The equivalent line impedance can be regarded as infinite or large for the harmonics, whereas it should ideally be zero for the fundamental frequency component.

The disadvantages of this filter are as follows:

1. It must handle the rated full load current.
2. At best, it can only improve harmonic current distortion of 30–40 %.
3. It slightly reduces the displacement power factor.

4.13 OPERATION OF THYRISTOR-CONTROLLED SVCs

The operation of thyristor-controlled SVCs at various conduction angles can be used advantageously to correct unbalanced line currents and minimize the transmission line loss in a system. Due to low-cost maintenance, the most important static reactive power devices are TCR and capacitor tapped.

In this chapter, an algorithm to evaluate the required VARs of reactive compensators for optimum compensation was utilized. This algorithm is based on the minimization of harmonics injected into the system.

The equations were derived from calculating the optimal reactive power and firing angles for the thyristor to meet unbalanced line currents and the minimum harmonics generated by the reactive power compensator.

For a three-phase power system, the reactive power compensator is assumed to be capacitive or inductive, applied in delta connection as shown in Figure 4.18; the phasor diagram of voltage and current is shown in Figure 4.19. An optimization problem for three-phase reactive power compensation can be written to minimize an objective function described as the summation of the squares of the r.m.s. values of line current of the three phases. This objective function is written as:

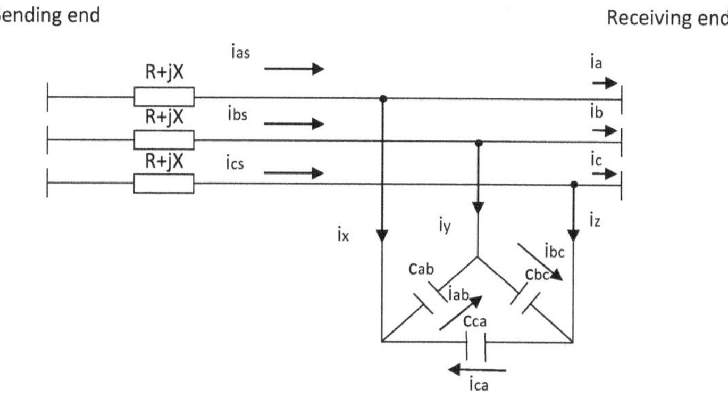

FIGURE 4.18 Diagram of reactive power compensator.

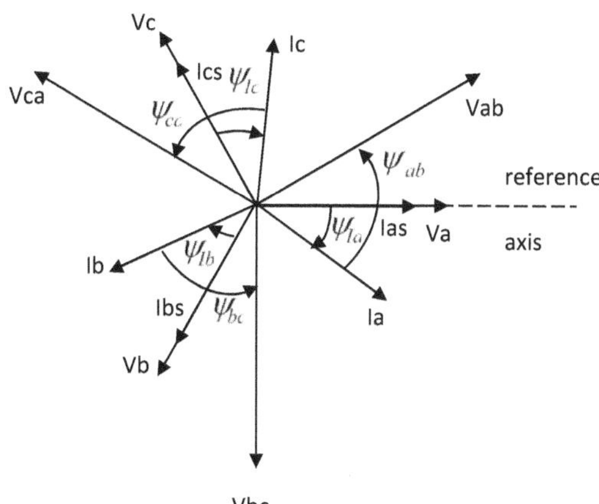

FIGURE 4.19 Phasor diagram of voltage and current in a three-phase system.

$$J_1(c) = \frac{1}{T}\int_0^T \left(i_{as}^{\ 2} + i_{bs}^{\ 2} + i_{cs}^{\ 2}\right)\mathrm{dt} \tag{4.28}$$

where

$$T = \frac{2\pi}{\omega} \tag{4.29}$$

Subject to an equality constraint

$$|I_{as}| = |I_{bs}| = |I_{cs}| \tag{4.30}$$

In Equation 4.5, I_{as}, I_{bs}, and I_{cs} are the phasor of peak values of the compensated line currents i_{as}, i_{bs}, i_{cs} for respective phase after installing the reactive power compensator.

The compensated line currents, ias, ibs, ics, can be expressed as:

$$i_{as} = i_a + C_{ab}dv_{ab}/dt - C_{ca}dv_{ca}/dt \tag{4.31}$$

$$i_{bs} = i_b + \frac{C_{bc}dv_{bc}}{dt} - \frac{C_{ab}dv_{ab}}{dt} \tag{4.32}$$

$$i_{cs} = i_c + C_{ca}dv_{ca}/dt - C_{bc}dv_{bc}/dt \tag{4.33}$$

Assume that the uncompensated line current i_a, i_b, i_c and line voltages v_{ab}, v_{bc}, v_{ca} are pure sinusoids with peak values I_a, I_b, I_c, and V_{ab}, V_{bc}, V_{ca}, respectively. In the power systems, the supply voltages are assumed to be equal in quantity as:

$$V_{ab} = V_{bc} = V_{ca} = V.$$

$$\begin{aligned} J_1(c) = \frac{1}{2}[I_a^2 + I_b^2 + I_c^2 + \omega^2 V^2 (2C_{ab}^2 \\ + 2C_{bc}^2 + 2C_{ca}^2 + C_{ab}C_{bc} + C_{bc}C_{ca} \\ + C_{ca}C_{ab}) + 2\omega V(y_{ab}C_{ab} \\ + y_{bc}C_{bc} + y_{ca}C_{ca})] \end{aligned} \tag{4.34}$$

where

$$y_{ab} = -I_a \sin\psi_{ab} + I_b \sin\left(2\pi/3 + \psi_{bc}\right) \tag{4.35}$$

$$y_{bc} = -I_b \sin\psi_{bc} + I_c \sin\left(2\pi/3 + \psi_{ca}\right) \tag{4.36}$$

$$y_{ca} = -I_c \sin\psi_{ca} + I_a \sin\left(2\pi/3 + \psi_{ab}\right) \tag{4.37}$$

where ψ_{ab} is the phase angle between V_{ab} and i_a, similarly for ψ_{bc} and ψ_{ca} and w is the angular frequency of the power systems.

The optimal conditions for the objective function $J_1(c)$ can be defined by its partial derivatives concerning C_{ab}, C_{bc}, and C_{ca} separately, and setting the resulting equations to zero as follows:

$$\omega V\left(4C_{ab}+C_{bc}+C_{ca}\right)+2y_{ab}=0 \tag{4.38}$$

$$\omega V\left(C_{ab}+4C_{bc}+C_{ca}\right)+2y_{bc}=0 \tag{4.39}$$

$$\omega V\left(C_{ab}+C_{bc}+4C_{ca}\right)+2y_{ca}=0 \tag{4.40}$$

$$\begin{aligned} C_{ab}^{*}=1/\left(2\omega V\right)\big[&I_a(f_3\sin\psi_{ab}+f_1 I_b \\ &\times\cos\psi_{ab})+\left(f_2\sin\psi_{bc}-f_3\cos\psi_{bc}\right) \\ &+I_c\left(-f_2\sin\psi_{ca}+f_1\cos\psi_{ca}\right)\big] \end{aligned} \tag{4.41}$$

$$\begin{aligned} C_{bc}^{*}=1/\left(2\omega V\right)\big[&I_a\left(-f_2\sin\psi_{ab}+f_1\cos\psi_{ab}\right) \\ &+I_b\left(f_3\sin\psi_{bc}+f_1\cos\psi_{bc}\right) \\ &+I_c\left(f_2\sin\psi_{ca}-f_3\cos\psi_{ca}\right)\big] \end{aligned} \tag{4.42}$$

$$\begin{aligned} C_{ca}^{*}=1/\left(2\omega V\right)\big[&I_a\left(f_2\sin\psi_{ab}-f_3\cos\psi_{ab}\right) \\ &+I_b\left(-f_2\sin\psi_{bc}+f_1\cos\psi_{bc}\right) \\ &+I_c\left(f_3\sin\psi_{ca}+f_1\cos\psi_{ca}\right)\big] \end{aligned} \tag{4.43}$$

Where constants $f_1=1/\left(3\sqrt{3}\right)$, $f_2=1/3$, and $f_3=5/\left(3\sqrt{3}\right)$. The "*" sign means the optimal terms for general cases; if any solution from the equations above is negative, it indicates that a compensating inductor should be installed instead of a compensating capacitor.

4.14 SVC PARAMETERS CALCULATION

The parameters of SCR are as follows.

4.14.1 Static VAR Compensator Configurations

The block diagram arrangement of a typical SVC consists of a tap capacitor, and thyristor-Controller Reactor (TCR) is shown in Figure 4.20.

These compensators essentially function as variable reactance (capacitive and inductive impedance). The three phases of TCR are usually delta connected to prevent triplet harmonics' injection into the system.

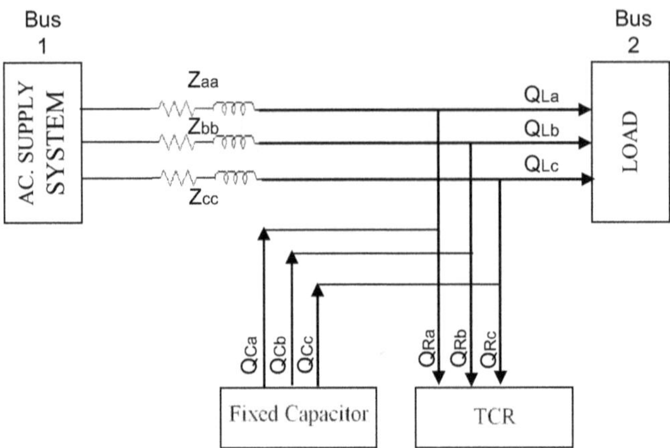

FIGURE 4.20 Block diagram of a typical industrial power supply system with load compensation.The block diagram arrangement of a typical SVC consists of a tap capacitor, and thyristor-Controller Reactor (TCR).

Considering the compensator as variable delta-connected reactances as shown in Figure 4.21, Q_{Ra}, Q_{Rb}, and Q_{Rc} are absorbed in the unsymmetrical reactances, x_{ab}, $x_{bc,}$ and x_{ca} of the TCR.

The TCR type compensator consists of a variable inductor by employing phase control of the thyristor switch. A Variable reactive power can be achieved as the TCR controls inductance via various firing angles.

4.14.2 Calculation of the TCR Firing Angle

The voltage and the resulting inductor current waveforms for a firing angle α are shown in Figure 4.22. The relationship between the inductor current and the firing angle can be expressed as follows:

$$i_L \begin{cases} i_{L1} = (V/\omega L)\left[\sin(\omega t) - \sin\alpha\right] & \alpha \le \omega t \le \pi - \alpha \\ i_{L2} = (V/\omega L)\left[\sin(\omega t) + \sin\alpha\right] & \pi + \alpha \le \omega t \le 2\pi - \alpha \end{cases} \tag{4.44}$$

where L is the inductance of the inductor current I_L, calculated from:

$$I_L = (V/\omega L)\left[(1/\pi)\left((\pi/2 - \alpha)(1 + 2\sin^2\alpha) - (3/2)\sin(2\alpha)\right)\right]^{1/2} \tag{4.45}$$

The SVC is used to provide a variable reactive power Q_{SVC}. The variable reactive power varies the capacitive reactive power Q_C and inductive reactive power Q_R.

$$Q_{SVC} = Q_C + Q_R \tag{4.46}$$

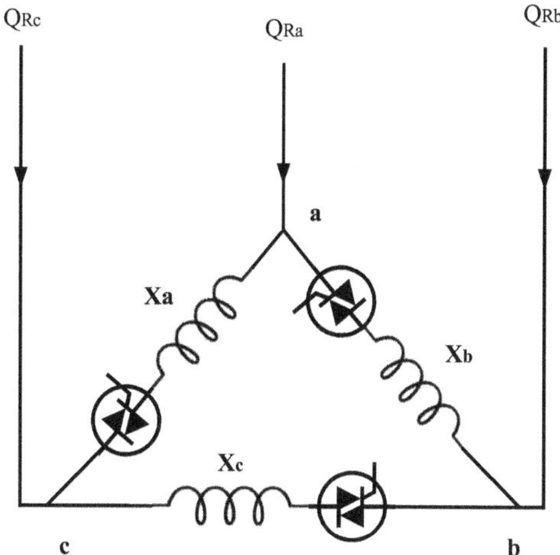

FIGURE 4.21 Delta-Connected Thyristor-Controlled Reactor.

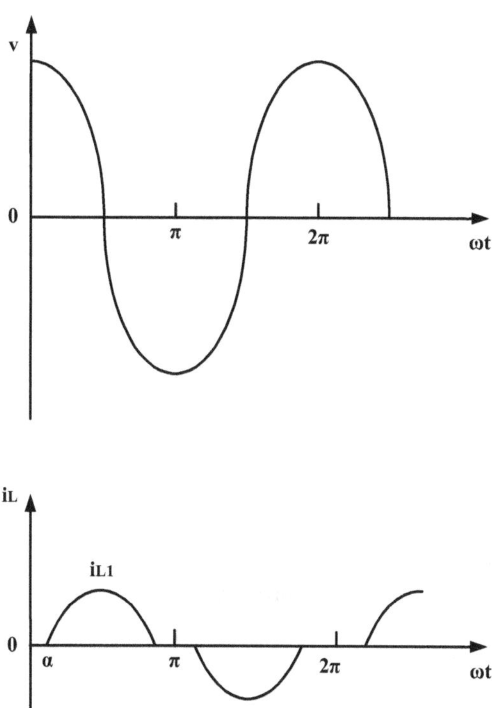

FIGURE 4.22 The voltage and resulting inductor current waveforms for a TCR compensator.

The capacitive and inductive reactive powers are expressed as follows:

$$Q_C = \omega C V^2 / 2 \tag{4.47}$$

$$Q_R = -I_L^2 \omega L \tag{4.48}$$

$$= -\left(V^2 / \pi\omega L\right)\left[\left(\pi / 2 - \alpha\right)\left(1 + 2\sin^2\alpha\right) - \left(3/2\right)\sin\left(2\alpha\right)\right] \tag{4.49}$$

where C is the capacitance of the SVC type compensator. The required optimal compensating reactive power is

$$Q_C^* = \omega C^* V^2 / 2 \tag{4.50}$$

The reactive power provided by the SVC must be equal to the required optimal compensating reactive power, that is

$$Q_{SVC} = Q_C^* \tag{4.51}$$

and

$$\left(C - C^*\right)\omega^2 L = S \tag{4.52}$$

where

$$S = \left(1/\pi\right)\left[\left(\pi - 2\alpha\right)\left(1 + 2\sin^2\alpha\right) - 3\sin\left(2\alpha\right)\right] \tag{4.53}$$

The above equations are derived based on a single-phase scheme for the SVC. For general three-phase compensators,

$$\left(C_i - C_i^*\right)\omega^2 L_i = S_i\, i = ab, bc, ca \tag{4.54}$$

and

$$S_i = \left(1/\pi\right)\left[\left(\pi - 2\alpha_i\right)\left(1 + 2\sin^2\alpha_i\right) - 3\sin\left(2\alpha_i\right)\right] i = ab, bc, ca \tag{4.55}$$

C_i and L_i are the capacitance and inductance of SVC for phase i, and C_i^* is the optimal compensation for phase i.

The coefficient S_i is in the range $0 \le S_i \le 1$. If $S_i < 0$, then $C_i < C_i^*$, which means that the required optimal compensating capacitive reactive power is larger than the maximum capacitive reactive power provided by the SVC, the thyristor-controlled reactor being fully conduction, that is, $\alpha_i = \pi/2$. On the other hand, if $S_i > 1$ it means that the required optimal compensating inductive reactive power is larger than the maximum inductive reactive power provided by the SVC, the thyristor-controlled reactor being fully opened, that is $\alpha_i = 0$.

The relationship between coefficient S and firing angle α is shown in Table 4.1. Figure 4.23 shows the relationship between coefficient S and the firing angle.

TABLE 4.1
Relationship between Coefficient S and Firing Angle α (Deg)

S	α	S	α	S	α
1	0	0.160193652	31	0.005369971	61
0.956164777	1	0.148097932	32	0.004520902	62
0.913547039	2	0.136695221	33	0.003781117	63
0.872144938	3	0.125960454	34	0.003140824	64
0.831954360	4	0.115869604	35	0.002589325	65
0.792970836	5	0.106397651	36	0.002117207	66
0.755187809	6	0.097520754	37	0.001715743	67
0.718597829	7	0.089214489	38	0.001377497	68
0.683191597	8	0.081454404	39	0.001094576	69
0.648959696	9	0.074217819	40	0.000859465	70
0.615890324	10	0.067480676	41	0.000666664	71
0.583971202	11	0.061219744	42	0.000510063	72
0.553188920	12	0.055412412	43	0.000384236	73
0.523529232	13	0.050036825	44	0.000284212	74
0.494976401	14	0.045070522	45	0.000205968	75
0.467513919	15	0.040491637	46	0.000146280	76
0.441124916	16	0.036280144	47	0.000100897	77
0.415790766	17	0.032414939	48	0.000067847	78
0.391492873	18	0.028875913	49	0.000043751	79
0.368211538	19	0.025643488	50	0.000027700	80
0.345926344	20	0.022699526	51	0.000016298	81
0.324616194	21	0.020025203	52	0.000009278	82
0.304260015	22	0.017602308	53	0.000004648	83
0.284835339	23	0.015414826	54	0.000001897	84
0.266319841	24	0.013445455	55	0.000000816	85
0.248690367	25	0.011678257	56	0.000000408	86
0.231923699	26	0.010098130	57	0.000000266	87
0.215995923	27	0.008691263	58	0.000000085	88
0.200883448	28	0.007443009	59	0.000000009	89
0.186561748	29	0.006339857	60	0.000000000	90
0.173006713	30				

4.15 HARMONICS DUE TO SVC OPERATION

Exercising control on the conduction intervals of the thyristor switch generates current harmonic components within the delta-connected reactances of the compensator and in the AC supply system line currents. If the line-to-line supply voltages at the compensator bus form a balanced set,

$$V_{ab} = V_m \cos \omega t \tag{4.56}$$

$$V_{bc} = V_m \cos\left(\omega t - 2\pi/3\right) \tag{4.57}$$

$$V_{ca} = V_m \cos\left(\omega t + 2\pi/3\right) \tag{4.58}$$

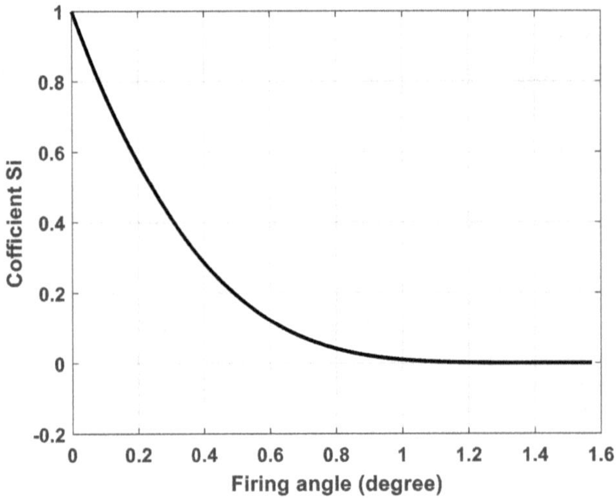

FIGURE 4.23 Relationship between coefficient S and firing angle.

where:

V_m is the peak value of the line to line voltage.
V_{ab}, V_{bc}, and V_{ca} are the voltages applied across the individual branches of the delta-connected reactance.

Fourier analysis has been applied to the discontinuous current waveforms to find the magnitudes of the fundamental and harmonic components. The branch currents of the delta-connected reactance is given by:

$$i_{ab} = \frac{V_m}{X_{ab}}\left[\left(1-\frac{2\alpha_1}{\pi}-\frac{\sin 2\alpha_1}{\pi}\right)\sin\omega t + \left(\sum_h \frac{(-1)^h-1}{\pi}\left(\frac{2\sin\alpha_1\cos\alpha_1}{h}-\frac{\sin(h+1)}{(h+1)}+\frac{\sin(h-1)\alpha_1}{(h-1)}\right)\right)\times\sin\omega t\right] \quad (4.59)$$

$$i_{bc} = \frac{V_m}{X_{bc}}\left[\left(1-\frac{2\alpha_2}{\pi}-\frac{\sin 2\alpha_2}{\pi}\right)\times\sin\left(\omega t-\frac{2\pi}{3}\right) + \left(\sum_h \frac{(-1)^h-1}{\pi}\left(\frac{2\sin\alpha_2\cos\alpha_2}{h}-\frac{\sin(h+1)}{(h+1)}+\frac{\sin(h-1)\alpha_2}{(h-1)}\right)\right) \times\sin h\left(\omega t-\frac{2\pi}{3}\right)\right] \quad (4.60)$$

$$i_{ca} = \frac{V_m}{X_{ca}}\left[\left(1 - \frac{2\alpha_3}{\pi} - \frac{\sin 2\alpha_3}{\pi}\right) \times \sin\left(\omega t + \frac{2\pi}{3}\right) + \left(\sum_h \frac{(-1)^h - 1}{\pi}\left(\frac{2\sin\alpha_3\cos\alpha_3}{h} - \frac{\sin(h+1)}{(h+1)} + \frac{\sin(h-1)\alpha_3}{(h-1)}\right)\right) \times \sin h\left(\omega t + \frac{2\pi}{3}\right)\right] \tag{4.61}$$

where $h = 3^{rd}, 5^{th}, 7^{th}, \ldots$ is the order of harmonics. The line currents are evaluated as the difference of two branch currents, thus

$$i_a = i_{ab} - i_{ca} \tag{4.62}$$

$$i_b = i_{bc} - i_{ab} \tag{4.63}$$

$$i_c = i_{ca} - i_{bc} \tag{4.64}$$

The line current harmonics are also obtained as the difference of the corresponding branch current harmonics. The fundamental and the harmonics components of the line currents are given by

$$i_f = \frac{V_m}{2\pi\omega L}\left[G_f^2 + H_f^2\right]^{\frac{1}{2}} \sin\left(\omega t - \varphi - \theta_f\right) \tag{4.65}$$

$$i_h = \frac{2V_m}{\pi\omega L}\left[G_h^2 + H_h^2\right]^{\frac{1}{2}} \sin\left(h\left(\omega t - \varphi\right) - \theta_h\right) \tag{4.66}$$

where:

i_f = instantaneous value of fundamental line current,
i_h = instantaneous value of harmonic line current,
ω = fundamental frequency (rad/sec),
L = inductance of each delta-connected reactance (Henries).

$$G_f = 3\pi - 4\gamma - 2\sin(2\gamma) - 2\beta - 2\sin(2\beta) \tag{4.67}$$

$$H_f = \sqrt{3}\left(\pi - 2\beta - \sin(2\beta)\right) \tag{4.68}$$

$$G_h = \left(\frac{\sin\left[(h+1)\gamma\right]}{h+1} - \frac{\sin\left[(h-1)\gamma\right]}{h-1} - \frac{2\sin\gamma\sin h\gamma}{h}\right) + \frac{1}{2}\left(\frac{\sin\left[(h+1)\beta\right]}{h+1} - \frac{\sin\left[(h-1)\beta\right]}{h-1} - \frac{2\sin\beta\sin h\beta}{h}\right) \tag{4.69}$$

$$H_h = \pm\frac{\sqrt{3}}{2}\left(\frac{\sin\left[(h+1)\beta\right]}{h+1} - \frac{\sin\left[(h-1)\beta\right]}{h-1} - \frac{2\sin\beta\sin h\beta}{h}\right) \tag{4.70}$$

$$\theta_f = \tan^{-1}\left(H_f/G_f\right) \tag{4.71}$$

$$\theta_h = \tan^{-1}\left(H_h/G_h\right) \tag{4.72}$$

$\phi = 0, \gamma = \alpha_1, \beta = \alpha_3$ for line current i_a
$\phi = 2\pi/3, \gamma = \alpha_2, \beta = \alpha_1$ for line current i_b
$\phi = 4\pi/3, \gamma = \alpha_3, \beta = \alpha_2$ for line current i_c
h = harmonic order, $(6k \pm 1)$, $k = 1, 2, 3\ldots$
+ sign for harmonics of order $(6k + 1)$
– sign for harmonics of order $(6k - 1)$.

For triple harmonics (3rd, 9th, etc.)

$$G_h = \frac{(h+1)}{4}\left[\left(\frac{\sin\left[(h+1)\gamma\right]}{h+1} - \frac{\sin\left[(h-1)\gamma\right]}{h-1} - \frac{2\sin\gamma\sin h\gamma}{h}\right) - \left(\frac{\sin\left[(h+1)\beta\right]}{h+1} - \frac{\sin\left[(h-1)\beta\right]}{h-1} - \frac{2\sin\beta\sin h\beta}{h}\right)\right] \tag{4.73}$$

$$H_h = 0. \tag{4.74}$$

PROBLEMS

4.1 What are the merits and demerits of the synchronous compensator?
4.2 What are the various methods of voltage control in the transmission system?
4.3 What are the various functions of the excitation system?
4.4 Show the voltage control and reactive power control are interrelated.
4.5 Give the function of AVR.
4.6 State difference between P-f and Q-V controls.
4.7 What are the methods to improve the voltage profile in the power system.
4.8 What is the need for the compensator in the AVR loop.
4.9 What is meant by stability compensation?
4.10 Draw the root loci for zero compensated loops.
4.11 Compare shunt and series capacitor.
4.12 Discuss the effect of compensation on the maximum power transfer in a transmission line.
4.13 What is SVC? Why is it used?
4.14 Draw the phasor diagram of SVC.
4.15 How is voltage control obtained by using tap changing transformer?
4.16 What are the methods of voltage control?
4.17 What are the different types of SVC?
4.18 What is a booster transformer? Where is it used?

4.19 What is the difference between ON load tap changing transformer and OFF load tap changing transformer?
4.20 What is meant by reactive power compensation?
4.21 Describe various methods of voltage control and explain any three in detail.
4.22 (i) Name the generators and consumers of reactive power in a power system. (ii) What are static VAR compensators? State the advantages of SVS.
4.23 Explain the following methods of voltage control: (i) Tap changing transformers, (ii) shunt reactors, (iii) synchronous phase modifiers, (iv) shunt capacitors, (v) series capacitors.
4.24 Draw the circuit for a typical excitation system, derive the transfer function model, and draw the block diagram.
4.25 What are the different methods of voltage control? Explain any two methods in detail.
4.26 (i) Develop a typical excitation arrangement to control an alternative's voltage and explain it. (ii) Briefly explain the role of tap changing transformer in voltage control?
4.27 What is a static VAR compensator? Where is it used? Explain its operation. Also, state the merits of static VAR compensator over the other voltage control methods.
4.28 (i) Briefly discuss the generation and absorption of reactive power. (ii) Derive the relations between voltage, power, and reactive power at a node for power system control applications.
4.29 Describe the various reactive power control methods and explain any two in detail.
4.30 Discuss static and dynamic analysis of AVR.

5 Power System Optimization

Power systems are very large and complex, and many unexpected events can influence them. This makes power system optimization problems difficult to solve. Hence methods for solving these problems ought to be an active research topic. This review presents an overview of important mathematical optimization methods: unconstrained optimization approaches, non-linear programming (NLP), and linear programming (L.P.), to be familiar with principles of power system utilization, especially from the commercial viewpoint. The fundamentals of optimization and problem formulation are discussed in this chapter.

5.1 OPTIMIZATION PROBLEM

An optimization problem includes finding the minimum or maximum value of a function of multiple variables.

Main components of an optimization problem:

1. Decision Variables
   The n variables to obtain their values at the function's optimum point (min or max).
   Decision variables

$$x_1, x_2, \cdots x_n \rightarrow x = \begin{bmatrix} x_1 \\ x_1 \\ \vdots \\ x_n \end{bmatrix} \tag{5.1}$$

   where x is the vector of the decision variable.
2. The objective function is required to find its optimum value. In other words, it describes a relation between decision variables that need minimizing or maximizing.
3. Problem constraints.
   m constraints as = or $\leq$ or $\geq$ valid at the optimum point.

DOI: 10.1201/9781003293965-5

Example 5.1

Consider an objective function of $f(x) = (n - 1)^2$. Minimize $f(x)$ for the problem constraints subject to $2 \leq n \leq 3$.

Solution

For $n = 2 \rightarrow f^* = (2 - 1)^2 = 1$, min value of $f(x)$ considering constraints.

In general, an objective function equation is shown in Figure 5.1 (between decision variables) can be a non-linear equation (e.g., second order, third order, sinusoidal, etc.). In the simplest form of the optimization problem, the objective function describes a linear relationship between decision variables. If both objective function and problem constraints are linear, it is known as L.P.

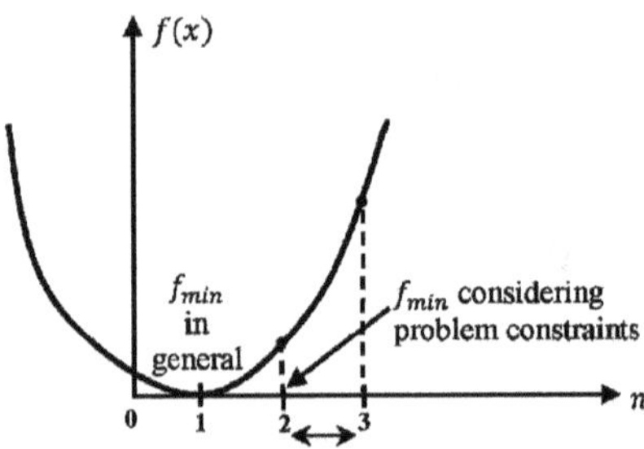

FIGURE 5.1 Objective function of Example 5.1.

Example 5.2

A plant produces three different types of products, as shown in Table 5.1, by applying various operations. Determine how many units of each product should be generated to maximize total plants profit.

TABLE 5.1
Data of Example 5.2

Operation	Operation Time to Generate One Unit of Each Product (Minutes)			Daily Operation Capacity (Minutes)
	Product 1 (x_1)	Product 2 (x_2)	Product 3 (x_3)	
1	1	2	1	450
2	3	0	2	480
3	1	4	0	440
Profit for generation of one unit of each product	3	2	5	

Solution

The first step of problem formulation is to convert the problem to a mathematical problem by writing a proper mathematical formula that describes the original problem completely).

1. Decision variables
   x_1 = number of generated products 1
   x_2 = number of generated products 2
   x_3 = number of generated products 3
2. Objective function
   Total plants profit is $f(x) = 3x_1 + 2x_2 + 5x_3$

   where $f(x) = f(x_1, x_2, x_3)$ and $x = \begin{bmatrix} x_1 \\ x_2 \\ x_3 \end{bmatrix}$
3. Problem's constraints
   For operation 1: $x_1 + 2x_2 + x_3 \leq 450$
   For operation 2: $3x_1 + 0x_2 + 2x_3 \leq 480$
   For operation 3: $x_1 + 4x_2 + 0x_3 \leq 440$
   The other inherent constraints are x_1, x_2, $x_3 \geq 0$ and integer number.

The mathematical form of the problem can be written as:

$$\max f(x) = f(x_1, x_2, x_3) = 3x_1 + 2x_2 + 5x_3$$

$$\text{Operating constraints}\, x_1 + 2x_2 + x_3 \leq 450$$

$$3x_1 + 2x_3 \leq 480$$

$$x_1 + 4x_2 + \leq 440$$

Since x_1, x_2, $x_3 \geq 0$ are an integer.

There is an analytical solution method for L.P. problems known as the "simplex method," which can be used to calculate the definitive answer of the problem.

5.2 FORM CHANGING OF OPTIMIZATION PROBLEM

5.2.1 Conventional Form

An optimization problem in conventional form conditions:

1. All decision variables are non-negative (zero or positive).
2. All constraints are in type less or equal.
3. The objective function is in maximum type.

Any optimization problem can be converted to a conventional form. For example, if the objective function is in minimum type (condition three not met):

$$\min f(x) \equiv \max g(x) at\, g(x) \triangleq -f(x) \tag{5.2}$$

If condition two is not met:

$$\text{e.g.,} k(x) \geq a \equiv -k \leq -a \tag{5.3}$$

$$H(x) = b \equiv H(x) \leq b \,\&\, H(x) \geq b \equiv H(x) \leq b \,\&\, H(x) \leq -b \tag{5.4}$$

Note that the only way to validate both above constraints is $H(x) = b$. If condition one is not met; for example, if a variable like x_1 without sign constraints, if replace it by two auxiliary variables like x_1^+ and x_1^- with both being non-negative:

$$x_1 \triangleq x_1^+ - x_1^- \tag{5.5}$$

where x_1 without sign constraint and $x_1^+, x_1^- \geq 0$ with sign constraint.

Example 5.3

Convert the following problem to conventional form:

$$\min z_o = 3x_1 - 3x_2 + 7x_3$$

Operating constraints $x_1 + x_2 + 3x_3 \leq 60$

$$x_1 + 2x_2 - 7x_3 \geq 70$$

$$5x_1 + 3x_2 = 40$$

$x_1^+, x_1^- \geq 0$, x_3 has no sign of constraints.

Solution

x_3^+ and x_3^- are considered as

$x_3 \triangleq x_3^+ - x_3^-$ and $x_3^+, x_3^- \geq 0$

For objective function $x - 1$

$$\max z_o = -3x_1 + 3x_2 - 7\left(x_3^+ - x_3^-\right)$$

$$\text{S.T } x_1 + x_2 + 3\left(x_3^+ - x_3^-\right) \leq 60$$

$$-x_1 - 2x_2 + 7\left(x_3^+ - x_3^-\right) \geq -70$$

$$5x_1 + 3x_2 \leq 40$$

$$-5x_1 - 3x_2 \leq -40$$

$$x_1, x_2, x_3^+, x_3^- \geq 0.$$

5.2.2 Standard Form

1. All constraints should be in an equality form (=).
2. The right element of each constraint should be non-negative.
3. All variables should be non-negative.

If the first condition is not met:

$$\text{e.g.,}\, k(x) \le a \tag{5.6}$$

$$H(x) \ge b \tag{5.7}$$

Consider two auxiliary variables S_i and S_j and add them to the problem as:

$$S_i, S_j \ge 0 \tag{5.8}$$

$$\left(k(x) \le a\right) \equiv \left(k(x) + S_i = a\right) \tag{5.9}$$

$$\left(k(x) \ge b\right) \equiv \left(H(x) + S_j = b\right) \tag{5.10}$$

Usually, constraints are numbered and auxiliary variables S_1, S_2, …, etc., are used for more clarity (the variable number is selected equal to corresponding constrained number).

If the second condition is not met, the constraint should be multiplied by −1.

If the third condition is not met, the same way as conventional form can be done (e.g. x_i is replaced with $x_i = x_i^+ - x_i^-$).

Example 5.4

Convert the following problem to standard form.

$$\min z_o = 3x_1 - 3x_2 + 7x_3$$

Operating constraints $x_1 + x_2 + 3x_3 \le 65$

$$x_1 + 2x_2 - 7x_3 \ge 75$$

$$5x_2 - 3x_3 = -45$$

$$-6x_1 + 4x_2 \le -40$$

x_1, x_2, ≥ 0, x_3 has no sign of constraints.

Solution

Similar to the previous example, x_3 is considered as (based on condition three):

$$x_3 \triangleq x_3^+ - x_3^- \text{ and } x_3^+, x_3^- \geq 0$$

For the first constraint, based on condition one, variables S_1 is added to the problem:

$$x_1 + x_2 + 3x_3 + S_1 = 65\; S_1 \geq 0$$

For second constraint:

$$x_1 + 2x_2 - 7x_3 - S_2 = 75$$

For third constraint (based on condition two):

$$-5x_2 + 3x_3 = 45$$

For the fourth constraint (based on conditions of one and two):

$$-6x_1 + 4x_2 \leq -40$$

Thus,

$$6x_1 - 4x_2 \leq 40$$

and

$$6x_1 - 4x_2 - S_4 = 40, S_4 \geq 0$$

Find result:

$$\min z_o = 3x_1 - 3x_2 + 7\left(x_3^+ - x_3^-\right)$$

Operating constraints $x_1 + x_2 + 3\left(x_3^+ - x_3^-\right) + S_1 = 65$

$$x_1 + 2x_2 - 7\left(x_3^+ - x_3^-\right) - S_2 = 75$$

$$-5x_2 + 3\left(x_3^+ - x_3^-\right) = 45$$

$$6x_1 - 4x_2 - S_4 = 40$$

$$x_1, x_2, x_3^+, x_3^-, S_1, S_2, S_4 \geq 0.$$

5.3 ECONOMIC LOAD DISPATCH

The Economic Load Dispatch (ELD) is allocating generation among the available generating units to minimize the total generation cost while meeting the equality and inequality constraints. The different types of power plants are as follows:

i. Thermal steam and gas power generation are the main objectives for ELD and other economic analysis.
ii. Hydroelectric power generation has no fuel cost; therefore, any generation is economical. However, hydroelectric power generation units have constraints based on the limitations of water storage. Therefore, coordinating with thermal units is necessary.

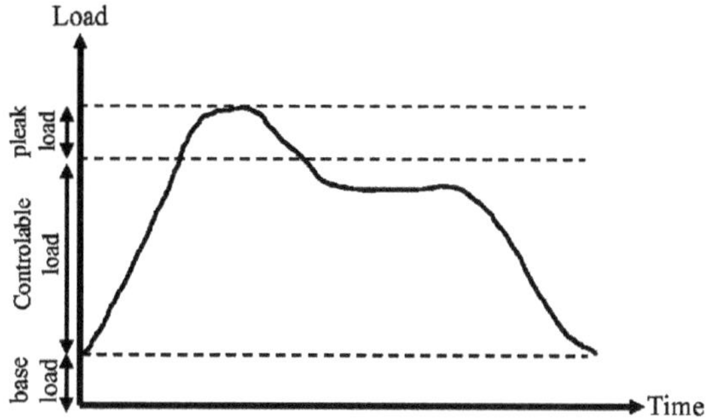

FIGURE 5.2 Load curve.

iii. Nuclear power loading is fixed and non-changeable. Therefore, it can be used for baseload and N.T. considered in economic calculations.
iv. Distribution generation and renewable energies are small and not considered in large-scale economic calculations, but calculations are done separately.

The load curve in Figure 5.2 shows that the peak load requires all generation units to supply the load, and a controllable load is subject to ELD calculations. However, the minimum demand is always needed and considered as a baseload.

5.4 THE SUBJECT OF ECONOMIC LOAD DISPATCH

Determination of each in-circuit generation unit share (generation amount) for demand-supply to minimize generation cost (especially fuel cost), which can be done by grid management (dispatching unit).

Since, as discussed before, economic analysis is done mostly for thermal power generation units, it is necessary to be familiar with the required thermal unit characteristics.

5.5 THERMAL UNITS CHARACTERISTICS

A typical electrical power generation unit is shown in Figure 5.3:

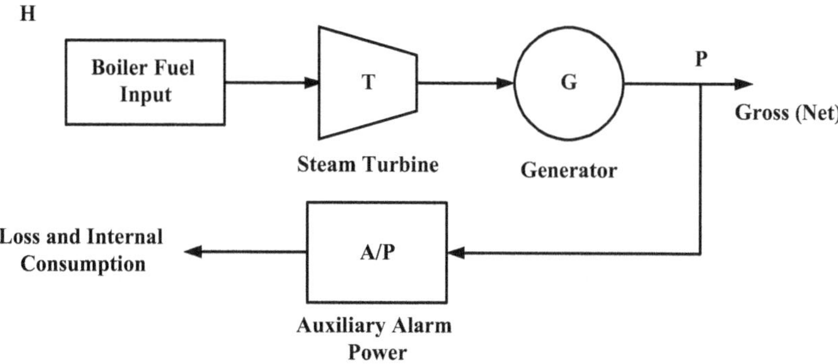

FIGURE 5.3 Electrical power generation unit.A typical electrical power generation unit.

Net output: Is the net electrical output power of the generation unit after a loss and internal consumption elimination.
Gross input: The input fuel to the unit in terms of fuel units per hour (e.g., a cubic meter of gas per hour m^3/hour). H gross input (heavy value of fuel per hour) is Btu/hour or MBtu/hour (3.192 MBtu = 1MWh).

5.5.1 Input–Output Characteristic

The curve of required input variations versus output power (H vs. P) is given in Figure 5.4.

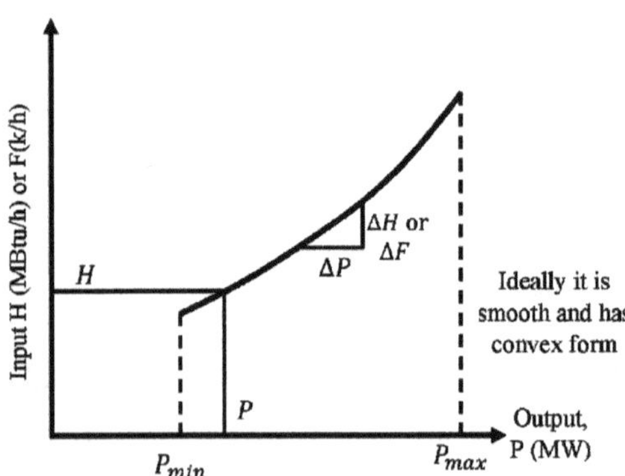

FIGURE 5.4 Input variations versus output power.The curve of required input variations versus output power (H vs. P).

where, P_{min} is the minimum generation based on thermodynamic issues. Fuel Cost (C): Fuel cost in terms of $/MBtu (for example) time-variant. Generation Cost Function (F): Obtained by multiplying fuel cost (C) to H in terms of $/h. The curve of the cost function is like the input–output function (H(P) because H is multiplied by a constant number (C). The cost function is usually considered a second-order function.

5.5.2 Incremental Cost Incremental Cost

Derivative of input–output or cost function:

It should be noted that ($\Delta F/\Delta P$) is more applicable. If the input–output function is considered a second-order function, the *I.C.* function will be linear (first order) (see Figure 5.5).

Net heat rate characteristics: *H/P* versus *P*, where *H/P* is considered as inverse of efficiency and P_{rated} is the nominal power at maximum efficiency (see Figure 5.6).

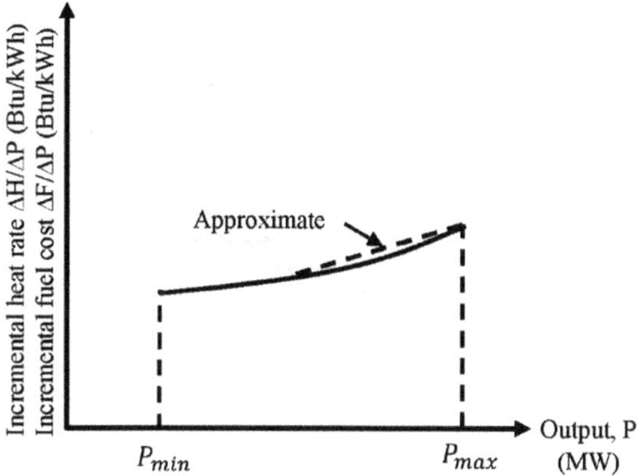

FIGURE 5.5 Incremental fuel cost and heat rate.

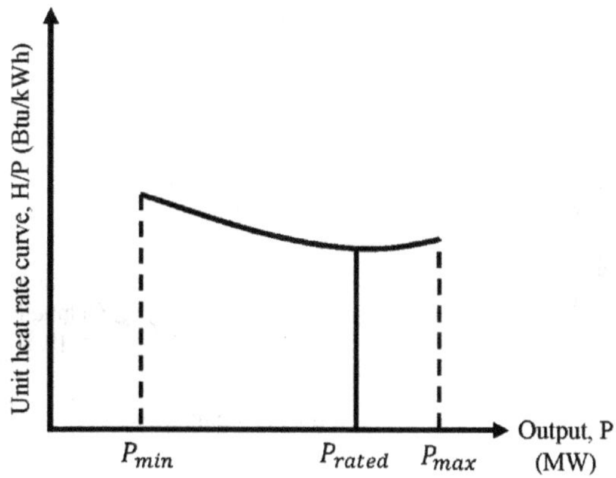

FIGURE 5.6 Heat rate characteristics: *H/P* versus *P*.

In practical cases, the input–output and cost function are not always smooth and differentiable because of the following:

1. Effect of multiple steam valves.
   The size of a control valve for a steam application can be difficult. This module attempts to highlight the subject by using first principles to explain the relationship between flow and pressure drop. It uses a simple nozzle to explain the phenomenon of critical pressure and how this can be predicted for steam flow through a control valve. Figure 5.7 shows multiple steam valves characteristics that affect the incremental heat rate of the multiple steam

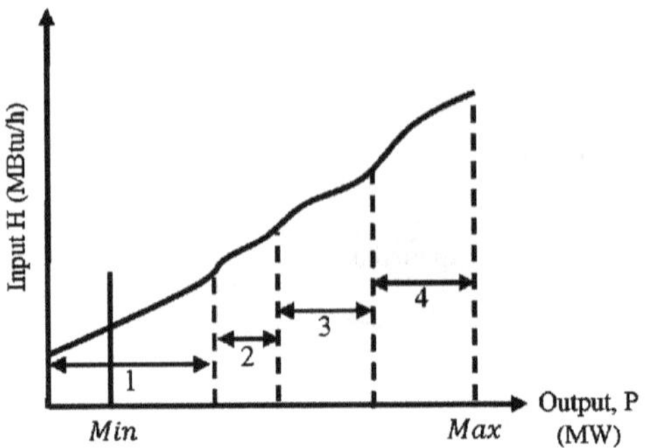

FIGURE 5.7 Multiple steam valves characteristics.

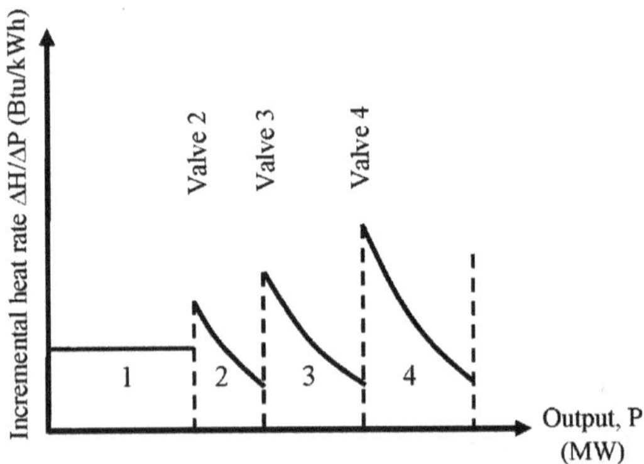

FIGURE 5.8 Incremental heat rate of the multiple steam valves characteristics.

valves characteristics as shown in Figure 5.8. Some factors, such as noise, erosion, and how steam is dried or superheated as it passes through a valve, affect the properties.

2. Common-header steam.

With the multiple low-pressure steam boilers connected to a common header, an idle boiler will act as a radiator and condense steam from the active boiler(s). Other than putting a check valve in the steam supply line from the boiler, is there a way to ensure that the idle boiler does not flood and the water treatment levels don't go out of control. The multiple boilers are installed and serve a common header, and each boiler should have what is known as a "boiler stop valve" in the main steam line, leaving the boiler before connecting to the header. This valve acts as a stop and checks valve. It prevents flow from an

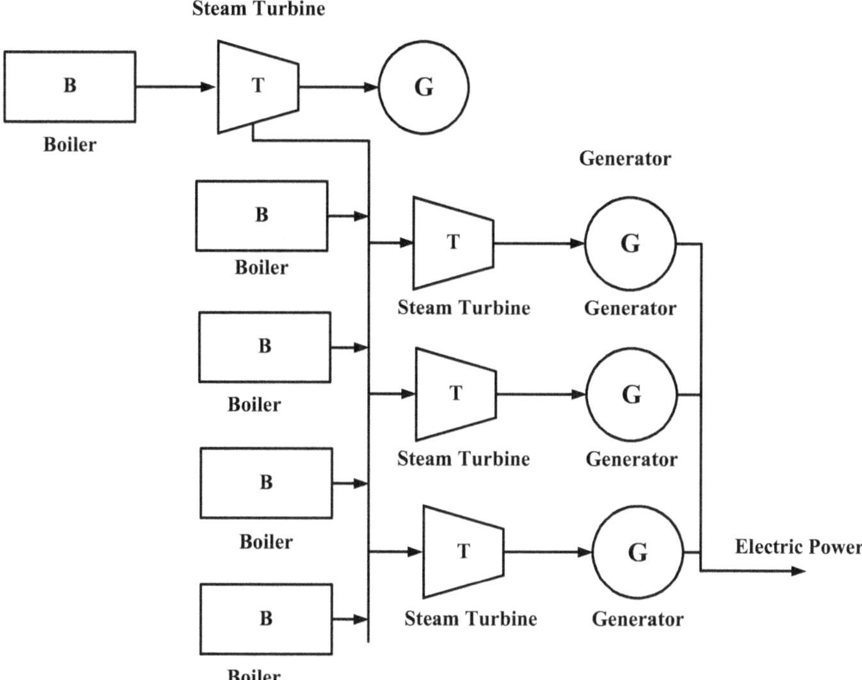

FIGURE 5.9 Common-header steam.

active boiler from entering and pressurizing an inactive boiler. The major valve manufacturers all list this type of valve. Figure 5.9 shows a common-header steam configuration.

3. Combined cycle power plant.

   A combined-cycle plant works to produce electricity and gets waste heat from the gas turbine to increase the efficiency of the generation system and increase the electrical output power. The gas turbine compresses air and mixes it with the fuel heated to a very high temperature. The hot air-fuel mixture moves through the gas turbine blades, making them spin. The spinning turbine drives a generator that converts a portion of the spinning energy into electrical power.

   Figure 5.10 shows a combined cycle power plant. A heat recovery steam generator (HRSG) takes exhaust heat from the gas turbine that would otherwise be released through the exhaust stack. The HRSG creates steam from the gas turbine exhaust heat and delivers it to the steam turbine. The steam turbine sends its energy to the generator drive shaft, converted into additional electricity.

   All the above-mentioned special cases have special solution methods. However, assume that the cost function is smooth and differentiable.

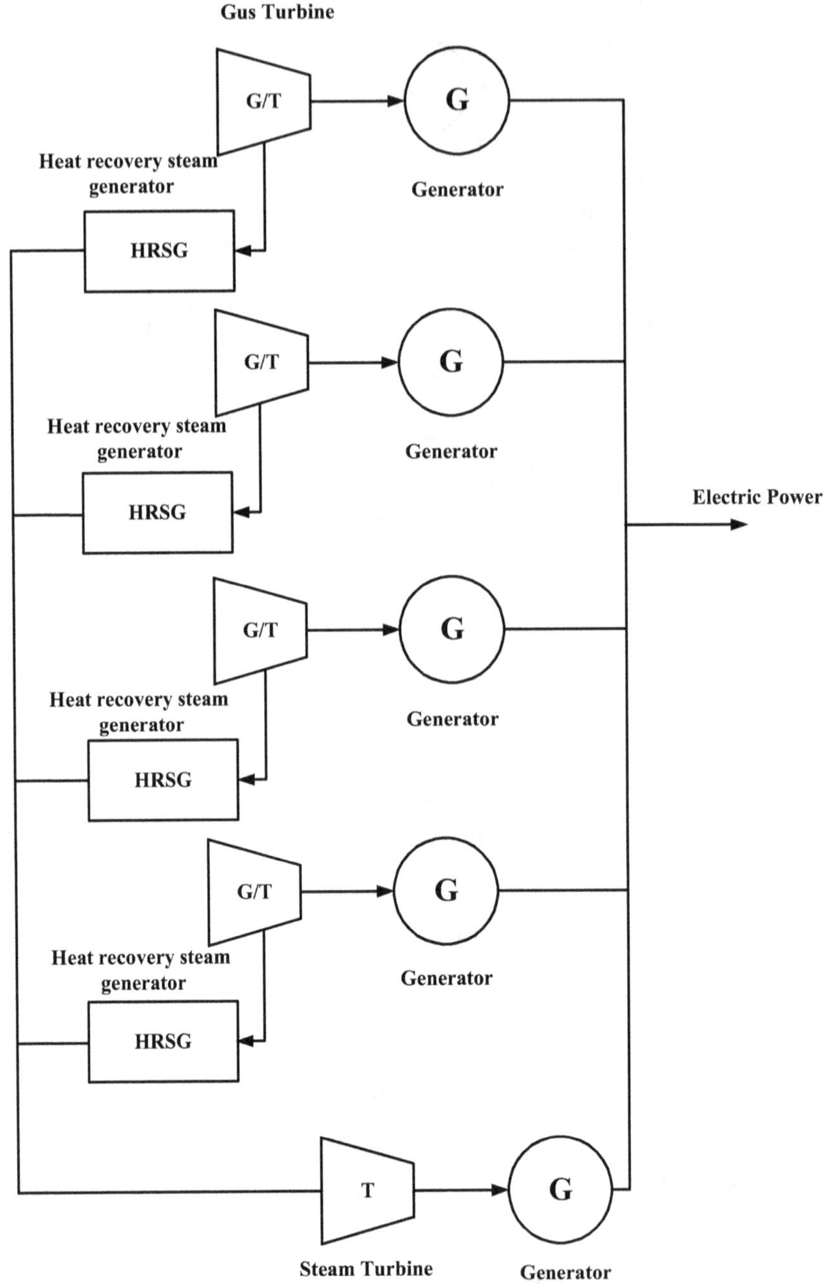

FIGURE 5.10 Combined cycle power plant.

5.6 ECONOMIC LOAD DISPATCH PROBLEM FORMULATION

Consider that N working generation units are in the circuit and should supply total load demand of grid equal to "Demand." The unit i has output power equal to P_i. To determine the share of each unit in supplying load, the total generation cost in the grid is minimized.

$$\text{Decision Vars} \rightarrow P_1, P_1, \ldots, P_N$$

$$\text{Objective Function} \rightarrow \text{Cost} = F_1(P_1) + F_2(P_1) + \ldots + F_N(P_N) \tag{5.11}$$

$$\text{Cost} = \sum_{i=1}^{N} F_i(P_i) \tag{5.12}$$

Optimization Problem:

$$\min \text{Cost} = \sum_{i=1}^{N} F_i(P_i) \tag{5.13}$$

Operating constraint $P_i^{\min} \leq P_i \leq P_i^{\min}$ $i = 1, 2, 3 \ldots, N$ (two constraints for each unit)

$$Demand\ power = P_1 + P_2 + \ldots + P_N = \sum_{i=1}^{N} P_i \tag{5.14}$$

The total number of constraints = $2N + 1$

Because $F_i(P_i)$ usually is a second-order function, and the ELD problem is a non-linear optimization problem. Firstly, should be familiar with the solution method of a non-linear optimization problem in general.

5.7 NON-LINEAR OPTIMIZATION PROBLEM USING LAGRANGE METHOD

Firstly, consider a problem without any constraints:

$$\min f(x_1, x_1, \ldots, x_n)$$

As may be familiar, the following conditions should be met at the optimum point.

$$\frac{\partial f}{\partial x_i} = 0 \Rightarrow \begin{array}{c} \dfrac{\partial f}{\partial x_1} = 0 \\ \dfrac{\partial f}{\partial x_2} = 0 \\ \vdots \\ \dfrac{\partial f}{\partial x_n} = 0 \end{array} \tag{5.15}$$

- All the above n equations should be met at the optimum point.
- These conditions only show the optimum point (max or min cannot be detected).

Example 5.5

Find the *Min* $f(x_1, x_1) = x_1^2 + 4x_2^2 - 4x_1^2 + 8$ using Lagrange coordinations.

Solution

$$\frac{\partial f}{\partial x_1} = 0\Delta 2x_1 - 4 = 0\Delta x_1 = \frac{4}{2} = 2$$

$$\frac{\partial f}{\partial x_2} = 0\Delta 8x_2 = 0\Delta x_2 = 0$$

Now, the values of x_1 and x_2 are considered as the optimum point

If the problem includes one equation constraint:

$$\min f\left(x_1, x_1, \ldots, x_n\right)$$

Operating constraints $g(x_1, x_1, \ldots, x_n) = 0$ (Note that any equation constraint can be converted to this form)

To solve this problem, the "Lagrange function" is defined as:

$$\mathcal{L} = f - \lambda g$$

where $\mathcal{L}$ is the Lagrange function, and λ is the Lagrange coefficient.

A coefficient (an auxiliary variable) adds the problem constraint to the objective function.

$$d\mathcal{L} \equiv 0$$

$$\Rightarrow \frac{\partial f}{\partial x_i} = 0 \; i = 1, 2, \ldots, n$$

$$\Rightarrow \frac{\partial f}{\partial x_i} - \lambda \frac{\partial f}{\partial x_i} = 0 \; i = 1, 2, \ldots, n$$

Another condition for optimum point is the constraint of the problem.

$$g\left(x_1, x_1, \ldots, x_n\right) = 0 \left(\equiv \frac{\partial f}{\partial \lambda} = 0 \right)$$

Consequently, at the optimum point

$$\frac{\partial f}{\partial x_i} - \lambda \frac{\partial f}{\partial x_i} = 0 \; i = 1, 2, \ldots, (n = \text{equations})$$

$$g\left(x_1, x_1, \ldots, x_n\right) = 0$$

Totally, $(n + 1)$ conditions

The above equations are called "coordination equations."

Example 5.6

Find the Min $f(x_1, x_1) = x_1^2 + 4x_2^2 - 4x_1^2 + 8$

Operating constraints $g = x_1 + 2x_2 - 3 = 0$ using Lagrange coordinations.

Solution

$$d\mathcal{L} = f - \lambda g = x_1^2 + 4x_2^2 - 4x_1^2 + 8 - \lambda(x_1 + 2x_2 - 3)$$

$$\frac{\partial \mathcal{L}}{\partial x_1} = 2x_1 - 4 - \lambda = 0$$

$$\frac{\partial \mathcal{L}}{\partial x_2} = 8x_2 - 2\lambda = 0$$

$$g(x_1, x_2) = x_1 + 2x_2 - 3 = 0$$

By solving this set of equations

$$x_1^* = 2.5$$

$$x_2^* = 0.25$$

$$\lambda = 1$$

Note that the obtained answer is located on constraint $g = 0$.

5.8 ELD PROBLEM SOLUTION REGARDLESS OF INEQUALITY CONSTRAINTS

$$\min Cost = \sum_{i=1}^{N} F_i(P_i) \tag{5.16}$$

Operating constraint

$$\sum_{i=1}^{N} P_i = Demand\ power \Rightarrow \sum_{i=1}^{N} P_i - Demand\ power = 0 \tag{5.17}$$

Lagrange function:

$$\mathcal{L} = \sum_{i=1}^{N} F_i(P_i) - \lambda\left(\sum_{i=1}^{N} P_i - Demand\ power\right) \tag{5.18}$$

Coordination equations:

$$\frac{\partial \mathcal{L}}{\partial P_k} = \frac{\partial F_k}{\partial P_k} - \lambda = 0 \, k = 1, 2, \ldots, N \tag{5.19}$$

$$\sum_{i=1}^{N} P_i - Demand\ power = 0 \tag{5.20}$$

$$\frac{\partial F_k}{\partial P_k} = (IC)_k \Rightarrow (IC)_k = \lambda\ k = 1, 2, \ldots, N \tag{5.21}$$

And

$$Demand\ power = \sum_{i=1}^{N} P_i \tag{5.22}$$

$$\begin{aligned} (IC)_1 &= \lambda \\ (IC)_2 &= \lambda \\ (IC)_N &= \lambda \end{aligned} \tag{5.23}$$

$$Demand\ power = \sum_{i=1}^{N} P_i \tag{5.24}$$

At the optimum point, all units have the same incremental cost or (*IC*) value (=λ). As stated before, the cost function is usually considered a second-order function. Therefore, the incremental cost function (*IC*) is in the form of a first-order (linear) function.

There is an infinite solution that gives the same (*IC*) value for all units. Among these solutions, only one is acceptable which satisfies the constraint of the problem (equality of generation and demand, $Demand\ Power = \sum_{i=1}^{N} P_i$).

Example 5.7

For Figure 5.11, if the power of four units *of* P_1, P_2, P_3, and P_4 are connected in parallel to supply a load, and the power of these units are validate the constraint?

Solution

The power demand equals the sum of the generation power,

$$Power\ demand = \sum_{i=1}^{4} P_i = P_1 + P_2 + P_3 + P_4$$

and since the λ line passes the incremental cost curve for each unit, so the power output from each generator validates the constraint.

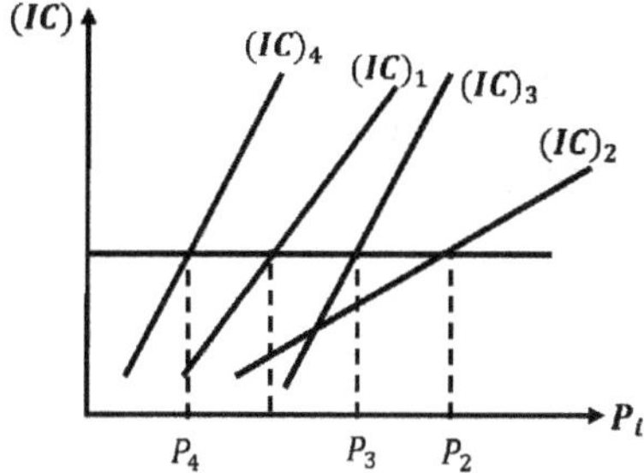

FIGURE 5.11 Fuel incremental cost of Example 5.7.

Example 5.8

Consider three units with characteristics presented below:

$$F_1(P_1) = 561 + 7.92P_1 + 0.001562P_1^2 \; 150\,\text{MW} \le P_1 \le 600\text{MW}$$

$$F_2(P_2) = 310 + 7.85P_2 + 0.00194P_2^2 \; 100\,\text{MW} \le P_2 \le 400\,\text{MW}$$

$$F_3(P_3) = 78 + 7.97P_3 + 0.00482P_3^2 \; 50\,\text{MW} \le P_3 \le 200\text{MW}$$

If the demand power = 850 MW, find these units' ELD.

Solution

At first, do not consider the inequality constraints. In other words, consider that all units are inside their allowed range.

Coordinate equations:

$$(IC)_1 = \frac{\partial F_1}{\partial P_1} = 7.92 + 0.003124P_1 = \lambda$$

$$(IC)_2 = \frac{\partial F_2}{\partial P_2} = 7.85 + 0.00388P_2 = \lambda$$

$$(IC)_3 = \frac{\partial F_3}{\partial P_3} = 7.97 + 0.00964P_3 = \lambda$$

$$\sum_{i=1}^{3} P_i = P_1 + P_2 + P_3 = Demand\ power = 850\text{MW}$$

By solving the above set of equations, the solution

$P_1 = 393.2$ MW, $P_2 = 334.6$ MW, $P_3 = 122.2$ MW, $\lambda = 9.148$

As can be seen, the obtained answers for all units are in their allowed ranges. Therefore, the solution is acceptable.

Example 5.9

Suppose that because of a change in fuel cost, the cost function for unit 1 becomes $F_1(P) = 459 + 6.48P_1 + 0.00128P_1^2$. Resolve Example 5.8 for these new conditions.

Solution

$$(IC)_1 = \frac{\partial F_1}{\partial P_1} = 6.48 + 0.00256P_1 = \lambda$$

Substituting the above equation coordination equations set and solving the equation's results in

$$P_1 = 704.6 \text{ MW}, P_2 = 118.8 \text{ MW}, P_3 = 32.6 \text{ MW}, \lambda = 8.284$$

As can be seen, units 1 and 3 are not within their limits. Therefore, unit 1 can be set to its maximum output and unit 3 to its minimum output. In this case, dispatch becomes

$$P_1 = P_1^{max} = 600\,\text{MW} \;\; P_3 = P_3^{min} = 50\text{MW}$$

$$P_2 = Demand\ power - (P_1 + P_3) = 850 - (600 + 50) = 200\text{MW}$$

P_2 is obtained within its limits. The common incremental cost of the system (λ_{sys}) is obtained from unit 2.

$$\lambda_{sys} = (IC)_2 = (\frac{\partial F_2}{\partial P_2}|_{P_2=200} = 8.626$$

Now, it is necessary to check Kuhn-Tucker conditions for fixed units:

$(IC)_1|_{P1=600} = 8.016 \leq \lambda_{sys}$, this is correct.
$(IC)_3|_{P3=50} = 8.16 \not\geq \lambda_{sys}$, this does not seem right.

Unit 3 does not meet the Kuhn-Tucker conditions. Considering this result and the fact that in initial results, $P_3 = 32.6$ MW is a near limit $P_3 = P_3^{min} = 50$ MW in compared with $P_1 = 704.6$ MW, which is far from $P_1 = P_1^{max} = 600$ MW, one can be obtained is only by fixing P_1 on P_1^{max} Probably enough to reach the final answer. Decreasing P_1 on P_1^{max} cause an increase in P_2 and P_3 and can cause P_3 to return to its range.

New coordination equations:

$$P_1 = 600\text{MW}$$

$$(IC)_2 = 7.85 + 0.00388P_2 = \lambda_{sys}$$

$$(IC)_3 = 7.97 + 0.00964P_3 = \lambda_{sys}$$

$$600 + P_2 + P_3 = 850 \Rightarrow P_2 + P_3 = 250\text{MW}$$

Solving the above equations set, results in

$$P_2 = 187.1\text{MW}\ P_3 = 62.9\text{MW}\ \lambda_{sys} = 8.576$$

P_2 and P_3 are obtained within their allowed ranges. Because λ_{sys} is changed, it is necessary to check Kuhn-Tucker conditions for unit 1.

$$(IC)_1\big|_{P_1=600} = 8.016 \le \lambda_{sys}\,(=8.576)$$

Therefore, the solution is acceptable, and the final optimum answer is obtained.

5.9 MEMORIZE KUHN–TUCKER CONDITIONS FOR ELD PROBLEMS

Consider various units, for example, and their typical incremental cost (IC) curves given in Figure 5.12 as:

For non-fixed units (3 and 4):

$$(IC)_3 = (IC)_4 = \lambda_{sys} \tag{5.25}$$

For unit 1, which is fixed on $P_1^{\min}$:

$$(IC)_1 = \lambda_1 \ge \lambda_{sys} \tag{5.26}$$

For unit 2, which is fixed on $P_2^{\max}$:

$$(IC)_1 = \lambda_2 \le \lambda_{sys} \tag{5.27}$$

As can be observed from the previous example, finding the solution of the ELD problem by direct solving coordinating equations set and repeatedly checking the Kuhn-Tucker conditions is a time-consuming and challenging process significantly when the number of units is highly increased in real grids. Therefore,

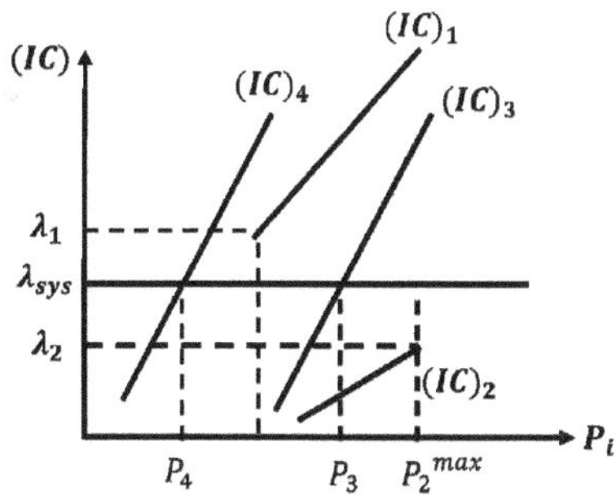

FIGURE 5.12 Incremental fuel cost for different units in different cases.

it is necessary to find faster methods for solving ELD problems. In the next sections, different methods are introduced for solving ELD problems faster and easier.

5.10 THE LAMBDA-ITERATION METHOD

Consider, for example, four units with incremental cost functions similar to Figure 5.13.

Because at the optimum point, all units have the same incremental cost value (equal to λ), start with an arbitrary value for λ (λ_o for example) and then calculate generation share for each unit (P_i values) using equations of incremental cost functions. Then calculate the total generation $\left(\sum_{i=1}^{N} P_i\right)$ and compare it with demand:

$$If \sum_{i=1}^{N} P_i < Demand\ power \rightarrow increase\ \lambda\ a\ bit$$

$$If \sum_{i=1}^{N} P_i > Demand\ power \rightarrow decrease\ \lambda\ a\ bit$$

In the next step, P_i values calculated for new λ and two above conditions checked again. This process is repeated until $\sum_{i=1}^{N} P_i \cong Demand\ power$. At this state, the final answer has been obtained. If one of the units has reached its minimum or maximum in this process, limit (P_i^{min} or P_i^{max}), its output power (P_i) is fixed on that limit. Because changing of λ value is small in each iteration (step), overtime at most, one unit has reached its limit and is fixed. Therefore, the Kuhn-Tucker conditions remain valid.

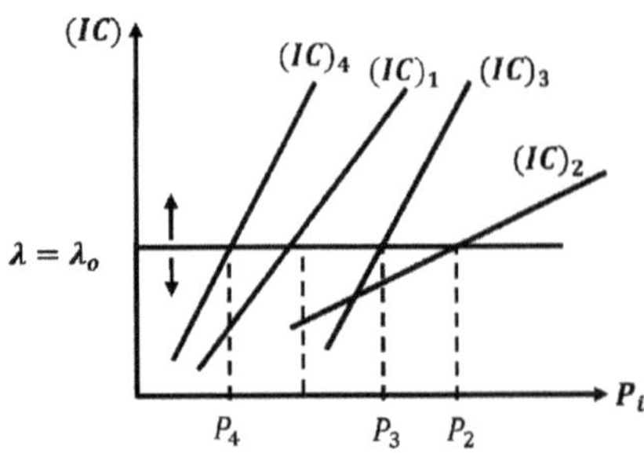

FIGURE 5.13 Incremental fuel cost for a different unit at economic considerations.

Example 5.10

Consider three-generation units with third-order cost functions as:

$$F(P_i) = A_i + B_iP_i + C_iP_i^2 + D_iP_i^3$$

$$320\,\text{MW} \le P_1 \le 800\,\text{MW}$$

$$300\,\text{MW} \le P_2 \le 1200\,\text{MW}$$

$$275\,\text{MW} \le P_3 \le 1100\,\text{MW}$$

Find the ELD for Demand = 2500 MW.

Solution

The initial value of λ is arbitrary. However, if it is selected near the final answer, the total required interactions decrease, and the speed of convergence increases. One simple way to a smart selection of λ initial value is consideration of a constant term of incremental cost functions (B coefficient in this example) and selecting initial λ a little larger (considering the effect of other terms). In this example $\lambda_o = 8$ is selected, and change in λ is ±10% for each iteration. Results for different iterations are listed in Table 5.3.

The Lambda-iteration method still has a low speed. It is necessary to find intelligent methods in selecting the optimum point that each iteration intelligently advances the answer toward the optimum point.

TABLE 5.2
Constants Parameters of Example 5.10

Unit	A	B	C	D
Unit 1	749.55	6.95	9.68×10^{-4}	1.27×10^{-7}
Unit 2	1285.0	7.051	7.375×10^{-4}	6.453×10^{-8}
Unit 3	1531.0	6.531	1.04×10^{-3}	9.98×10^{-8}

TABLE 5.3
Results of Example 5.10

Iteration	λ	Total Generation (M.W.)	P_1	P_2	P_3
1	8.0000	1731.6	494.3	596.7	640.6
2	8.8000	2795.0	800.0	1043.0	952.0
3	8.5781	2526.0	734.7	923.4	867.9
4	8.5566	2497.5	726.1	911.7	859.7
5	8.5586	2500.0	726.9	912.7	860.4

5.11 FIRST-ORDER GRADIENT SEARCH

Consider the objective function of the ELD problem:

$$Cost = \sum_{i=1}^{N} F_i(P_i) = F_1(P_1) + F_2(P_1) + \ldots + F_N(P_N) \tag{5.28}$$

Taylor series expansion of cost function is:

$$\begin{aligned} Cost + \Delta Cost &= F_1(P_1) + F_2(P_1) + \ldots \\ &+ F_N(P_N) + \frac{\partial F_1}{\partial P_1}\Big|_{P_1} \Delta P_1 \\ &+ \frac{\partial F_2}{\partial P_2}\Big|_{P_2} \Delta P_2 + \ldots + \frac{\partial F_N}{\partial P_N}\Big|_{P_N} \Delta P_N \end{aligned} \tag{5.29}$$

where, $F_1(P_1) + F_2(P_1) + \cdots + F_N(P_N)$ is the cost function. The second and higher-order derivations are neglected.

$$\Delta Cost = \frac{\partial F_1}{\partial P_1}\Big|_{P_1} \Delta P_1 + \frac{\partial F_2}{\partial P_2}\Big|_{P_2} \Delta P_2 + \ldots + \frac{\partial F_N}{\partial P_N}\Big|_{P_N} \Delta P_N \tag{5.30}$$

The Taylor series is written at a point that equality constraint $\sum_{i=1}^{N} P_i = Demand\ power$ is met but not necessarily the optimum point. In other words, despite the previous method, starting from a point in which equality constraint is valid and then moving toward the solution to get optimum point.

In moving the process toward an optimum point, the validity of the equality constraint must be kept in each iteration. Therefore, if the generation input is reduced, the generation share of another unit should be increased with the same value and vice versa. In other words, in each iteration, the following condition should be met:

$$\sum_{i=1}^{N} \Delta P_i = 0 \tag{5.31}$$

A unit can be arbitrarily chosen as a dependent unit (unit x). Some amount of generation power of the dependent unit is decreased (or increased) and is added (or subtracted) to the most effective unit in $\Delta Cost$. The most effective unit can be found based on $\Delta Cost$ equation. Assume that change in the generation unit of the dependent unit is ΔP_x, therefore it is necessary that:

$$\sum_{i=1}^{N} \Delta P_i = 0 \Rightarrow \Delta P_x = -\sum_{\substack{i=1 \\ i \neq x}}^{N} \Delta P_i = -(\Delta P_1 + \cdots + \Delta P_N) \tag{5.32}$$

Substituting the above equation in $\Delta Cost$ Equation 5.29:

$$\Delta Cost = \frac{\partial F_1}{\partial P_1}\Delta P_1 + \cdots - \frac{\partial F_x}{\partial P_x}\left(\Delta P_1 + \cdots + \Delta P_N\right) + \cdots + \frac{\partial F_N}{\partial P_N}\Delta P_N \tag{5.33}$$

$$\Delta Cost = \left(\frac{\partial F_1}{\partial P_1} - \frac{\partial F_x}{\partial P_x}\right)\Delta P_1 + \left(\frac{\partial F_2}{\partial P_2} - \frac{\partial F_x}{\partial P_x}\right)\Delta P_2 + \cdots + \left(\frac{\partial F_N}{\partial P_N} - \frac{\partial F_x}{\partial P_x}\right)\Delta P_N \tag{5.34}$$

$$\Delta Cost = \sum_{\substack{i=1 \\ i \neq x}}^{N} \left(\frac{\partial F_i}{\partial P_i} - \frac{\partial F_x}{\partial P_x}\right)\Delta P_i \tag{5.35}$$

The unit with the largest value of the coefficient $\left|\frac{\partial F_i}{\partial P_i} - \frac{\partial F_x}{\partial P_x}\right|$ (absolute value) is a most effective unit on $\Delta Cost$ with a specified value for ΔP. This unit is known as an effective cost unit.

After selection of an effective unit, based on the sign of the effective unit coefficient $\left|\frac{\partial F_i}{\partial P_i} - \frac{\partial F_x}{\partial P_x}\right|$, the sign of its power change ΔP_i is selected so that $\Delta Cost$ becomes negative (cost decreased in each iteration). The sign of ΔP_x selected despite the ΔP_i as stated before.

By applying ΔP_x to the dependent unit and $\Delta P_i (= -\Delta P_x)$ to the affected unit, new values for P_i and P_x is calculated for the first iteration (generation of other units remained constant). Now using new values for the generation of units, $\Delta Cost$ Equation 5.35 is recalculated, and by selecting a new effective unit, the above process is done again for the next iteration. This process is repeated repeatedly until a final answer is obtained. If, during iterations, any of the dependent or effective units reach to its generation limit (P^{min} or P^{max}), that unit is fixed to that limit, and another unit is selected.

The iteration process is repeated until all of $\left(\frac{\partial F_i}{\partial P_i} - \frac{\partial F_x}{\partial P_x}\right)$ coefficients become approximately zero. In this case, because of zero coefficients, change in units generations (ΔP_i) does not affect Cost reduction.

On the other hand,

$$\left(\frac{\partial F_i}{\partial P_i} - \frac{\partial F_x}{\partial P_x}\right) = 0, i = 1, 2, \ldots, N \Rightarrow \begin{cases} \frac{\partial F_1}{\partial P_1} - \frac{\partial F_x}{\partial P_x} \\ \vdots \\ \frac{\partial F_N}{\partial P_N} - \frac{\partial F_x}{\partial P_x} \end{cases} \tag{5.36}$$

$$\frac{\partial F_1}{\partial P_1} = \frac{\partial F_2}{\partial P_2} = \cdots = \frac{\partial F_N}{\partial P_N} \tag{5.37}$$

$$(IC)_1 = (IC)_2 = \cdots = (IC)_N \text{ (except fixed units)} \tag{5.38}$$

which is equivalent to validation of coordination equations.

Example 5.11

Consider three units with characteristics presented below. Find the ELD solution for *Demand* = 850 MW using the first-order gradient search method.

$$(IC)_1 = 7.92 + 0.003124P_1 150\text{MW} \le P_1 \le 600\,\text{MW}$$

$$(IC)_2 = 7.85 + 0.003884P_2 100\text{MW} \le P_2 \le 400\,\text{MW}$$

$$(IC)_3 = 7.97 + 0.00964\,P_3\ 50\text{MW} \le P_3 \le 300\,\text{MW}$$

Solution

Start with an arbitrary answer which meets the equality constraint

$$\sum_{i=1}^{N} P_i = Demand\ power$$

for example:

$$P_1 = 400\ \text{MW},\ P_2 = 300\ \text{MW},\ P_3 = 150\ \text{MW}$$

Unit 3 is selected as a dependent unit (arbitrarily)

$$\frac{\partial F_i}{\partial P_i} - \frac{\partial F_x}{\partial P_x} = \frac{\partial F_i}{\partial P_i} - \frac{\partial F_3}{\partial P_3} = (IC)_i - (IC)_3$$

$$(IC)_i - (IC)_3 = \begin{cases} (IC)_1 |_{P_1=400} - (IC)_3 |_{P_3=150} = -0.2464 \\ (IC)_2 |_{P_2=300} - (IC)_3 |_{P_3=150} = -0.4020\, maximum\ absolute\ value \end{cases}$$

$$\Delta Cost = \left[(IC)_1 - (IC)_3\right]\Delta P_1 + \left[(IC)_2 - (IC)_3\right]\Delta P_2$$

$|(IC)_2 - (IC)_3|$ has the maximum absolute value. Therefore, unit 2 is selected as an effective unit. Because the sign of $(IC)_2 - (IC)_3$ is negative, the sign of ΔP_2 should be positive to obtain negative value for $\Delta Cost$ (equivalent to Cost reduction). Therefore, some amount (e.g., 50MW) is subtracted from P_3 (dependent unit) and added to P_2 ($\Delta P_3 = -\Delta P_2$) consequently

$$P_1 = 400\,\text{MW}, P_2 = 350\,\text{MW}, P_3 = 100\,\text{MW}$$

Same as first iteration coefficient of $\Delta Cost$ are recalculated.
(Keeping unit 3 as a dependent unit)

$$((IC)_1 |_{P_1=400} - ((IC)_3 |_{P_3=100} = 0.2356$$
$$((IC)_2 |_{P_2=350} - ((IC)_3 |_{P_3=100} = 0.2740$$

$$\Delta Cost = 0.2356\Delta P_1 + 0.2740\Delta P_2$$

Still, unit 2 has a higher coefficient and is selected as an effective unit. But now its coefficient is positive. Therefore, it is necessary to subtract some amount from P_2 and add it to P_3 (obviously, ΔP should be smaller than before, e.g. $\Delta P_2 = -25$ MW and $\Delta P_3 = 25$MW chosen).

Consequently, new values for generation is

$$P_1 = 400\,\text{MW}, P_2 = 325\,\text{MW}, P_3 = 125\,\text{MW}$$

The calculation can be continued until reaching the final answer, which are

$$P_1 = 393.2\,\text{MW}, P_2 = 334.6\,\text{MW}, P_3 = 122.2\,\text{MW}$$

Using the above values to calculate $\Delta Cost$ results in

$$\Delta Cost = 0.00035\ \Delta P_1 + 0.00024\ \Delta P_2$$

Because all of the coefficients in $\Delta Cost$ are very small, the final answer is obtained, and because all constraints, including generation limits, are met, the answer is acceptable.

5.12 SECOND-ORDER SEARCH

5.12.1 Second-Order Search Formulation

Taylor series expansion of cost function considering second-order terms and neglecting third and higher orders:

$$\begin{aligned} Cost + \Delta Cost = {} & F_1(P_1) + F_2(P_2) \\ & + \cdots + F_N(P_N) + \frac{\partial F_1}{\partial P_1}\Delta \times P_1 + \frac{\partial F_2}{\partial P_2}\Delta P_2 + \cdots \\ & + \frac{\partial F_N}{\partial P_N}\Delta P_N + \frac{1}{2}\left[\frac{\partial^2 F_1}{\partial P_1^2}(\Delta P_1)^2 \right. \\ & \left. + \frac{\partial^2 F_2}{\partial P_2^2}(\Delta P_2)^2 + \cdots + \frac{\partial^2 F_N}{\partial P_N^2}(\Delta P_N)^2\right] \end{aligned} \tag{5.39}$$

Where,

$$Cost = F_1(P_1) + F_2(P_2) + \ldots + F_N(P_N) \tag{5.40}$$

and

$$\Delta Cost = \frac{\partial F_1}{\partial P_1}\Delta P_1 + \frac{\partial F_2}{\partial P_2}\Delta P_2 + \ldots + \frac{\partial F_N}{\partial P_N}\Delta P_N \tag{5.41}$$

Again starting from a point that will meet the equality constraint $\left(\sum_{i=1}^{N} P_i = Demand\ power\right)$.

$$\sum_{i=1}^{N}\Delta P_i = 0 \Rightarrow \Delta P_x = -\sum_{\substack{i=1\\ i\neq x}}^{N}\Delta P_i = -\Delta P_1 - \Delta P_2 - \cdots_{\neq x} - \Delta P_N \tag{5.42}$$

Where x is the dependent unit,

$$\begin{aligned}(\Delta P_x)^2 &= (-\Delta P_1 - \Delta P_2 - \cdots_{\neq x} - \Delta P_N)^2 \\ &= \Delta P_1^2 + \Delta P_2^2 - \cdots_{\neq x} - \Delta P_N^2 + 2\Delta P_1\Delta P_2 + 2\Delta P_1\Delta P_3 \\ &\quad + \cdots_{\neq x} + 2\Delta P_1\Delta P_N + 2\Delta P_2\Delta P_3 + \ldots\end{aligned} \tag{5.43}$$

Combining the above equations:

$$\begin{aligned}\Delta Cost = \sum_{i=1}^{N}\left(\frac{\partial F_i}{\partial P_i} - \frac{\partial F_x}{\partial P_x}\right)\Delta P + \frac{1}{2}\Bigg[\frac{\partial^2 F_1}{\partial P_1^2}(\Delta P_1)^2_{\ i} \\ + \frac{\partial^2 F_2}{\partial P_2^2}(\Delta P_2)^2 + \cdots + \frac{\partial^2 F_x}{\partial P_x^2}(\Delta P_1^2 + \Delta P_2^2 + \cdots \\ + \Delta P_N^2 + 2\Delta P_1\Delta P_2 + 2\Delta P_1\Delta \\ \times P_3 + \cdots) + \cdots + \frac{\partial^2 F_N}{\partial P_N^2}(\Delta P_N)^2\Bigg]\end{aligned} \tag{5.44}$$

At the optimum point, it is necessary that,

$$\frac{\partial \Delta Cost}{\partial \Delta P_i} = 0 \text{ for all } i = 1, 2, \ldots, N \text{ and } i \neq x \tag{5.45}$$

$$\frac{\partial \Delta Cost}{\partial \Delta P_1} = \left(\frac{\partial F_1}{\partial P_1} - \frac{\partial F_x}{\partial P_x}\right) + \frac{\partial^2 F_1}{\partial P_1^2}\Delta P_1 + \frac{\partial^2 F_x}{\partial P_x^2}(\Delta P_1 + \Delta P_2 + \ldots + \Delta P_N) = 0 \tag{5.46}$$

$$\frac{\partial \Delta Cost}{\partial \Delta P_2} = \left(\frac{\partial F_2}{\partial P_2} - \frac{\partial F_x}{\partial P_x} \right) + \frac{\partial^2 F_2}{\partial {P_2}^2} \Delta P_2 + \frac{\partial^2 F_x}{\partial {P_x}^2} \sum_{\substack{i=1 \\ i \neq x}}^{N} \Delta P_i = 0 \tag{5.47}$$

$$\vdots$$

$$\frac{\partial \Delta Cost}{\partial \Delta P_N} = \left(\frac{\partial F_N}{\partial P_N} - \frac{\partial F_x}{\partial P_x} \right) + \frac{\partial^2 F_N}{\partial {P_N}^2} \Delta P_N + \frac{\partial^2 F_x}{\partial {P_x}^2} \sum_{\substack{i=1 \\ i \neq x}}^{N} \Delta P_i = 0 \tag{5.48}$$

Where,

$$\sum_{\substack{i=1 \\ i \neq x}}^{N} \Delta P_i = \Delta P_1 + \Delta P_2 + \ldots + \Delta P_N \tag{5.49}$$

For simplicity and more clarity, define F_i' and F_i'' as,

$$F_i' \triangleq \frac{\partial F_i}{\partial P_i} \text{ and } F_i'' \triangleq \frac{\partial^2 F_i}{\partial P_i^2} \tag{5.50}$$

By stating equation set (***) in matrix form and using the above definitions:

$$\begin{bmatrix} & & & \cdots & \\ F_1'' + F_x'' & F_x'' & F_x'' & \cdots & \\ F_x'' & F_2' + F_x'' & F_x'' & \cdots & F_x'' \\ F_x'' & F_x'' & F_3' + F_x'' & \ddots & F_x'' \\ \vdots & \vdots & \vdots & \cdots & F_x'' \\ F_x'' & F_x'' & F_x'' & & \vdots \\ & & & & F_N' + F_x'' \end{bmatrix} \begin{bmatrix} \Delta P_1 \\ \Delta P_2 \\ \Delta P_3 \\ \vdots \\ \Delta P_N \end{bmatrix}_{\neq x} = \begin{bmatrix} F_x' + F_1' \\ F_x' + F_2' \\ F_x' + F_3' \\ \vdots \\ F_x' + F_N' \end{bmatrix} \tag{5.51}$$

Where,

$$\bar{A} = \begin{bmatrix} & & & \cdots & \\ F_1'' + F_x'' & F_x'' & F_x'' & \cdots & \\ F_x'' & F_x'' + F_x'' & F_x'' & \cdots & F_x'' \\ F_x'' & F_x'' & F_3' + F_x'' & \ddots & F_x'' \\ \vdots & \vdots & \vdots & \cdots & F_x'' \\ F_x'' & F_x'' & F_x'' & & \vdots \\ & & & & F_N' + F_x'' \end{bmatrix} \tag{5.52}$$

$$\overline{\Delta P} = \begin{bmatrix} \Delta P_1 \\ \Delta P_2 \\ \Delta P_3 \\ \vdots \\ \Delta P_N \end{bmatrix} \text{ and } \overline{B} = \begin{bmatrix} F_x' + F_1' \\ F_x' + F_2' \\ F_x' + F_3' \\ \vdots \\ F_x' + F_N' \end{bmatrix} \tag{5.53}$$

Also,

$$\overline{A}.\overline{\Delta P} = \overline{B} \tag{5.54}$$

$$\overline{\Delta P} = \overline{A}^{-1}.\overline{B}. \tag{5.55}$$

5.12.2 Second-Order Search Algorithm

The second-order gradient search has high-speed convergence. Usually, the final answer is obtained in one or two iterations. The main disadvantage of this method is the difficulty of considering the generation limits of units.

An initial guess is chosen such that

$$\sum_{i=1}^{N} P_i = Demand\ power$$

A dependent unit is chosen arbitrarily.

F_1' and F_1'' values calculated at chosen work point. Matrices $\overline{A}$ and $\overline{B}$ are established and ΔP_i values are obtained from $\overline{\Delta P} = \overline{A}^{-1}.\overline{B}$. ΔP_x also obtained from $\Delta P_x = \sum_{i=1}^{N} \Delta P_i$. Consequently, new P_i values are calculated.

By calculating new F_1' and F_1'' values using new P_i values, the process mentioned above is repeated for the next iteration.

The process is repeated until $\overline{B} \cong 0$. In this case $\overline{\Delta P} \cong 0$. It means no more change in units generations in the next iteration, and the final answer is obtained. On the other hand;

$$\overline{B} = 0 \Rightarrow \begin{cases} F_x' + F_1' = 0 \\ F_x' + F_2' = 0 \\ \vdots \\ F_x' + F_N' = 0 \end{cases} \Rightarrow \begin{matrix} F_1' = F_2' = \ldots = F_N' = \lambda \\ (IC)_1 = (IC)_2 = \ldots = (IC)_N = \lambda \end{matrix} \tag{5.56}$$

Example 5.12

Solve the previous example using the second-order gradient search method.

Solution

The initial guess is selected the same as before:

$$P_1 = 400\,\text{MW}, P_2 = 300\,\text{MW}, P_3 = 150\,\text{MW}$$

Unit 3 is chosen as a dependent unit.

$$F_1|_{P_1=400} = 9.1696 \qquad F_1'' = 0.003124$$

$$F_1|_{P_2=300} = 9.0140 \qquad F_2'' = 0.00388$$

$$F_1|_{P_3=150} = 9.4160 \qquad F_3'' = 0.00964$$

$$\begin{bmatrix} 0.003124 + 0.00964 & 0.00964 \\ 0.00964 & 0.00388 + 0.00964 \end{bmatrix} \begin{bmatrix} \Delta P_1 \\ \Delta P_2 \end{bmatrix} = \begin{bmatrix} 9.4160 - 9.1696 \\ 9.4160 - 9.0140 \end{bmatrix}$$

$$\begin{bmatrix} \Delta P_1 \\ \Delta P_2 \end{bmatrix} = \begin{bmatrix} -6.83 \\ 34.603 \end{bmatrix} \Rightarrow \begin{cases} P_1^{new} = P_1^{old} + \Delta P_1 = 400 - 6.83 = 393.17\,MW \\ P_2^{new} = P_2^{old} + \Delta P_2 = 300 + 34.603 = 334.603\,MW \\ P_3^{new} = Demand\ power - \left(P_1^{new} + P_2^{new}\right) = 122.23\,MW \end{cases}$$

$$\Delta P_3 = -\left(\Delta P_1 + \Delta P_2\right)$$

The obtained values are within the allowed ranges of units. In the next iteration, the F_1' and F_1'' values should be recalculated using new P_i values. Because Cost functions (F_i) are modeled with second-order functions, F_1'' values are not changed. Therefore, no need to recalculate A^{-1}. But B should be recalculated:

$$(F_3'|_{P_3=122.23} - (F_1'|_{P_1=393.17} \cong -0.000034$$

$$(F_3'|_{P_3=122.23} - (F_2'|_{P_2=334.603} \cong -0.000049$$

From the above results, the value of $\bar{B} \cong 0$. Therefore, the final answer was obtained in the first iteration. As can be seen, the answer is the same as the previous example (and even more accurate).

5.13 BASEPOINT AND PARTICIPATION FACTORS METHOD

This method assumes that the economic dispatch problem must be solved repeatedly by moving the units from one optimum point to another as the load (demand) changes by a reasonably small amount. This is usually the case because the ELD problem should be solved periodically to respond to load changes in a real power system (e.g., every one hour or every half hour). Therefore, the previous solution is available, and the new demand is usually near the previous one because of the short time between two consecutive solutions.

Starting from a given acceptable solution (the base point) is possible. In the next step, the scheduler considers the demand change and investigates how much each generating unit needs to be moved (i.e., participate in the demand change) to serve the new load at the most economical operating point.

Considering the increased cost curve of the i-th unit in Figure 5.14, as the unit load is changed by the amount of ΔP_i, the incremental system cost moves from λ^o to $\lambda^o + \Delta\lambda$. For a slight change in power output on this single unit given by

$$\frac{\Delta\lambda}{\Delta P_i} = \frac{d}{dP_i}(IC)_i = \frac{dF_i'}{dP_i} = F_i'' \Rightarrow \Delta P_i = \frac{\Delta\lambda}{F_i''} \tag{5.57}$$

This is true for each of the N-units on the system so that

$$\begin{aligned} \Delta P_1 &= \frac{\Delta\lambda}{F_1''} \\ \Delta P_2 &= \frac{\Delta\lambda}{F_2''} \\ &\vdots \\ \Delta P_N &= \frac{\Delta\lambda}{F_N''} \end{aligned} \tag{5.58}$$

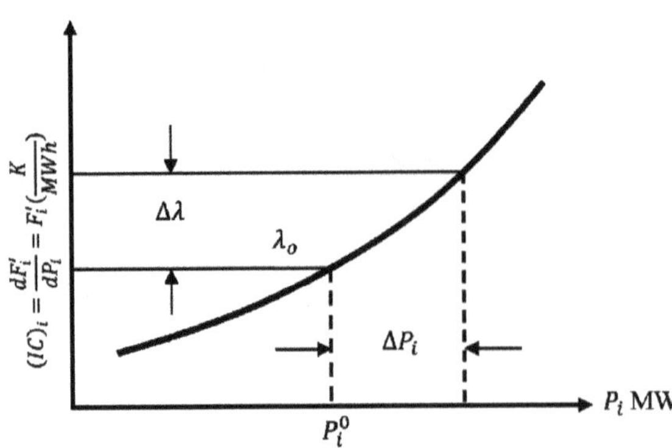

FIGURE 5.14 Increased cost curve of the i-th unit basepoint and participation factors.

The summation of Equation 5.58 results in

$$(\Delta P_1 + \Delta P_2 + \cdots + \Delta P_N = \Delta\lambda\left(\frac{1}{F_1''} + \frac{1}{F_2''} + \cdots \frac{1}{F_N''}\right) \tag{5.59}$$

$$\sum_{i=1}^{N} \Delta P_i = \Delta\lambda \sum_{i=1}^{N} \frac{1}{F_i''} \tag{5.60}$$

On the other hand,

$$\Delta Demand\ power = Demand\ power^{new} - Demand\ power^{old}$$

$$= \sum_{i=1}^{N} P_i^{new} - \sum_{i=1}^{N} P_i^{old}$$

$$= \sum_{i=1}^{N} \left(P_i^{new} - P_i^{old}\right) \tag{5.61}$$

Using (5.61) in (5.60) results in

$$\Delta Demand\ power = \Delta\lambda \sum_{i=1}^{N} \frac{1}{F_i''}$$

Dividing Equation (5.58) to the above equation:

$$\frac{\Delta P_i}{\Delta Demand\ power} = \frac{\frac{1}{F_i''}}{\sum_{i=1}^{N} \frac{1}{F_i''}} = \alpha_i$$

Where α_i is known as the participation factor, and the above equation can be used to determine each generation unit for new demand:

$$\Delta P_i = \alpha_i \Delta Demand\ power$$

$$\Delta P_i = P_i^{new} - P_i^{old}$$

Or

$$P_i^{new} = P_i^{old} + \Delta P_i$$

Example 5.13

A power system has P_1^{old} = 393.2 MW, P_2^{old} = 334.6 MW, P_3^{old} = 122.2 MW, and *Demand power*^{old} = 850 MW. Find the economic dispatch for Demand = 900 MW.

Solution

$$\Delta Demand\ power = Demand\ power^{new} - Demand\ power^{old}$$
$$= 900 - 850 = 50\,\text{MW}$$

Using data from a previous example

$$\alpha_1 = \frac{\frac{1}{F_1''}}{\sum_{i=1}^{N}\frac{1}{F_i''}} = \frac{(0.003124)^{-1}}{(0.003124)^{-1} + (0.00388) + (0.00964)} = 0.47$$

$$\alpha_2 = \frac{\frac{1}{F_2''}}{\sum_{i=1}^{N}\frac{1}{F_i''}} = \frac{(0.00388)^{-1}}{\sum_{i=1}^{N}\frac{1}{F_i''}} = 0.38$$

$$\alpha_3 = 1 - (\alpha_1 + \alpha_2) = 1 - (047 + 0.38) = 0.15$$

Note that based on the definition and concept of participation factor,

$$\sum_{i=1}^{N}\alpha_i = 1$$

$$P_1^{new} = P_1^{old} + \Delta P_1$$
$$= P_1^{old}\ \alpha_i \Delta Demand\ power$$
$$= 393.2 + (0.47)50 = 416.7\,\text{MW}$$
$$P_2^{new} = 334.6 + (0.38)(50) = 353.6\text{MW}$$
$$P_3^{new} = 122.2 + (0.15)(50)$$
$$= 900 - (P_1^{new} + P_2^{new}) = 129.7\ \text{MW}$$

It is evident that if cost functions are modeled with second-order equations (or I.C. functions are linear), a small change in demand is not required to use this method.

PROBLEMS

5.1 Solve Example 5.2 using MATLAB® software's L.P. command (linprog)?

5.2 Consider an objective function of $f(x) = (n - 2)^2$. Minimize $f(x)$ for the problem constraints subject to $2 \leq n \leq 4$.

5.3 A plant produces three different types of products, as shown in Table 5.1, by applying various operations. Determine how many units of each product should be generated to maximize total plants profit.

5.4 Convert the following problem to conventional form:

$$\min z_o = 3x_1 + 3x_2 + 7x_3$$

Operating constraints $x_1 + x_2 + 3x_3 \leq 40$

$$x_1 + 2x_2 - 7x_3 \geq 30$$

$$5x_1 + 3x_2 = 50$$

$x_1^+, x_1^- \geq 0$, x_3 has not to sign constraint

5.5 Convert the following problem to standard form:

$$\min z_o = 3x_1 - 5x_2 + 7x_3$$

Operating constraints $x_1 + x_2 + 3x_3 \leq 70$

$$x_1 + 2x_2 \mp 3x_3 \geq 55$$

$$5x_2 - 3x_3 = -35$$

$$-6x_1 + 4x_2 \leq -20$$

$x_1, x_2, \geq 0$, x_3 has not to sign constraint

5.6 Find the $Min\, f(x_1, x_1) = x_1^2 + 2x_2^2 - 2x_1^2 + 6$ using Lagrange coordinations.

5.7 Find the $Min\, f(x_1, x_1) = 2x_1^2 + 2x_2^2 - 2x_1^2 + 4$

Operating constraints $g = 5x_1 + 2x_2 - 4 = 0$ using Lagrange coordinations.

TABLE 5.4
Data of Example 5.2

Operation	Operation Time to Generate One Unit of Each Product (Minutes)			Daily Operation Capacity (Minutes)
	Product 1 (x_1)	Product 2 (x_2)	Product 3 (x_3)	
1	2	2	2	550
2	3	0	2	580
3	2	4	0	540
Profit for generation of one unit of each product	4	5	5	

TABLE 5.5
Constants Parameters of Example 5.10

Unit	A	B	C	D
Unit 1	770	7.06	9.68×10^{-4}	1.27×10^{-7}
Unit 2	1300	7.08	7.375×10^{-4}	6.453×10^{-8}
Unit 3	1550	6.56	1.04×10^{-3}	9.98×10^{-8}

5.8 Consider three units with characteristics presented below:

$$F_1(P_1) = 580 + 7.92P_1 + 0.00157P_1^2 \; 150\,\text{MW} \le P_1 \le 640\text{MW}$$

$$F_2(P_2) = 410 + 7.85P_2 + 0.002P_2^2 \; 100\,\text{MW} \le P_2 \le 440\,\text{MW}$$

$$F_3(P_3) = 178 + 7.97P_3 + 0.00482P_3^2 \; 50\,\text{MW} \le P_3 \le 240\text{MW}$$

If demand = 950 MW, find these units' ELD.

5.9 Consider three-generation units with third-order cost functions as:

$$F(P_i) = A_i + B_iP_i + C_iP_i^2 + D_iP_i^3$$

$$330\,\text{MW} \le P_1 \le 800\,\text{MW}$$

$$320\,\text{MW} \le P_2 \le 1200\,\text{MW}$$

$$285\,\text{MW} \le P_3 \le 1100\,\text{MW}$$

Find the ELD for Demand = 2700 MW

5.10 Consider three units with characteristics presented below. Find the ELD solution for *Demand* = 900 MW using the first-order gradient search method.

$$(IC)_1 = 7.92 + 0.003124P_1 \; 150\text{MW} \le P_1 \le 600\,\text{MW}$$

$$(IC)_2 = 7.85 + 0.003884P_2 \; 100\text{MW} \le P_2 \le 400\,\text{MW}$$

$$(IC)_3 = 7.97 + 0.00964P_3 \; 50\text{MW} \le P_3 \le 300\,\text{MW}$$

5.11 Solve the Problem 5.9 using the second-order gradient search method.

5.12 A power system has $P_1^{old} = 393.2$ MW, $P_2^{old} = 334.6$ MW, $P_3^{old} = 122.2$ MW, and Demand power^{old} = 900 MW for the system described in Problem 5.9. Find the economic dispatch for *Demand power* = 950MW.

6 Economic Dispatch

Economic dispatch is a typical energy management system feature. Its role is to distribute the non-base load over a collection of dispatchable units. The necessary power is produced at the least cost, and a minimum reserve capacity is provided. The minimum reserve capacity is typically based on an operational requirement that adequate reserve is capable of responding to a generation contingency and a sudden rise in load within a given period. Although recognizing that this function concept has served a large number of energy management systems, its position in providing the necessary protection level should be examined regarding the two contingencies: Demand contingencies because of sudden power demand rise and generation contingency of loss of one generating unit.

This chapter will focus on how to know the relation between the external electric energy of the machine and its voltage. To make network operation with minimum cost and higher power generation as possible and how to reduce the losses in the transmission line. This is done by reducing voltage drop and the ability to control the manner of the multi-machine power system through controlling the limits of generator' voltages and reactive network power.

6.1 ECONOMIC DISPATCH IN POWER SYSTEM NETWORKS

Economic dispatch offers a definite reserve ability for contingencies of demand. If a contingency generation happens, the reserve ability depends on the contingency itself. Since the standby unit could have contributed to the reserve requirement, there is no guarantee of how much reserve is available. It seems that this obvious weakness of the standard economic dispatch has not been of particular concern to utilities with a large number of units because every single unit's reserve contribution is minimal.

In particular, the economic dispatch function must be executed in such a way that the following conditions are met: the impact of an outage on any single generating unit is limited and available reserves on other online units cover the loss of a generation of any single unit.

The first criterion is established by the so-called impact factor, which defines the maximum percentage of total device load assignable to any unit in the system. In other words, the impact factor limits the maximum percentage of load required if the worst single-generation contingency occurs. The corresponding cover factor determines the second criterion, the least percentage of any unit generation protected by reserves on the other units in the system. For example, suppose a 100°/0 cover factor is applied. In that case, the second requirement ensures that any single online unit failure can be fully compensated within the reserved time by redistributing the system load among the online units available.

DOI: 10.1201/9781003293965-6

An engineer is always concerned with the cost of products and services. Proper operation is very important for a power system to return a profit on the capital invested. Rates fixed by regulatory bodies and the importance of conservation of fuel place extreme pressure on power companies to achieve maximum efficiency of operation and to improve efficiency continually to maintain a reasonable relation between the cost of a kilowatt-hour to a consumer and the cost to the company of delivering a kilowatt-hour in the face of constantly rising prices for fuel, labor, supplies, and maintenance.

Engineers have been very successful in increasing the efficiency of boilers, turbines, and generators so continuously that each new unit added to the system's generating plants operates more efficiently than any older unit on the system. In operating the system for any load condition, each plant's contribution and from each unit within a plant must be determined to minimize the cost of delivered power. How the engineer has met and solved this challenging problem is the subject of this project.

An early method of minimizing the cost of delivered power called for supplying power from only the most efficient plant at light loads. As the load increased, power would be supplied by the most efficient plant until the maximum efficiency of that plant was reached. Then for further increase in load, the next most efficient plant would start to feed power to the system, and a third plant would not be called upon until the point of maximum efficiency of the second plant was reached. Even with transmission losses neglected, this method fails to minimize cost.

We shall first study the most economic distribution of a plant's output between the generators or units within the plant. Since system generation is often expanded by adding existing plants, the various plant units usually have different characteristics. The method that will be developed is also applicable to the economic scheduling of plant outputs for a given loading of the system without considering transmission losses. We shall develop a method of expressing transmission loss as a function of the various plants' outputs. Then we shall determine how the output of each of the plants of a system is scheduled to achieve a minimum cost of power delivered to the load.

The aspects of economic load dispatch are the following:

i Load forecasting – both short term and long term.
ii Economic scheduling of generators (unit commitment).
iii Economic loading of the units.

Operating constraints:

Several operating restrictions must be taken into account. These would include the following:

i. Capacity restrictions of individual generators
   are determined mainly by thermal restrictions and boiler capability.
ii. Reserve requirements for system security. This accounts for such emergencies as an outage of line generators or transformers. Excess spinning reserve is required as a margin to account for such forced analogs in addition to possible forecast errors.

Distribution of load between generations:
i. Lossless case.

This assumes that units considered are either within the same station or close enough together that the effect of transmission line losses may be neglected (urban area). The objective is to distribute plant load between units in such a way as to minimize the station fuel cost. In general, the important factors to minimize the cost of electrical energy are as follows:

i. Operating efficiencies of the generators.
ii. Fuel cost and another cost as maintenance.
iii. Transmission losses.

The input–output curve of the thermal plant is important in describing the plant efficiency. A curve is a plot of fuel input in(Btu/hr) versus electrical power output in (kW or MW), as shown in Figure 6.1.

The coordinates of the curve are converted to money units per hour versus power output, as shown in Figure 6.2 and given as a function of the power generated as:

The cost function for unit 1

$$F_1 = f(P_1) \tag{6.1}$$

The cost function for unit 2

$$F_2 = f(P_2) \tag{6.2}$$

The cost function for unit n

$$F_n = f(P_n) \tag{6.3}$$

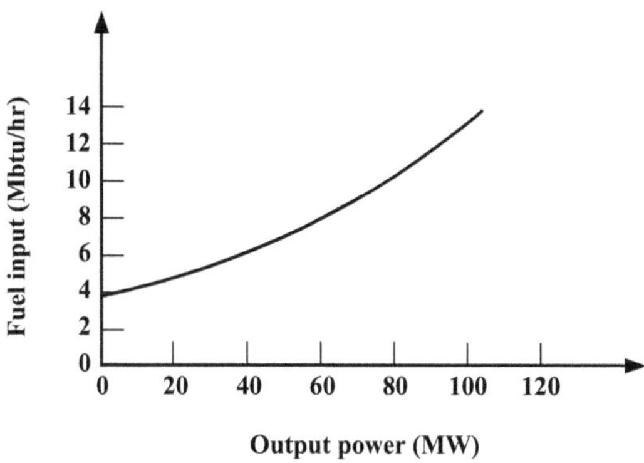

FIGURE 6.1 Input–output curve.

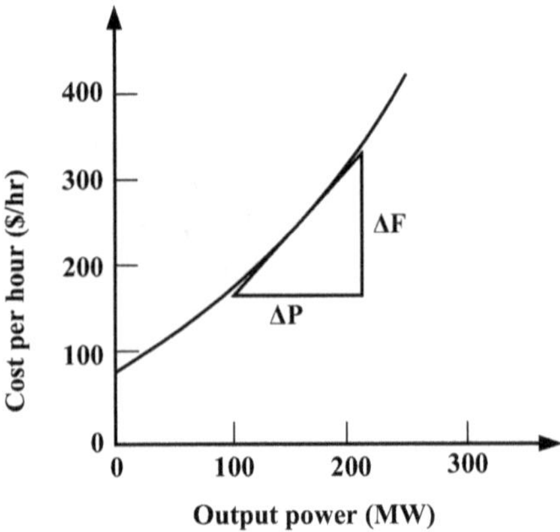

FIGURE 6.2 Money units per hour versus power output curve.

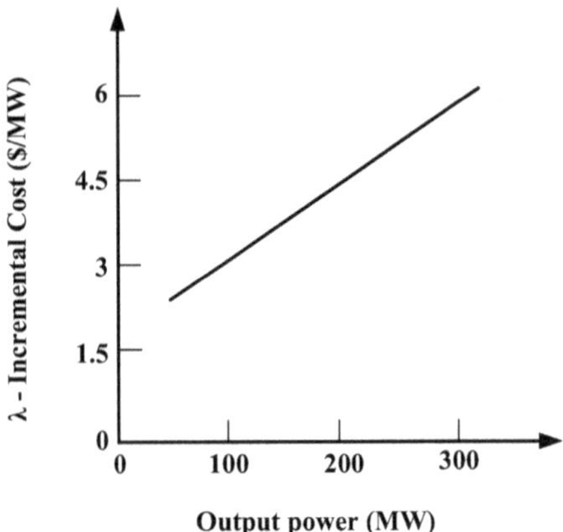

FIGURE 6.3 The incremental cost λ versus power curve.

The important term is the slope $\frac{\Delta F}{\Delta P}$ for various outputs. A plot of this slope is λ versus power will yield an incremental cost curve as shown in Figure 6.3.

The incremental fuel cost curve measures how costly it will be to produce the net increment of power. Let us assume that two or more generators within the same plant will be economically operated together, as shown in Figure 6.4. The basic criteria for such operation is that each unit operates at the same value of incremental fuel cost λ i.e.

$$\lambda = \frac{dF_1}{dP_1} = \frac{dF_2}{dP_2} = \ldots = \frac{dF_n}{dP_n} \tag{6.4}$$

For nth units, the total power is given:

$$P_{total} = P_1 + P_2 + \ldots + P_n \tag{6.5}$$

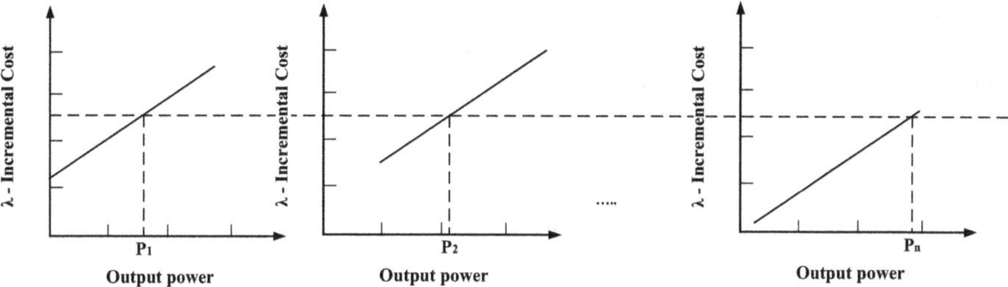

FIGURE 6.4 The nth units economically operated together.

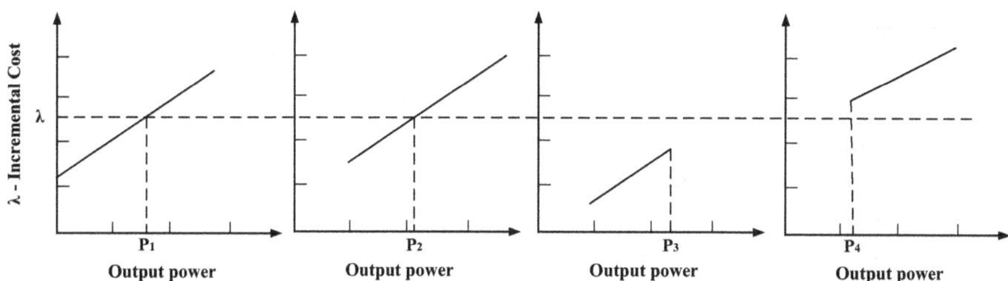

FIGURE 6.5 Four units operated economically.

For four units (n = 4), Figure 6.5 shows the principle for four units, and the total power is given:

$$P_{total} = P_1 + P_2 + P_3 + P_4 \tag{6.6}$$

Unit 3 should be fully loaded before the other; unit 4 might be used as a reserve.

6.2 FUEL TYPES AND COST

Let C be the cost, expressed, for example, in dollars per hour, if the number of the generator of N, producing energy in the generator unit *i*. The total controllable system production cost, therefore, will be

$$C = \sum_{i=1}^{N} Ci \frac{\$}{\text{h}} \tag{6.7}$$

Increasing the prime mover torque increases individual actual generations, requiring increased fuel expenditure. The reactive generations Q_{Gi} have no observable effect on cost, as varying field currents control them. Therefore, the individual generator unit output costs are given by

$$C = \sum_{i=1}^{N} C_i\left(P_{Gi}\right) = C_1\left(P_{G_1}\right) + \ldots\ldots + C_N\left(P_{G_N}\right)\frac{\$}{\text{h}} \tag{6.8}$$

The cost function C can be written as a sum of terms that depends only on one independent variable, so the C is separable. The fuel costs, of course, constitute the major portions as operation and maintenance, etc.

6.3 INCREMENTAL FUEL COST

If we express the ordinates of the input–output curve in dollars per hour and let:

$$F_n = input\ to\ unit\ n, dollars\ per\ hour$$

$$P = output\ of\ unit\ n, MW$$

The incremental fuel cost of the unit in \$/MW is dF_n/dP_n.

Incremental fuel cost for a unit for any given power output is the limit of the ratio of the increase in fuel input cost in dollars per hour to the spending increase in power output in megawatts as the power output approaches zero. The incremental fuel cost could be obtained by determining the increased fuel cost for a definite time interval during which a small amount increases the power output. The approximate output is the additional cost in dollars per hour to increase the output by 1 MW.

Incremental fuel cost for a unit for any given power output is the limit of the ratio of the increase in fuel input cost in \$/MW. hr to the corresponding increase in power output in MW as power output increase approaches zero. Incremental fuel cost could be obtained by calculating the increased fuel cost over a certain time when a small amount increases power output.

6.4 OPTIMIZATION TECHNIQUES

The major cost in plant operation is fuel, while the other costs may be added to the *fuel cost* and the sum referred to as the fuel cost. The fuel cost is \$/h and is a function of the power generation in MW. The fuel cost curve is usually assumed to be parabolic of the form:

$$C_i = \alpha_i + \beta_i P_i + \gamma_i P_i^2 \tag{6.9}$$

where α_i, β_i, and γ_i are constant.

The *incremental fuel-cost* curve is the slope of the fuel-cost characteristics curve defined by

$$\frac{dC_i}{dP_i} = 2\gamma_i P_i + \beta_i \tag{6.10}$$

The incremental fuel-cost curve indicates how expensive it will be to generate the next power increment. Units: C_i is in *\$/h* and $\left(\frac{dC_i}{dP_i}\right)$ is in \$/MW. hr.

6.4.1 Economic Dispatch Neglecting Losses and Generator Limits

The simplest economic allocation is that of power generators on the same bus. Suppose we have n generators connected to a bus and supplying a load demand P_D. We would like to minimize the total cost of generation.

$$C_t = \sum_{i=1}^{ng} C_i = \sum_{i=1}^{n} \left(\alpha_i + \beta_i P_i + \gamma_i P_i^2 \right) \tag{6.11}$$

Subject to the constraint

$$\sum_{i=1}^{ng} P_i = P_D \tag{6.12}$$

This situation is shown in Figure 6.6. It is noted that the number of "dispatchable" generators is n_g, while the total number of generators is n. This means n_g generators are not subject to economic allocation (they may be on a flat load, such as the case of nuclear generators, or they may have reached a maximum or minimum limit of generation).

As usual, we now form the function $\mathcal{L}$:

$$\mathcal{L} = C_t + \lambda \left(P_D - \sum_{i=1}^{ng} P_i \right) \tag{6.13}$$

The necessary conditions for a minimum are:

$$\frac{\partial \mathcal{L}}{\partial P_i} = 0 \tag{6.14}$$

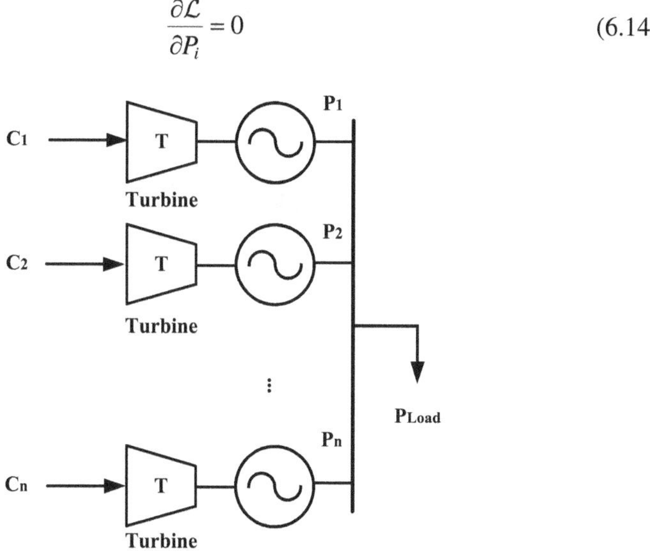

FIGURE 6.6 Plants connected to the common bus.

$$\frac{\partial \mathcal{L}}{\partial \lambda} = 0 \tag{6.15}$$

The first condition results in

$$\frac{\partial C_t}{\partial P_i} + \lambda(0-1) = 0 \tag{6.16}$$

Also, we know that $C_t = C_1 + C_2 + C_3 + \ldots + C_{ng}$, therefore

$$\frac{\partial C_t}{\partial P_i} = \frac{dC_t}{dP_i} = \lambda \tag{6.17}$$

Thus, the condition for optimal dispatch is

$$\frac{dC_t}{dP_i} = \lambda \, i = 1,2,\ldots,ng \tag{6.18}$$

$$\lambda = 2\gamma_i P_i + \beta_i \tag{6.19}$$

And the second condition above simply states

$$\sum_{i=1}^{ng} P_i = P_D \tag{6.20}$$

Assuming no limits on generation, and no losses, the most economical state of operation is when all the plants have the same incremental cost λ, and their total generation equals the load demand, so the power at optimum operation is given by

$$P_i = \frac{\lambda - \beta_i}{2\gamma_i} \tag{6.21}$$

The relations given by this Equation are called the *coordination equations.* Using this Equation in the constraint equation, resulted in

$$\sum_{i=1}^{ng} \frac{\lambda - \beta_i}{2\gamma_i} = P_D \tag{6.22}$$

and solving for λ we have

$$\lambda = \frac{P_D + \sum_{i=1}^{ng} \frac{\beta_i}{2\gamma_i}}{\sum_{i=1}^{ng} \frac{1}{2\gamma_i}} \tag{6.23}$$

Knowing the value of λ, we can compute the power using (6.21).

In the case of lossless power systems (i.e., generators on a common bus), it is easy to solve the optimal dispatch problem analytically. The solution cannot be found analytically in the general case where losses are significant and the generators are geographically dispersed. In such cases, an iterative solution is employed. Such a solution is detailed below for the case of lossless systems.

6.4.2 Economic Dispatch Neglecting Losses and Including Generator Limits

In the real world, generators have limits on their generation, both minimum and maximum generation. Thus, we have the inequality constraints

$$P_{i(\min)} \le P_i \le P_{i(\max)}\, i = 1, 2, \ldots, ng \tag{6.24}$$

which are in addition to the equality constraint

$$\sum_{i=1}^{ng} P_i = P_D \tag{6.25}$$

The necessary conditions for optimal dispatch, according to Lagrange conditions in lossless systems with included generator limits, become

$$\frac{dC_i}{dP_i} = \lambda \; for \; P_{i(\min)} \le P_i \le P_{i(\max)} \tag{6.26}$$

$$\frac{dC_i}{dP_i} \le \lambda \; for \; P_i = P_{i(\max)}$$

$$\frac{dC_i}{dP_i} \ge \lambda \; for \; P_i = P_{i(\min)}$$

Thus, for an estimated lambda value, the powers are found for each participating generator. Suppose the power is outside the limits of a generator. In that case, that generator is "pegged" at that limit, either P_{min} or P_{max}, and the generator no longer participates in optimizing dispatched power.

6.4.3 Economic Dispatch Including Losses

The transmission losses may be neglected when the loads and generators are in a small geographical region. The optimal generation dispatch is achieved with all plants operating at equal incremental production cost. However, transmission losses become significant in a large system over long distances and need to be considered. using *Kron's loss formula,* which is

$$P_L = \sum_{i=1}^{ng}\sum_{j=1}^{ng} P_i B_{ij} P_j + \sum_{i=1}^{ng} B_{0i} P_i + B_{00} \tag{6.27}$$

$$C_t = \sum_{i=1}^{ng} C_i$$

$$= \sum_{i=1}^{n} \left(\alpha_i + \beta_i P_i + \gamma_i P_i^2 \right) \tag{6.28}$$

This description is to be performed subject to the constraint that the total generation must equal the demand plus the losses, thus

$$\sum_{i=1}^{ng} P_i = P_D + P_L \tag{6.29}$$

satisfying the inequality constraints:

$$P_{i(\min)} \le P_i \le P_{i(\max)}\; i = 1,2,\ldots,ng$$

where $P_{i(\min)}$ and $P_{i(\max)}$ are the minimum and maximum generating limits, respectively, for plant i.

Using the Lagrange multiplier and adding additional terms to include the inequality constraints, we have

$$\mathcal{L} = C_t + \lambda\left(P_D + P_D - \sum_{i=1}^{ng} P_i \right) + \sum_{i=1}^{ng} \mu_{i(\max)}\left(P_i - P_{i(\max)}\right) + \sum_{i=1}^{ng} \mu_{i(\min)}\left(P_i - P_{i(\min)}\right) \tag{6.30}$$

The constraints should be understood to mean that $\mu_{i(\max)} = 0$, when

$P_i < P_{i(\max)}$, and $\mu_{i(\min)} = 0$, when $P_i > P_{i(\min)}$. Thus, if the power is not beyond the limits, both $\mu_{i(\max)} = 0$ and $\mu_{i(\min)} = 0$; thus, L has only the equality constraint. Suppose a generator's power is beyond one of the limits. In that case, that limit is the power to which the generator is "pegged." The remaining generators participate in optimizing the load dispatch. These statements are clear from the necessary conditions for a minimum, which are as follows:

$$\frac{\partial \mathcal{L}}{\partial P_i} = 0 \tag{6.31}$$

$$\frac{\partial \mathcal{L}}{\partial \lambda} = 0 \tag{6.32}$$

$$\frac{\partial \mathcal{L}}{\partial \mu_{i(\max)}} = P_i - P_{i(\max)} = 0 \tag{6.33a}$$

$$\frac{\partial \mathcal{L}}{\partial \mu_{i(\min)}} = P_i - P_{i(\min)} = 0 \tag{6.33b}$$

The last two equations imply that the power should not go beyond its limit, and when it is within its limits, then $\mu_{i(\max)} = \mu_{i(\min)} = 0$, and the Kuhn–Tucker function becomes the same as the Lagrange one.

The first equation results in

$$\frac{\partial C_t}{\partial P_i} + \lambda\left(0 + \frac{\partial P_L}{\partial P_i} - 1\right) = 0$$

Since $C_t = C_1 + C_2 + C_3 + \ldots + C_{ng,}$ the above Equation becomes

$$\frac{\partial C_i}{\partial P_i} + \lambda \frac{\partial P_L}{\partial P_i} = \lambda\, i = 1, 2, \ldots, ng \tag{6.34}$$

The term $\frac{\partial P_L}{\partial P_i}$ is known as the *incremental transmission loss.* The second condition is given by

$$\sum_{i=1}^{ng} P_i = P_D + P_L \tag{6.35}$$

The classical form of the Equation 6.35 is

$$\left(\frac{1}{1 - \frac{\partial P_L}{\partial P_i}}\right)\frac{\partial C_i}{\partial P_i} = \lambda\, i = 1, 2, \ldots, ng \tag{6.36}$$

$$L_i \frac{\partial C_i}{\partial P_i} = \lambda\, i = 1, 2, \ldots, ng \tag{6.37}$$

where L is known as the penalty factor of plant i and is given by

$$L_i = \frac{1}{1 - \frac{\partial P_L}{\partial P_i}}. \tag{6.38}$$

6.4.4 The B-Coefficient and Algorithms

Thus, transmission losses result in adding a penalty factor with a value that depends on the plant's location for the loads. It is obvious in Equation 6.36 that the most economical dispatch cost is obtained when the incremental cost compounded by the corresponding penalty factor is equal for all participating plants. The incremental transmission loss can be found from the loss equation

$$\frac{\partial P_L}{\partial P_i} = 2\sum_{j=1}^{ng} B_{ij}\, P_j + B_{0i} \tag{6.39}$$

and the incremental production cost is given by

$$\frac{dC_i}{dP_i} = 2\gamma_i P_i + \beta_i \; i = 1,2,\ldots,ng \tag{6.40}$$

Using these two equations in the Equation (reproduced below for convenience)

$$\frac{\partial C_i}{\partial P_i} + \lambda \frac{\partial P_L}{\partial P_i} = \lambda \; i = 1,\, 2,\ldots,ng \tag{6.41}$$

we have

$$\beta_i + 2\gamma_i P_i + 2\lambda \sum_{j=1}^{ng} B_{ij} P_j + B_{0i}\lambda = \lambda$$

or (divide by 2λ and pull out B_{ii} from the summation)

$$\left(\frac{\gamma_i}{\lambda} + B_{ii}\right) P_i + \sum_{j=1, j\neq i}^{ng} B_{ij} P_j = \frac{1}{2}\left(1 - B_{0i} - \frac{\beta_i}{\lambda}\right) \tag{6.42}$$

Expanding the above Equation for all plants and detailing it into full matrix form, we have

$$\begin{bmatrix} \frac{\gamma_1}{\lambda} + B_{11} & B_{12} & \cdots & B_{1ng} \\ B_{21} & \frac{\gamma_i}{\lambda} + B_{ii} & \cdots & B_{2ng} \\ \vdots & \vdots & \ddots & \vdots \\ B_{ng1} & B_{ng2} & \cdots & \frac{\gamma_{ng}}{\lambda} + B_{ngng} \end{bmatrix} \begin{bmatrix} P_1 \\ P_2 \\ \vdots \\ P_{ng} \end{bmatrix} = \frac{1}{2} \begin{bmatrix} 1 - B_{01} - \frac{\beta_1}{\lambda} \\ 1 - B_{02} - \frac{\beta_2}{\lambda} \\ \vdots \\ 1 - B_{0ng} - \frac{\beta_{ng}}{\lambda} \end{bmatrix}. \tag{6.43}$$

6.5 MATHEMATIC FORMULATION

We employ the gradient method to solve the optimization problem by iteration. The constraint equation is written as follows:

$$f(\lambda) = \sum_{i=1}^{ng} \frac{\lambda - \beta_i}{2\gamma_i} = P_D \tag{6.44}$$

Taking the first-order terms of the Taylor series expansion about the point $\lambda^{(k)}$ gives

$$f(\lambda)^{(k)} + \left(\frac{df(\lambda)}{d\lambda}\right)^{(k)} \Delta\lambda^{(k)} = P_D \tag{6.45}$$

or

$$\Delta\lambda^{(k)} = \frac{P_D - f(\lambda)^{(k)}}{\left(\frac{df(\lambda)}{d\lambda}\right)^{(k)}}$$

$$\Delta\lambda^{(k)} = \frac{P_D - \sum_{i=1}^{ng} P_i^{(k)}}{\left(\frac{df(\lambda)}{d\lambda}\right)^{(k)}} = \frac{\Delta P^{(k)}}{\left(\frac{df(\lambda)}{d\lambda}\right)^{(k)}}$$

$$= \frac{\Delta P^{(k)}}{\sum_{i=1}^{ng} \left(\frac{dP_i}{d\lambda}\right)^{(k)}} \tag{6.46}$$

or

$$\Delta\lambda^{(k)} = \frac{\Delta P^{(k)}}{\sum_{i=1}^{ng} \left(\frac{1}{2\gamma_i}\right)^{(k)}} \tag{6.47}$$

Thus, the correction is

$$\lambda^{(k+1)} = \lambda^{(k)} + \Delta\lambda^{(k)} \tag{6.48}$$

where

$$\Delta P^{(k)} = P_D - \sum_{i=1}^{ng} P_i^{(k)} \tag{6.49}$$

The iteration is continued till $\Delta P^{(k)}$is smaller than a specified accuracy. Of course, as usual, a starting value $\lambda^{(1)}$ is used to start the iteration. The steps in this process are outlined below:

Step 1. Assume an initial value for lambda, say $\lambda^{(1)}$.

Step 2. Using the Equation

$$P_i^{(1)} = \frac{\lambda^{(1)} - \beta_i}{2\gamma_i}$$

Compute all the powers for the first iteration.

Step 3. Check if the sum of powers equals the demand, and find the difference, thus

$$\Delta P_i^{(1)} = P_D - \sum_{i=1}^{ng} P_i^{(1)}$$

A solution is reached if this difference is smaller than the specified accuracy. If not, continue.

Step 4. Compute the change in lambda:

$$\Delta\lambda^{(1)} = \frac{\Delta P^{(1)}}{\sum_{i=1}^{ng}\left(\frac{1}{2\gamma_i}\right)^{(1)}}$$

and compute the second estimate for lambda, namely:

$$\lambda^{(2)} = \lambda^{(1)} + \Delta\lambda^{(1)}$$

The process is repeated till step 3 indicates convergence.

For the system with loss consideration: To find the optimal dispatch, first, an estimated guess of λ, say $\lambda^{(1)}$ is made. The simultaneous linear equations are solved using MATLAB®. The iterative process is continued using the gradient method. Thus, from Equation 6.49 above, we have P_i at the kth iteration expressed as:

$$\Delta\lambda^{(k)} = \frac{P_D - \sum_{i=1}^{ng} P_i^{(k)}}{\left(\frac{df(\lambda)}{d\lambda}\right)^{(k)}} = \frac{\Delta P_i^{(k)}}{\left(\frac{df(\lambda)}{d\lambda}\right)^{(k)}}$$

$$P_i^{(k)} = \frac{\lambda^{(k)}(1 - B_{0i}) - \beta_i - 2\lambda^{(k)}\sum_{j\neq i} B_{ij}P_j^{(k)}}{2\left(\gamma_i + \lambda^{(k)}B_{ii}\right)} \tag{6.50a}$$

And using this Equation in the equality constraint gives

$$\sum_{i=1}^{ng} P_i^{(k)} = \sum_{i=1}^{ng} \frac{\lambda^{(k)}(1-B_{0i}) - \beta_i - 2\lambda^{(k)} \sum_{j \neq i} B_{ij} P_j^{(k)}}{2\left(\gamma_i + \lambda^{(k)} B_{ii}\right)} = P_D + P_L^{(k)} \tag{6.50b}$$

or

$$f(\lambda)^{(k)} = P_D + P_L^{(k)} \tag{6.51}$$

Expanding the left side above in a Taylor series and keeping only first-order terms results in

$$f(\lambda)^{(k)} + \left(\frac{df(\lambda)}{d\lambda}\right)^{(k)} \Delta\lambda^{(k)} = P_D + P_L^{(k)} \tag{6.52}$$

or

$$\Delta\lambda^{(k)} = \frac{P_D + P_L^{(k)} - \sum_{i=1}^{ng} P_i^{(k)}}{\left(\frac{df(\lambda)}{d\lambda}\right)^{(k)}} = \frac{\Delta P^{(k)}}{\left(\frac{df(\lambda)}{d\lambda}\right)^{(k)}}$$

$$\Delta\lambda^{(k)} = \frac{\Delta P^{(k)}}{\sum_{i=1}^{ng} \left(\frac{1}{2\gamma_i}\right)^{(k)}} \tag{6.53}$$

where

$$\Delta P^{(k)} = P_D + P_L^{(k)} - \sum_{i=1}^{ng} P_i^{(k)} \tag{6.54}$$

and

$$\sum_{i=1}^{ng} \left(\frac{\partial P_i}{\partial \lambda}\right)^{(k)} = \sum_{i=1}^{ng} \frac{\gamma_i(1-B_{0i}) - B_{ii}\beta_i - 2\gamma_i \sum_{j \neq i} B_{ij} P_j^{(k)}}{2\left(\gamma_i + \lambda^{(k)} B_{ii}\right)^2} \tag{6.55}$$

$$\lambda^{(k+1)} = \lambda^{(k)} + \Delta\lambda^{(k)} \tag{6.56}$$

The process is repeated till $\Delta P(k)$ is less than a specified accuracy. If we simplify the loss equations and use the simpler expression

$$P_L = \sum_{i=1}^{ng} B_{ii} P_i^2 \tag{6.57}$$

(where - assuming $B_{ij} = 0$ and $B_{00} = 0$), the solution of the simultaneous equations simplifies to

$$P_i^{(k)} = \frac{\lambda^{(k)} - \beta_i}{2\left(\gamma_i + \lambda^{(k)} B_{ii}\right)} \tag{6.58}$$

which are known as the *coordination equations,* and Equation 6.55 simplifies to

$$\sum_{i=1}^{ng} \left(\frac{\partial P_i}{\partial \lambda}\right)^{(k)} = \sum_{i=1}^{ng} \frac{\gamma_i - B_{ii}\beta_i}{2\left(\gamma_i + \lambda^{(k)} B_{ii}\right)^2} \tag{6.59}$$

The iteration process remains the same using these simpler functions.

Example 6.1

The Load Flow Data of the 5-Bus Power System, Shown In Figure 6.7, is Given In Tables 6.1–6.4.

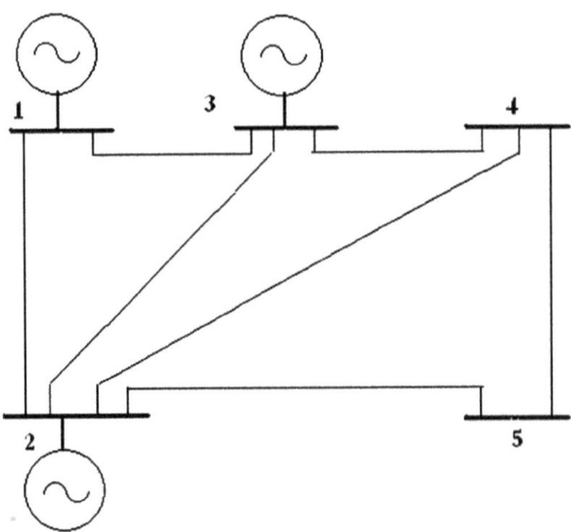

FIGURE 6.7 The 5-bus power system of Example 6.1.

TABLE 6.1
Line Data for the 5-Bus Power System (100MVA Base)

Line No.	Bus Number		Impedance (p.u.)		Capacitive Susceptance (1/2) B
	From	To	R	X	
1	1	2	0.02	0.06	0.03
2	1	3	0.08	0.24	0.025
3	2	3	0.06	0.18	0.02
4	2	4	0.06	0.18	0.02
5	2	5	0.04	0.12	0.015
6	3	4	0.01	0.03	0.01
7	4	5	0.08	0.24	0.025

TABLE 6.2
Initial Bus Data for the 6-Bus Power System

Bus No.	Voltage		Load		Generation	
	V PU	δ Deg.	P MW	Q MVAR	P MW	Q MVAR
1	1.06	0.0	0	0	0	0
2	1.045	0.0	20	10	40	30
3	1.03	0.0	20	15	30	10
4	1.00	0.0	50	30	0	0
5	1.00	0.0	60	40	0	0

TABLE 6.3
Generator Cost

Generators	α	β	γ
G1	0.008	7	200
G2	0.009	6.3	180
G3	0.007	6.8	140

TABLE 6.4
The 5-Bus Reactive Power Injected Limits

Bus	Qmax	Qmin
1	50	10
2	50	10
3	40	10

Table 6.5 shows the objective function result, and Table 6.6 shows the system summary. The bus data and branch data are given in Tables 6.7 and 6.8, respectively. Table 6.9 gives the initial and the new value of power demand, while Table 6.10 shows the output of each unit for various values of λ.

TABLE 6.5
The Objective Function Result

Iter.	f(x)
0	996.7
1	1633.92
2	1622.31
3	1620.43
4	1605.44
5	1626.98
6	1593.34
7	1595.6
8	1595.6

Converged in 6.41 seconds
Objective Function Value = 1595.60 $/hr

TABLE 6.6
System Summary

Item	Number	Item	P (MW)	Q (MVAr)
Buses	5	Total Gen. Capacity	320	30 to 140
Generators	3	Generation	152.1	85.9
Loads	4	Load	150	95
Branches	7	Losses (I^2 * Z)	2.09	6.27
Inter-Ties	5	Total Inter-Tie Flow	89.5	51.1

TABLE 6.7
Bus Data

Bus	Voltage		Generation		Load		λ ($/MVA.hr)	
	Mag (pu)	Ang (deg)	P (MW)	Q (MVAr)	P (P.U.MW)	Q (P.U.MVAr)	P	Q
1	1.050	−0.00	29.02	10	-	-	7.464	0
2	1.041	−0.521	67.94	35.87	20	10	7.523	0
3	1.035	−1.025	55.13	40	20	15	7.572	0
4	1.021	−1.662	-	-	50	30	7.645	0
5	1.026	−3.023	-	-	60	40	7.821	0
Total			152.09	85.87	150	95		

TABLE 6.8
Branch Data

Branch	From Bus	To Bus	From Bus Injection P (PU. MW)	From Bus Injection Q (PU. MVAr)	To Bus Injection P (MW)	To Bus Injection Q (MVAr)	Loss (I^2 * Z) P (MW)	Loss (I^2 * Z) Q (MVAr)
1	1	2	19.79	8	−19.7	−11.01	0.088	0.26
2	1	3	9.23	2	−9.16	−4.51	0.070	0.21
3	2	3	5.67	0.14	−5.65	−2.24	0.019	0.06
4	2	4	13.99	5.67	−13.85	−7.35	0.133	0.4
5	2	5	47.99	31.12	−46.76	−28.98	1.227	3.68
6	3	4	49.94	31.75	−49.61	−31.82	0.33	0.99
7	4	5	13.46	9.16	−13.24	−13.24	0.223	0.67
Total losses							2.09	6.27

TABLE 6.9
Initial and the New Value of Power Demand

Load Bus	Initial Demand(MW)	New Demand(MW)
Bus2	20	26
Bus3	20	26
Bus4	50	65
Bus5	60	78

Case(a). Increase in the power demand by (30) percentage. Table 6.10 shows these changes.

TABLE 6.10
Output of Each Unit for Various Values of λ

λ ($/MVA)	P_{g1}1 (MW)	P_{g2} (MW)	P_{g3} (MW)	P_g Total (MW)
6.36	5	5	5	15
6.5	5	11	5	21
6.8	5	27.78	5	37.78
7	5	38.89	14.29	58.18
7.3	18.75	55.56	35.71	110.02
7.51	31.88	67.3	50.8	149.98

So we conclude:

1. The iteration increased from 8 to 11.
2. The generators run to meet the load demand when increasing the power demand.

$$\sum_{i=1}^{ng} P_i = P_D$$

3. The cost increases from 1595.60 \$/h to 1948.53 \$/h. The increase in power generation is the reason for increasing the cost.

$$C_i = \alpha_i + \beta_i P_i + \gamma_i P_i^2$$

4. The Lagrange multiplier for equality constraint(λ) increases from 7.821\$/MVA to 8.182 \$/MVA.

$$\frac{dC_i}{dP_i} = 2\gamma_i P_i + \beta_i$$

5. Energy loss also increases from [2.09 MW, 6.27 MVAR] to[3.238 MW, 9.71 MVAR] due to an increase in the generation

$$P_L = \sum_{i=1}^{ng}\sum_{j=1}^{ng} P_i\, B_{ij}\, P_j + \sum_{i=1}^{ng} B_{0i}\, P_i + B_{00}$$

Solution
Beginning from the three units cost Equation:

$$C_1 = 0.008\, P_{g1}^2 + 7\, P_{g1} + 200$$

$$C_2 = 0.009\, P_{g2}^{\dagger} + 6.3\, P_{g2} + 180$$

$$C_3 = 0.007\, P_{g3}^{\dagger} + 6.8\, P_{g3} + 140$$

Assume the load varies from 15 MW to 150 MW.
Minimum load for each unit = (15/3) = 5 MW.

$$dC_1 / dPg_1 = 0.016\, P_{g1} + 7$$

$$dC_2 / dPg_2 = 0.018\, P_{g2} + 6.3$$

$$dC_3 / dPg_3 = 0.014\, P_{g3} + 6.8$$

At light load (5 MW) each unit, the incremental cost(I.C.) or (dCi/dPgi) or (λ) will be:

$$\lambda_1 = 7.08\$ / MVA$$

$$\lambda_2 = 6.36\$ / MVA$$

$$\lambda_3 = 6.87\$ / MVA$$

To get P_{g1}, P_{g2} and P_{g3} schedules, we will assume several $\lambda(dC/dP_g)$ values according to Figure 6.10.
Total load obtained optimally, at $\lambda = 7.51$ *\$/MVA*

To know the saving in fuel cost \$/h for the optimal scheduling of total load of 150 MW compared to equal distribution(150 MW/3) = 50 MW of the same load between the units:

For unit (1),

$$C1 = \int_{31.88}^{50} \left(dC_1 / dP_{g1}\right).dP_{g1}$$

$$= \left[0.008 * (50)^2 + 7 * (50) + 200\right] - \left[0.008 * (31.88)^2 + 7 * (31.88) + 200\right]$$

$$= 138.71\$ / h$$

For unit (2),

$$C_2 = \int_{67.3}^{50} \left(dC_2 / dP_{g2}\right).dP_{g2}$$

$$= \left[0.009 * (50)^2 + 6.3 * (50) + 180\right] - \left[0.009 * (67.3)^2 + 6.3 * (67.3) + 180\right]$$

$$= -126.5\$ / h$$

For unit (3),

$$C_3 = \int_{50.8}^{50} \left(dC_3 / dP_{g3}\right).dP_{g3}$$

$$= \left[0.007 * (50)^2 + 6.8 * (50) + 140\right] - \left[0.007 * (50.8)^2 + 6.8 * (50.8) + 140\right]$$

$$= -5.25\$ / h$$

Net saving caused by optimum scheduling is

$$138.71 - 126.5 - 5.25 = 7\,\$/\text{h}$$

$$\textit{Annually saving} = 61320\,\$/\text{year}$$

Example 6.2

The Total Output of a Two Generator Station is to Be 325 MW. Determine how this Load Should be Shared to Give the Most Economical Distribution. The Incremental Cost Curve Can be Represented Mathematically as a Straight Line as

$$\lambda_A = \frac{dF_A}{dP_A} = 0.005\,P_A + 2.4\,\frac{\$}{\text{MW.hr}}$$

$$\lambda_B = \frac{dF_A}{dP_A} = 0.008\,P_B + 2.25\,\frac{\$}{\text{MW.hr}}$$

Solution

Setting the incremental costs equal

$$\lambda_A = \lambda_B$$

$$0.005\,P_A + 2.4 = 0.008\,P_B + 2.25$$

$$P_A = 1.6\,P_B - 30$$

and

$$P_L = P_A + P_B = 325$$

Solving to get

$$P_A = 188.46\,\text{MW}$$

$$P_B = 136.53\,\text{MW}$$

$$\lambda_A = \lambda_B = 3.3423\,\frac{\$}{\text{MW.hr}}$$

Example 6.3

In a Power System Where Transmission Line Losses are Negligible, the Load of 10.0 p.u is Supplied from Two Generators.

The cost of the generators is given by:

$$C_1(P_{G1}) = 0.5 + P_{G1} + P_{G1}^2$$

$$C_2(P_{G2}) = 0.5 + 0.5P_{G2} + 2P_{G2}^2$$

Determine the optimum value of P_{G1} and P_{G2} such that the demand is met.

Solution

The requirement that the demand is met is given by

$$P_{G1} + P_{G2} = 10$$

The cost function is

$$C = C_1(P_{G1}) + C_2(P_{G2})$$

$$= 1.0 + P_{G1} + P_{G1}^2 + 0.5P_{G2} + 2P_{G2}^2$$

The Lagrange expression is

$$l = 1.0 + P_{G1} + P_{G1}^2 + 0.5P_{G2} + 2P_{G2}^2 - \lambda(10 - P_{G1} - P_{G2})$$

$$\frac{\partial l}{\partial P_{G1}} = 1.0 + 2P_{G1} - \lambda = 0$$

$$\frac{\partial l}{\partial P_{G2}} = 0.5 + 2P_{G2} - \lambda = 0$$

$$\frac{\partial l}{\partial \lambda} = 10 - P_{G1} - P_{G2} = 0$$

Solving to get $P_{G1} = 4.875\ p.\ u$

$$P_{G2} = 5.125\,p.u$$

$$\lambda = 10.75\,p.u.$$

Example 6.4

The Fuel Costs for the Two Units Supplying A Common Load Without Transmission Line Losses are.

$$C_1(P_{G1}) = 8.644\,P_{G1} + 0.10707\,P_{G1}^2\,MBtu/hr\,0 \le P_{G1} \le 50$$

$$C_2(P_{G2}) = 7.55P_{G2} + 0.01416\,P_{G2}^2\,\frac{MBtu}{hr}\,0 \le P_{G2} \le 100$$

Subject to the power flow constraint $P_{G1} + P_{G2} = P_L$, the two units are optimally operated. Determine the output of each unit as the common system load varies between 0 and 150 MW.

Solution

The total cost $C_T = C_1(P_{G1}) + C_2(P_{G2})$

$$\frac{dC_T}{dP_{G1}} = 0 = \frac{dC_1(P_{G1})}{dP_{G1}} + \frac{dC_2(P_{G2})}{dP_{G2}} \cdot \frac{dP_{G2}}{dP_{G1}}$$

$$P_{G1} + P_{G2} = P_L$$

$$\frac{dP_{G2}}{dP_{G1}} = -1$$

$$8.644 + 0.21408\,P_{G1} + \left(7.55 + 0.01416\left(P_{G1} - P_L\right)\right)(-1) = 0$$

$$P_{G1} = 0.0062\left(P_L - 38.56\right)$$

$$P_{G2} = 0.8832\,P_L + 4.504$$

And

$$\lambda = 7.6795 + 0.025\,P_L$$

Example 6.5

Assume Power System With Three-Generation Units.

$$F_1(P_1) = 561 + 7.92\,P_1 + 0.001562P_1^2\,150^{MW} \le P_1 \ge 600^{MW}$$

$$F_2(P_2) = 310 + 7.85\,P_2 + 0.00194\,P_2^2\,100^{MW} \le P_2 \ge 400^{MW}$$

$$F_3(P_3) = 78 + 7.97\,P_3 + 0.00482\,P_3^2\,50^{MW} \le P_3 \ge 200^{MW}$$

Total losses of the system are obtained from

$$P_{loss} = 0.00003P_1^2 + 0.00009P_2^2 + 0.00012P_3^2$$

Find the answer to the ELD problem for *Demand* = 850*MW*.

Solution

Establishment of coordination equations considering penalty factors and P_{loss} results in a set of non-linear equations:

$$L_1 = \frac{1}{1 - \frac{\partial P_{loss}}{\partial P_1}} = \frac{1}{1 - 0.00006P_1}$$

$$L_2 = \frac{1}{1 - \frac{\partial P_{loss}}{\partial P_2}} = \frac{1}{1 - 0.00018P_2}$$

$$L_3 = \frac{1}{1 - \frac{\partial P_{loss}}{\partial P_3}} = \frac{1}{1 - 0.00006P_3}$$

The coordination equations are written as:

$$\lambda = \left(\frac{1}{1 - 0.00006P_1}\right)(7.92 + 0.003124P_1)$$

$$(L_1)(IC)_1$$

$$\lambda = \left(\frac{1}{1 - 0.00018P_2}\right)(7.85 + 0.00388P_2)$$

$$(L_2)(IC)_2$$

$$\lambda = \left(\frac{1}{1 - 0.00006P_3}\right)(7.97 + 0.00964P_3)$$

$$(L_3)(IC)_3$$

$$P_1 + P_2 + P_3 - 850 - \left(0.00003P_1^2 + 0.00009P_2^2 + 0.00012P_3^2\right) = 0$$

$$(Demand)(P_{loss})$$

Solving these non-linear sets is difficult. Therefore, the following method is proposed:

a. Firstly, an initial guess which satisfies the equality constraint (neglecting losses) is selected.

$$P_1 = 400MW\ P_2 = 300MW\ P_3 = 150MW$$

b. P_{loss} and L_i values are calculated at the above working point after substituting the values of P_1, P_2, and P_3:

$$L_1 = \frac{1}{1-\frac{\partial P_{loss}}{\partial P_1}} = \frac{1}{1-0.00006P_1} = 1.03734$$

$$L_2 = \frac{1}{1-\frac{\partial P_{loss}}{\partial P_2}} = \frac{1}{1-0.00018P_2} = 1.05708$$

$$L_3 = \frac{1}{1-\frac{\partial P_{loss}}{\partial P_3}} = \frac{1}{1-0.00006P_3} = 1.03734$$

$$P_{loss} = 0.00003(600)^2 + 0.00009(300)^2 + 0.00012(150)^2 = 21.6MW$$

c. The coordination equations were established using the above constant values for L_1, L_2, L_3, and P_{loss}. These equations are linear and can be solved easily using solution methods such as Lambda iteration or gradient search. Here because of small-scale problems, the equations are set to be solv directly, and the coordination equations are given by:

$$\lambda = 1.03734(7.92 + 0.00312P_1)$$

$$(L_1)(IC)_1$$

$$\lambda = 1.05708(7.85 + 0.00388P_2)$$

$$(L_2)(IC)_2$$

$$\lambda = 1.03734(7.97 + 0.00904P_3)$$

$$(L_3)(IC)_3$$

$$P_1 + P_2 + P_3 - 850 - 21.6 = 0$$

$$(Demand)(P_{loss})$$

d. Again P_{loss} and L_i values can be recalculated using these new P_i values. The coordination equations were reestablished and resolved to obtain another set of P_i values.

$$\lambda = 9.5277$$

$$P_1 = 433.94$$

$$P_2 = 300.11$$

$$P_3 = 121.72$$

e. This process is repeated until the answer converges to the final answer (The last iteration is when the obtained answer is approximately equal to the previous iteration's answer).

In the general case, usually an explicit expression for P_{loss} is not available for a real power system. However, P_{loss} and L_i values can be estimated during load flow calculations. Therefore, an algorithm like the above example can solve the problem. The ELD and load flow problems are solved simultaneously based on this algorithm. The algorithm is constructed from the main loop and two inner loops. At first, an initial guess for P_i value is assumed. Then the load flow program is run using assumed P_i values. The values for P_{loss} and L_1, L_1, ..., L_N are calculated from given P_i values. Then coordination equations are established and solved using obtained constant P_{loss} and L_i values each of load flow and ELD solution programs has its internal loop (solved with multiple iterations). Using obtained P_i values from the ELD solution, the main loop runs again, starting with load flow calculations. The main steps of the applied algorithm are given in Figure 6.8.

6.6 OPTIMUM POWER FLOW

The Optimum Power Flow (OPF) formulation focuses on what constraints are considered, including voltage limits, generation capability, transformer tap setting, and line flow limits. The OPF problem is used to find an efficient way to solve the nonlinear optimization problem considering the power system's special characteristics. Many optimization methods have been applied in solving OPF problems, such as the gradient searching method, Newton method, interior-point method, and sequential linear/quadratic programming method. Some new optimization methods, such as the simulated annealing method, fuzzy logic, artificial neural network method, and genetic algorithm, are applied to solve the OPF problem. To extend OPF applications, it is important to obtain a practical and reasonable OPF solution by coordinating optimization techniques and OPF formulations.

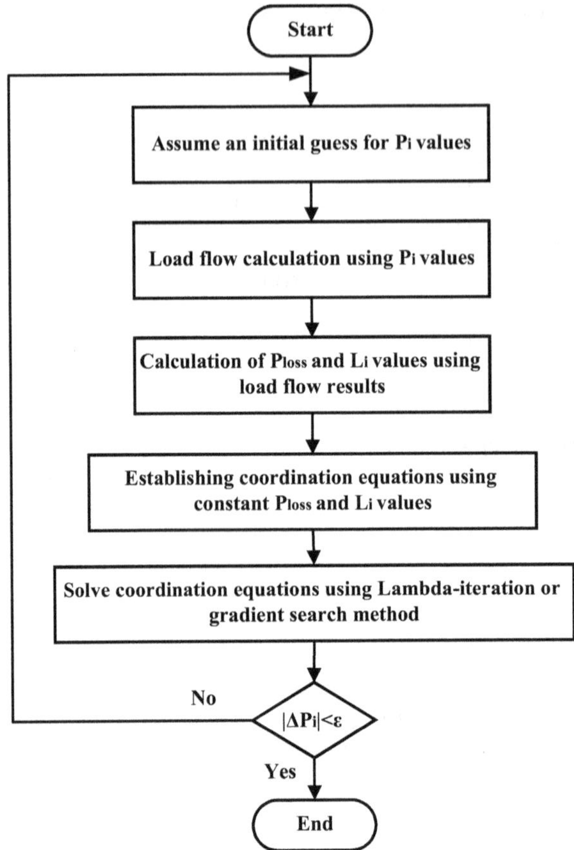
Start
Assume an initial guess for Pi values
Load flow calculation using Pi values
Calculation of Ploss and Li values using load flow results
Establishing coordination equations using constant Ploss and Li values
Solve coordination equations using Lambda-iteration or gradient search method
No
|ΔPi|<ε
Yes
End

FIGURE 6.8 ELD solution flow chart.

The OPF problem can be divided into two sub-problems: the real power optimization problem (MW dispatch) and the reactive power optimization problem (MVAR dispatch). The first problem's main objective is to minimize the system fuel cost, and the second problem is to minimize the power system losses. For both optimization sub-problems, acceptable system performance must be maintained in terms of limits on generator real and reactive outputs, transformers tap-settings, and bus voltages levels. In many cases, the optimal reactive power flow (ORPF) is considered independently, and in some others, it is combined with the real power dispatch. However, most real-time applications (ORPF) programs are run independently of the real power dispatch.

The changes in bus voltage in the system, resulting from the system configuration changes or in power demands, can be minimized by reallocating reactive power generations in the system by control devices (transformer tap-settings, generator voltage magnitudes, and switching of VAR sources). Also, it is possible to minimize system losses by reactive power redistribution in the system. In this respect, the (ORPF) problem has attracted much attention and has grown into a powerful power system operation and planning tool.

6.7 VOLTAGE STABILITY AND REACTIVE POWER FLOW PROBLEM

Voltage stability concerns are on power systems that are heavily loaded, failed, or lack reactive power. The essence of voltage stability can be studied by analyzing reactive power output, transmission, and distribution. It is known as a power system's capacity to maintain appropriate voltages at all buses in the system under normal operating conditions and after disturbance. Definitions of voltage stability are most presented in sources.

Power or load flow solution is closely related to voltage stability research. It is an important tool for voltage stability assessment. The load flow results are the voltages at each system bus that allow measuring all other system quantities such as actual and reactive power flow, current flows, voltage drops, and power losses. Newton-Raphson (N-R) is the most general and reliable algorithm for solving the power flow problem. His wording and description are standard articles in many textbooks.

Voltage levels are primarily dictated by reactive power balance and specifications. Because inductive line losses make transmitting large quantities of reactive power over long lines ineffective, lots of reactive power needed by loads must be supplied locally. Also, the reactive power generators are small, which can strongly affect voltage levels.

6.8 POWER LOSS AND POWER FLOW CONTROL

Load flow deals with electrical power flow from one or more sources to energy-consuming loads along available paths, typically seen in a single-line diagram. According to Kirchhoff's rules, electrical energy flows in a network separate branches according to their respective impedances with a voltage balance. The flow shifts as the circuit configuration changes, generation shifts, or load specifications change. Information on these changes is critical for industrial plants and electrical utility operators to ensure efficient operation, reduce losses, preserve service reliability, and coordinate protective relays for unforeseen emergencies.

Losses in the electrical network and actual and reactive electricity flow through load flow simulation for all equipment connecting the buses. Quantifying and minimizing losses is important because they decide the power system's economic activity. Different methods can assess active power losses. Simply computable as (I^2R), the power loss in a line can also be determined by taking the algebraic number of total power flows in either direction, or the total loss will be the sum of all line losses.

Generally, overall system losses cannot be minimized to the degree that power flow controller implementation is justified. Just reactive power flow losses are easily avoidable, typically very small. Reducing losses due to active power flow will require a decrease in line resistance, requiring drastic conductors.

Active and reactive power flow and losses in the transmission line depend on the transmission line's impedance, voltage magnitudes at sending and receiving ends, and phase angle between voltages. Generator voltage sensors, discrete VAR sources, and transformer tap settings will control these variables in real time and vary the transmitted power and losses depending on device conditions. Moreover, these controls must be calibrated to optimal values to eliminate total power losses in the power system.

Optimization has become an essential part of design in all major disciplines. These fields aren't just engineering. The key reason for this inclusion is to manufacture economically important goods or services with embedded efficiency. Using the word optimization usually means the best outcome for the given circumstances.

6.9 THE OPTIMIZATION PROBLEM

Optimization presupposes knowledge of design rules for a particular problem, specifically the ability to mathematically represent the design. These concepts include variables and priorities. Optimization variables represent the individual elements and it can be classified into two categories:

Established variables: Define constants that won't shift when different designs are compared. They have no role in deciding optimum design. Mathematically, these variables are numerical constants or parameters.

Unknown variables are fascinating variables; the aim is to find a collection of numeric values for these unknown variables to achieve optimality. In unknown variables, two major subcategories can be found:

Control-variables: These unknown variables represent individual elements that can be managed directly in the optimization process.

State-variables (or dependent variables): these unknown variables cannot be managed directly in the optimization process. Their value results from the control variable's preference and how the machine responds to its values.

6.10 MATHEMATICAL FORMULATION OF THE OPTIMIZATION PROBLEM

The general optimization problem formulation is summarized as follows:

$$MinimizeF\left(x\right)Objectivefunction$$

$$subjecttog_i\left(x\right)=0, i=1,2,\ldots,m\,Equalityconstraint \tag{6.60}$$

$$and\,h_j\left(x\right)\leq 0, j=1,2,\ldots,n\,Inequality\ contsraint$$

There are m equality constraints and n inequality constraints, and the number of variables is equal to the dimension of the vector x. The optimization problem of Equation 6.60 is to determine the set of values of the vector x for which all equality constraints $g^i\,(x)=0$ and all inequality constraints $h_j(x) \leq 0$ are satisfied and for which the objective function is at a strict local optimum.

6.11 OPTIMIZATION TECHNIQUES

The optimization techniques

6.11.1 Quadratic Programming

Quadratic programming (QP) is a branch of mathematics that finds extreme quadratic functions when linear equalities and inequalities constrain the variables.

The classic objective function of a QP problem is as follows:

$$Minimize\ F(x) = \frac{1}{2}x^t Hx + c^t x \tag{6.61}$$

Subject to

$$\left.\begin{aligned} A_1 x &= b_1 \\ A_2 x &\le b_2 \\ x &\ge 0 \end{aligned}\right\} \tag{6.62}$$

where:

x is the vector of unknowns, dim $[x]=d$;
c is the vector of cost coefficients, $dim\ [c]=d$;
H is a $(d.d)$ matrix;
A_1 is an $(m.d)$ matrix;
A_2 is an $(n.d)$ matrix;
b_1 is the vector specifying the right-hand sides of the equality constraints, $dim[b_1]=m$;
b_2 is the vector specifying the inequality constraints' right-hand sides, $dim\ [b_2]=n$.

All these matrices and vectors except x are numerically given. Also, the matrix H must be positive definite, and symmetric. With these conditions for H, the QP describes a convex problem.

6.11.2 Linear Programming

Linear programming (LP) is a branch of mathematics that finds extreme linear function values when linear equalities and inequalities constrain the variables.

Problem statement

The standard linear programming problem is defined as:

$$Minimize\ \ F = c^t x \tag{6.63}$$

Subject to:

$$A_1 x = b_1$$

$$A_2 x \le b_2 \tag{6.64}$$

$$x \ge 0$$

where:

x is the vector of unknowns, $dim\ [x]=d$;
c is the vector of cost coefficients, $dim\ [c]=d$;
A_1 is an $(m.d)$ matrix;
A_2 is an $(n.d)$ matrix;
b_1 is the vector specifying the right-hand sides of the equality constraints, $dim\ [b_1]=m$;
b_2 is the vector specifying the inequality constraints' right-hand sides, $dim[b_2]=n$.

6.11.3 *FMINCON* Function

The purpose of the *fmincon* is to find the minimum of a constrained non-linear multivariable function, stated as:

$$Minimize\ F(x)$$

$$\begin{aligned} &g(x)=0 \\ &h(x)\le 0 \\ &A_1.x=b_1 \\ &A_2.x\le b_2 \\ &lb\le x\le ub \end{aligned} \tag{6.65}$$

where x, b_1, b_2, lb, and ub are vectors. A_1 and A_2 are matrices, $g(x)$ and $h(x)$ are function that return vectors, and $F(x)$ is a function that returns a scalar. $F(x)$, $g(x)$, and $h(x)$ can be non-linear functions.

The *fmincon* function uses a sequential quadratic programming (SQP) method. In this method, a quadratic programming (QP) subproblem is solved at each iteration.

6.12 OPTIMAL POWER FLOW

The ever-increasing dependence of modern societies on a reliable electrical service makes the efficient and secure operation of the power grids the primary goal of power utilities. The optimal power flow (OPF) software helps utilities maintain a reliable electricity supply while maximizing profit and minimizing operation costs. Figure 6.9 displays the OPF main input and output parameters.

In traditional OPF algorithms, the goal is to run minimal cost or minimum MW losses. OPF defines the different quantities as parameters, control variables, and

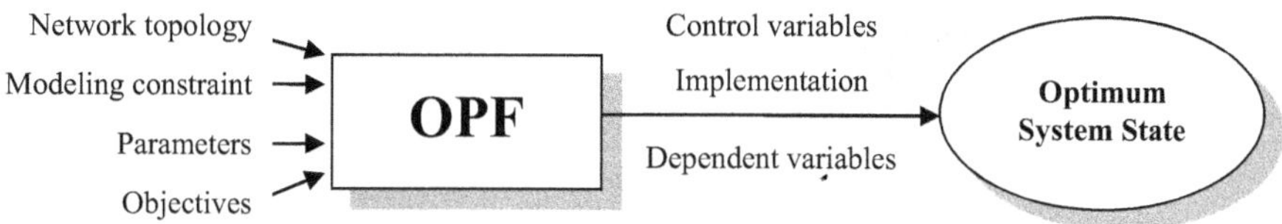

FIGURE 6.9 OPF program.

dependent variables for any network. The greater the portion of control variables, the greater the degree of freedom the optimization problem has. The most common control variables in power system studies and operation are generator voltages, power generators, VAR sources switching, and tap settings of the transformer.

OPF is a problem of non-linear optimization. The objective function (formulating losses or generation costs) is non-linear node voltage or generator MW power output. Most of the equality constraints are non-linear power flow equations. Inequality constraints are imposed by the built equipment's operating capabilities or various quality requirements.

Various mathematical techniques are proposed to solve the OPF mystery. Most of the techniques mentioned in the literature used one of five methods). Lambda iteration method is also called the equal incremental cost criterion (EICC) method. This method has its roots in the common method of economic dispatch used since the 1930s.

i. Gradient method.
ii. Newton's method.
iii. Linear programming method.
iv. Interior point method.

6.12.1 Mathematical Formulation of the OPF Problem

The OPF problem can be formulated as a non-linear optimization problem. The objective function F(x) is scalar and is a model of optimization target reflecting the power utility's economic and security-oriented interests. Objective functions most common are:

- Minimizing net fuel costs.
- Reducing active power loss.
- Reducing reactive power loss.
- Safe operation maintenance with minimal control settings variance.

Objective F(x) is usually non-linear. Vector variables x are divided into control variables and dependent variables. G(x) equality constraints are traditional power flow equations.

Operational restrictions. Most network-dependent variables cannot surpass those lower and upper limits. These restrictions are "soft" constraints, referring to limitations and specifications dependent on protection and power efficiency. Limiting the most common operational constraints:

- Load bus voltage magnitude.
- Reactive PV-generators.
- Present branch, MW/MVAR/MVA flows.
- Line angle/voltage magnitude decrease.

Limited control variables. No control variables surpass lower and upper limits. These can be "hard" constraints, particularly when corresponding to a physical operating range. Popular control variable constraints are as follows:

- Regulating transformer magnitude or tap angle.
- Active power generation.
- PV-bus voltage magnitude.
- Switched reactors and/or condensers.

Constraints vector consists of linear and non-linear functions. The set of variables selected defines the problem dimensions and the number of linear and non-linear constraints.

6.12.2 Classification of the OPF Algorithms Solution

The OPF algorithms will be discussed in two classes:

- **Class A:** Methods where optimization starts from a solved load flow (N-R load flow). The Jacobian and other sensitivity relations are used in the optimizing process; usually, LP or QP based. The process is iterative. After each LP or QP iteration, the load flow is solved again. The optimization is thus separated from the conventional power flow solution algorithm.
- **Class B:** Methods relying on the exact optimality conditions that contain the partial derivatives of the objective function and the constraints of the original problem. The load flow relations are integrated into the optimization algorithm as equality constraints.

This chapter's optimization method falls into class A so that the power flow equations are solved explicitly.

6.12.3 Comparison of the OPF Algorithms Solution Classes

Optimization methods are measured by their speed, flexibility, and robustness efficiency. No single optimization approach satisfies all requirements and can be considered the best non-linear problem-solving algorithm. Class-A and Class-B approaches have relative merits and perform well with either application. However, a method may show poor performance in any problem. In both class-A and class-B algorithms, the size of the linear inequality constraint set is similar, so there is no difference between both classes on this basis. Applied to the OPF problem for class-A algorithms; however, convergence problems can appear only linearly when approximating objective function when solving total loss and minimizing partial losses. The effort to obtain coefficients (especially the quadratic objective function matrix) is very large when using quadratic approximation and consumes much CPU time during the solution process. High efficiency can be obtained with the class-A algorithm if the overall operating cost is the objective function. Class-A algorithms should handle equality and inequality constraints well. Reducing the optimization problem to a reduced set of variables and equality constraints benefits class-A algorithms. Class-A methods are particularly appealing since the starting point is a solved load flow, which provides a feasible solution to the optimization problem's equality constraints. Initially, iterative solutions do not need to be very precise, so the total number of load flow iterations during the repeated execution of the power flow and QP-optimization model is not substantially greater than that for ordinary load flow, e.g., two to three times higher. Class-B approaches are desirable. They solve all objective function problems without variations during the solution process, which is a strong advantage for class-B methods. This is primarily because the class-B algorithms solve the optimal conditions of the original optimization problem, while class-A algorithms solve only optimal conditions of the approximate optimization problem. The downside of class-B algorithms is that the number of variables to be controlled is very high (much larger than in the class-A algorithms, where the number of variables is significantly reduced). Generally, it's unclear which class is better. Ultimately, the decision can be made based on the best combination of algorithmic robustness, computer code performance, and computer code maintenance.

The generation cost is represented as

$$C(i) = \gamma(i) + \beta(i)P(i) + \alpha(i)P^2(i) \tag{6.66}$$

Correspondingly, price curves for each generator in participating utilities are defined by a straight line with slope *m(i)*, as in Figure 6.10. The linear price curve introduces non-linearities in the problem. However, it is a more realistic price representation than a fixed price for power.

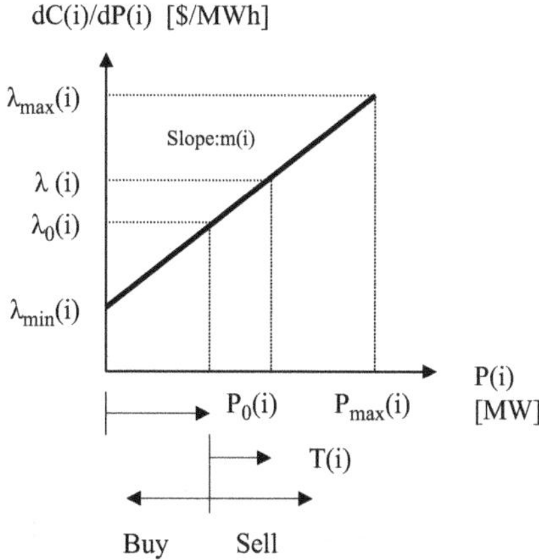

FIGURE 6.10 Incremental price cost of the generator.

Active and reactive power flow and losses in the transmission line depend on the transmission line impedance, voltages' magnitudes at sending and receiving ends, and the voltages' phase angle. The controls on generator voltages, discrete VAR sources, and transformer tap settings can control these variables in real time and, thus, vary the transmitted power and losses according to system conditions. Moreover, adjusting these controls at optimum values is necessary to minimize the power system's total power losses.

6.13 NON-LINEAR FUNCTION OPTIMIZATION

Non-linear function optimization is an important tool in computer-aided design and is a part of a broader non-linear programming class. The underlying theory and computational methods are discussed in many books. The basic goal is to minimize some non-linear objective cost functions subject to non-linear equality and inequality constraints.

The mathematical tools used to solve unconstrained parameter optimization problems come directly from multivariable calculus. The necessary condition to minimize the cost function

$$F\left(x1,\ x2,\ldots,\ xn\right) \tag{6.67}$$

is obtained by setting a derivative of f concerning the variables equal to zero, i.e.,

$$\frac{\partial f}{\partial xi} = 0\,i = 1,\ 2,\ \ldots,\ n \tag{6.68}$$

or

$$\nabla f = 0 \tag{6.69}$$

where

$$\nabla f = \left(\frac{\partial f}{\partial x1}, \frac{\partial f}{\partial x2}, \ldots\ldots, \frac{\partial f}{\partial xn}\right) \tag{6.70}$$

which is known as the gradient vector. The terms associated with second derivatives are given by

$$H = \frac{\partial^2 f}{\partial xi.\partial xj} \tag{6.71}$$

The above equation results in a symmetric matrix called the Hessian matrix of the function.

Once the derivative of f vanishes at local extrema $(x_1, x_2, \ldots, x_n)$, for f to have a relative minimum, the Hessian matrix evaluated at $(x_1, x_2, \ldots, x_n)$ must contain a positive definite matrix. This condition requires that all the Hessian matrix's eigenvalues evaluated at $(x_1, x_2, \ldots, x_n)$ be positive.

The unconstrained minimum is found by setting partial derivatives (concerning the parameters that may vary) equal to zero and solving the parameter values. Among the sets of parameter values obtained, the matrix of second partial derivatives of the cost function is positive definite are local minima. If there is a single local minimum, it is also the global minimum; otherwise, the cost function must be evaluated at each local minimum to determine which one is the global minimum.

Figure 6.11 shows the total power-cost curve of Example 6.6.

Example 6.6

Write a MATLAB Program to Determine the Optimum Power Generated From Two Units Generators; Each Has the Following Characteristics

The fuel inputs per hour of plants 1 and 2 are given as

$$F_1 = 0.0043P_1^2 + 6.4P_1 + 230\,\$/\text{hr}$$

$$F_2 = 0.0055P_2^2 + 7.3P^2 + 650\,\$/\text{hr}$$

Solution

Power limits for each generator

Unit	$\mathbf{P_{min}}$ **(MW)**	$\mathbf{P_{max}}$ **(MW)**
1	50	250
2	50	250

Demand load

$$P_D = 350\text{MW}$$

```
function [d] = data
% n c  b  a    min max
d=[1 230 6.4 0.0043 50  250;
 2 650 7.3 0.0055 50  250;
 ];
end
function [d] = load
d=350;
end

clc
clear all
D=load;
d=data;
b=d(:,3);
a=d(:,4);
Pl=d(:,5);
Ph=d(:,6);
```

```
dP=D;
x=max(b) %lambda
while abs(dP)>0.00001
   P=(x-b)./a/2
   P=min(P,Ph);
   P=max(P,Pl);
   dP=D-sum(P);
   x=x+dP*2/(sum(1./a));

end

for n=1:2
C=d(:,2)+b.*P+a.*P.*P    % Costs Equation

totalCost=sum(C);
display(totalCost);
table(d(:,1),P,C,'V',{'Unit' 'Power' 'Cost'})
end
% plot cost curves
figure
s=C(:,:)
plot(P,s)
grid
xlabel('Total Power (MW)')
ylabel('Cost $/hr')

 ans =

 lambda =
   7.3000
C =
  1.0e+03 *
   2.0336
   1.4996

totalCost =
    3.5332e+03

    Unit   Power     Cost
    ____   ______   ______
    1     242.35   2033.6
    2     107.65   1499.6

s =

   1.0e+03 *
    2.0336
    1.4996
```

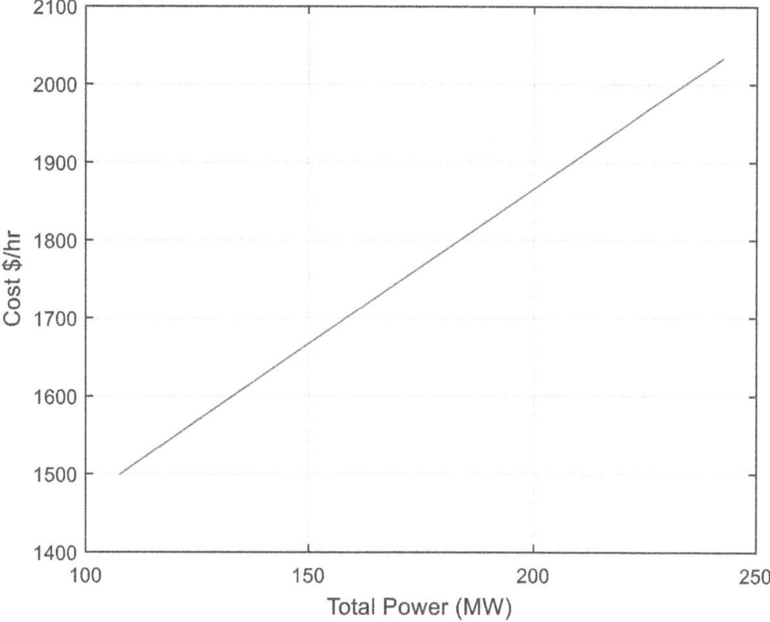

FIGURE 6.11 Total power-cost curve of Example 6.6.

Example 6.7

The Optimization Procedures Were Linear Programming and Sequential Quadratic Programming using MATLAB Code Program. The Above-Mentioned Methods and Procedures are Applied to a 6-Bus Ward and Hall system. The 6-Bus Power System, Shown in Figure 6.12, has been Studied for Comparing the Power and Energy Loss Minimization Methods. The Relevant Load Flow System Data are Given in Tables 6.11–6.14.

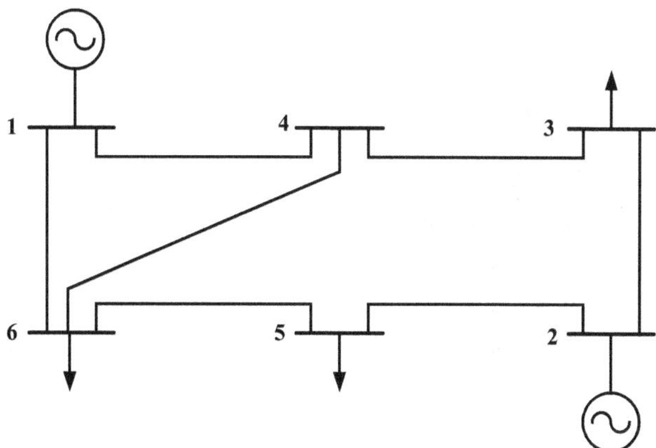

FIGURE 6.12 The 6-bus power system.

TABLE 6.11
Line Data for the 6-Bus Power System (100 MVA base)

Line No.	Bus Number From	To	Impedance (p.u.) R	X	Tap Setting
1	1	6	0.1223	0.518	1
2	1	4	0.080	0.370	1
3	4	6	0.097	0.407	1
4	6	5	0.000	0.300	1
5	5	2	0.282	0.640	1
6	2	3	0.723	1.050	1
7	4	3	0.000	0.133	1

TABLE 6.12
Initial Bus Data for the 6-Bus Power System

	Voltage		Generation		Load	
Bus No.	V PU.	δ Deg.	P MW.	Q MVAR	P MW	Q MVAR
1	1.05	0.0	--	--	--	--
2	1.10	--	50.0	--	0.0	0.0
3	1.00	--	--	--	55.0	13.0
4	1.00	--	--	--	0.0	0.0
5	1.00	--	--	--	30.0	18.0
6	1.00	--	--	--	50.0	10.0

TABLE 6.13
Generator Cost Constants

Generators	Α	β	γ
G1	0.022	0.25	0
G2	0.0175	0.25	0

TABLE 6.14
The 6-Bus Power System Variables Limits

Bus	P_g	Q_g	Q_{max}	Q_{min}	V_g	MVA	Status	P_{max}	P_{min}
1	0	0	150	−150	1.05	100	1	100	0
2	0.5	0	150	−15	1.1	100	1	120	0

TABLE 6.15
The Objective Function Result

Iter.	F(x)	Optimality Procedure
0	0.129375	Infeasible start point
1	$7.29112e^{-36}$	1.75
2	0.339594	5.38
3	0.281187	3.08
4	0.372644	5.94
5	0.312507	0.146
6	0.355378	0.323
7	0.355659	0.00167 Hessian modified

Converged in 7.09 seconds
Objective Function Value = 0.36 $/hr

TABLE 6.16
System Summary

Item	Number	P (MW)	Q (MVAr)
Buses	6	220.0	−165.0 to 300.0
Generators	2	220.0	−165.0 to 300.0
Loads	3 Load	1.4	0.5
Branches	6 Losses (I^2 * Z)	0.00	0.00
Inter-ties	4 Total Inter-tie Flow	1.4	0.5

TABLE 6.17
The Bus Data

Bus	Voltage		Generation		Load		Lambda($/MVA-hr)	
	Mag (pu)	Ang (deg)	P(pu MW)	Q(puMVAr)	P(pu MW)	Q(pu MVAr)	P(pu)	Q(pu)
1	1.027	0.000	0.61	0.36	-	-	0.277	0.000
2	1.028	0.089	0.74	0.13	-	-	0.276	–0.000
3	1.026	–0.075	-	-	0.55	0.13	0.277	0.000
4	1.027	–0.055	-	-	-	-	0.277	0.000
5	1.026	–0.051	-	-	0.30	0.18	0.277	0.000
6	1.026	–0.079	-	-	0.50	0.18	0.277	0.000
Total			1.35	0.49	1.35	0.49		

TABLE 6.18
Branch Data

Branch	From Bus	To Bus	From Bus Injection		To Bus Injection		Loss (I^2 * Z)	
			P (PU.MW)	Q (PU. MVAr)	P (MW)	Q (MVAr)	P (MW)	Q (MVAr)
1	1	6	0.33	0.21	–0.33	–0.21	0.000	0.00
2	1	4	0.28	0.15	–0.28	–0.15	0.000	0.00
3	6	5	–0.17	0.03	0.17	–0.03	–0.000	0.00
4	5	2	–0.47	–0.15	0.47	0.15	0.001	0.00
5	2	3	0.27	–0.02	–0.27	0.02	0.001	0.00
6	4	3	0.28	0.15	–0.28	–0.15	–0.000	0.00

Case(a). Increasing in the power demand by (30) percentage. Table 6.19 shows these changes;

TABLE 6.19
Initial and the New Value of Power Demand

Load Bus	Initial Demand (p.u)	New Demand (p.u)
Bus3	0.55	0.72
Bus5	0.30	0.39
Bus6	0.50	0.65

So we conclude:

1. The iteration increases from 7 to 9, and the running program's time increases from 0.36 to 8.05 sec.
2. When increasing the power demand, the generators run to meet the load demand given by

$$P_D = \sum_{i=1}^{ng} P_i$$

3. The cost increases from 0.36 \$/h to 0.47 \$/h. The reason for increase in the cost is an increase in power generation, according to

$$C(i) = \alpha(i) + \beta(i)P(i) + \gamma(i)P^2(i)$$

4. Lagrange multiplier for equality constraint(λ) increases from 0.277 \$/MVA to 0.285 \$/MVA, according to

$$\frac{dC(i)}{dP} = \beta(i) + 2\gamma(i)P(i)$$

5. Energy loss also increases due to an increase in the generation, and according to

$$P_L = \sum_{i=1}^{ng}\sum_{j=1}^{ng} P_i B_{ij} P_j + \sum_{i=1}^{ng} B_{0i} P_i + B_{00}$$

Case(b). The change in reactive power constraint limits.

Q limits are [150, −150] in the initial case. There is no change when increasing this limit or decreasing it due to a small system.

The main steps of the applied algorithm are given in Figure 6.13.

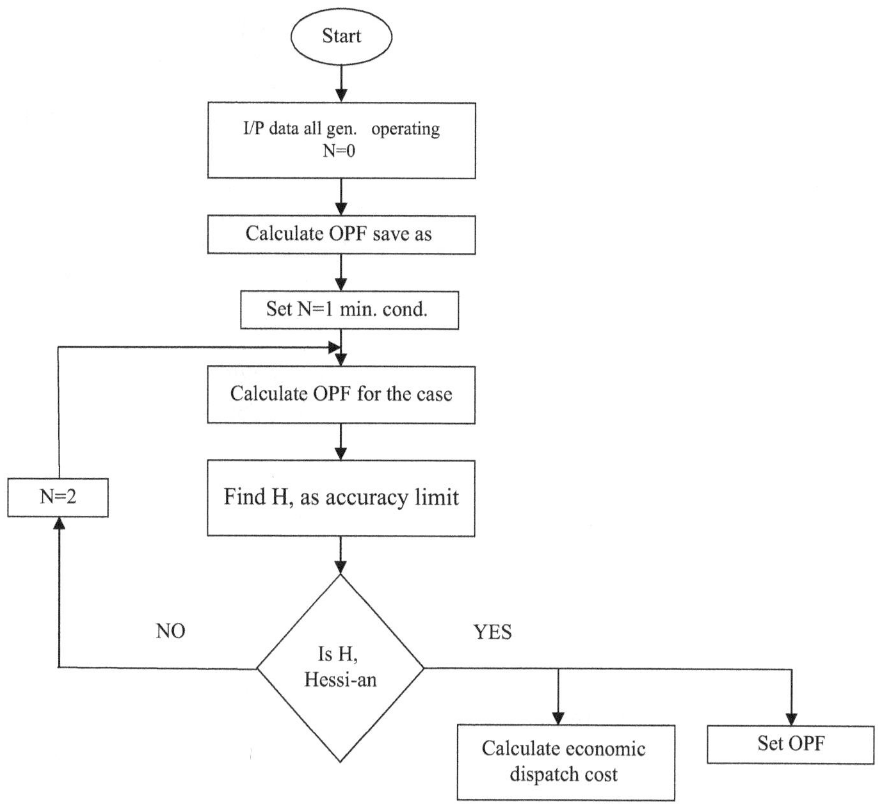

FIGURE 6.13 The main steps of the applied algorithm.

6.14 HYDROTHERMAL COORDINATION

The systematic coordination of the operation of a system of hydroelectric generation plants is usually more complex than the scheduling of an all-thermal generation system. The reason is both simple and important. That is, the hydroelectric plants may very well be coupled both electrically (i.e., they all serve the same load) and hydraulically (i.e., the water outflow from one plant may be a very significant portion of the inflow to one or more other, downstream plants).

No two hydroelectric systems in the world are alike. They are all different. The differences are the natural differences in the watersheds, the man-made storage and release elements used to control the water flows, and the many different types of natural and man-made constraints imposed on the operation of hydroelectric systems. River systems may be simple with relatively few tributaries (e.g., the Connecticut River), with dams in series (hydraulically) along the river. River systems may encompass thousands of acres, extend over vast multinational areas, and include many tributaries and complex arrangements of storage reservoirs (e.g., the Columbia River basin in the Pacific Northwest).

Reservoirs may be developed with a very large storage capacity with a few high-head plants along the river. Alternatively, the river may have been developed with more dams and reservoirs, each with a smaller storage capacity.

However, the one single aspect of hydroelectric plants that differentiates the coordination of their operation more than any other is the existence of the many and highly varied constraints. In many hydro systems, the generation of power is an adjunct to the control of floodwaters or the regular, scheduled release of water for irrigation. Recreation centers may have developed along the shores of a large reservoir so that only small surface water elevation changes are possible. Water release in a river may well have to be controlled so that the river is navigable at all times. With high-volume water releases, sudden changes may be prohibited because the release could result in a large wave traveling downstream with potentially damaging effects. Fish ladders may be needed. International treaties may dictate water releases.

6.15 DIFFERENT TYPES OF HYDRO-SCHEDULING

1. Short-Range (subject of this section).
2. Long-Range.

In a hydroelectric power system, three general categories of problems arise. These depend on the balance between hydroelectric generation, thermal generation, and load. In all hydroelectric systems, the scheduling could be done by simulating the water system and developing a schedule that leaves the reservoir levels with a maximum amount of stored energy.

Hydrothermal systems where the hydroelectric system is the largest component may be scheduled by economically scheduling the system to produce the minimum cost for the thermal system. These are basic problems in scheduling energy. A simple example is illustrated in the Section 6.16, where the hydroelectric system cannot produce sufficient energy to meet the expected load. The largest hydrothermal systems category includes those with a closer balance between the hydroelectric and thermal generation resources and those where the hydroelectric system is a small fraction of the total capacity. In these systems, the schedules are usually developed to minimize thermal-generation production costs, recognizing all the various hydraulic constraints that may exist.

6.16 SCHEDULING ENERGY

Suppose, as in Figure 6.14, we have two sources of electrical energy to supply a load, one hydro and another steam. The hydro plant can supply the load by itself for a limited time. That is, for any period j,

$$P_H^{\max} \geq P_{load\,j} \; j = 1, \ldots, j_{\max}$$

However, the energy available from the hydro plant is insufficient to meet the load:

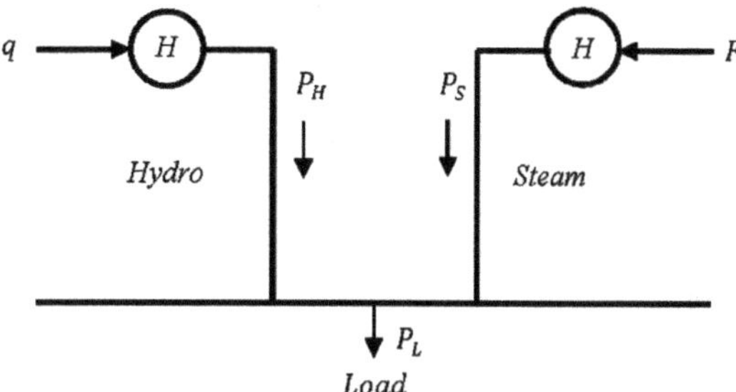

FIGURE 6.14 Hydro and steam electrical energy.

$$\sum_{j=1}^{j_{\max}} P_{Hj}\, n_j \le \sum_{j=1}^{j_{\max}} P_{load}\, n_j \; n = \text{number of hours in a period } j$$

$$\sum_{j=1}^{j_{\max}} n_j = T_{\max} = \text{total interval}$$

We would like to use up the entire amount of energy from the hydro plant so that the cost of running the steam plant is minimized. The steam-plant energy required is:

$$\sum_{j=1}^{j_{\max}} P_{load}\, n_j - \sum_{j=1}^{j_{\max}} P_{Hj}\, n_j = E$$

where

$$\sum_{j=1}^{j_{\max}} P_{load}\, n_j \text{is load energy}$$

$$\sum_{j=1}^{j_{\max}} P_{Hj}\, n_j \text{ is hydro} - \text{energy}$$

E is steam energy.

We will not require the steam unit to run for the entire interval of T_{max} hours. Therefore,

$$\sum_{j=1}^{N_s} P_{sj}\, n_j = E$$

N_s is the number of periods the steam plant is run
Then,

$$\sum_{j=1}^{N_s} n_j \le T_{\max}$$

The scheduling problem becomes:

$$\min F_T = \sum_{j=1}^{N_s} F(P_{sj}) n_j$$

Subject to

$$\sum_{j=1}^{N_s} P_{sj}\, n_j - E = 0$$

and the Lagrange function is

$$\mathcal{L} = \sum_{j=1}^{N_s} F(P_{sj}) n_j + \alpha\left(E - \sum_{j=1}^{N_s} P_{sj}\, n_j \right)$$

Then,

$$\frac{\partial \mathcal{L}}{\partial P_{sj}} = \frac{\partial F(P_{sj})}{\partial P_{sj}} - \alpha = 0 \text{ for } j = 1, 2, \ldots, N_s$$

Or

$$\frac{\partial F(P_{sj})}{\partial P_{sj}} = \alpha \text{ for } j = 1, 2, \ldots, N_s$$

This means that the steam plant should be run at constant incremental cost for the entire period. Let this optimum value of steam-generated power be P_s^* which is the same for all time intervals the steam unit is on. The total cost over the interval is

$$F_T \sum_{j=1}^{N_s} F(P_s^*) n_j = F(P_s^*) \sum_{j=1}^{N_s} n_j = F(P_s^*) T_s$$

where

$$T_s = \sum_{j=1}^{N_s} n_j = \text{total run time for the steam plant}$$

Let the steam-plant cost be expressed as:

$$F(P_s) = A + BP_s + CP_s^2$$

Then,

$$F_T = \left(A + BP_s^* + CP_s^{*2}\right)T_s$$

Also, note that

$$\sum_{j=1}^{N_s} P_{sj}\, n_j = \sum_{j=1}^{N_s} P_s^* n_j = P_s^* T_s = E$$

$$T_s = \frac{E}{P_s^*}$$

$$F_T = \left(A + BP_s^* + CP_s^{*2}\right)\frac{E}{P_s^*}$$

Now, we can establish the value of P_s^* by minimizing F_T:

$$\frac{\partial F_T}{\partial P_s^*} = \frac{-AE}{P_s^{*2}} + CE = 0$$

$$P_s^* = \sqrt{\frac{A}{C}}$$

This means the unit should be operated at its maximum efficiency point long enough to supply the energy (E) needed. Note, if

$$F(P_s) = A + BP_s + CP_s^2 = f_c\, xH(P_s)$$

where f_c is the fuel cost, then the heat rate is

$$\frac{H(P_s)}{P_s} = \frac{1}{f_c}\left(\frac{A}{P_s} + B + CP_s\right)$$

and the heat rate has a minimum when,

$$\frac{\partial}{\partial P_s}\left[\frac{H(P_s)}{P_s}\right] = 0 = -\frac{A}{P_s^2} + C$$

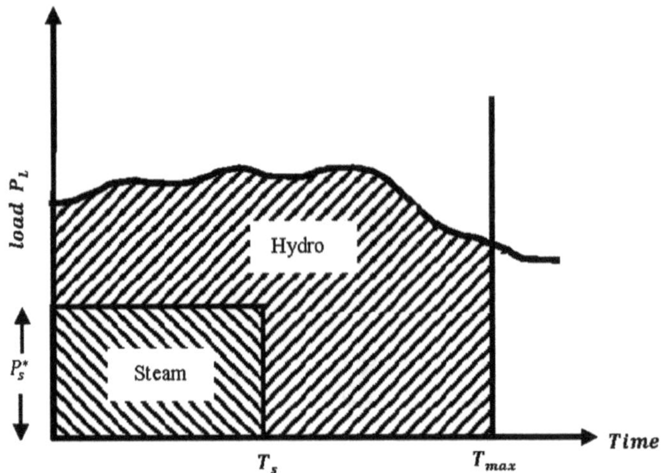

FIGURE 6.15 Load curve for the load connected to the hydro and steam power generator.

giving best efficiency at

$$P_s = \sqrt{\frac{A}{C}} = P_s^2$$

Example 6.8

A Hydro Plant and a Steam Plant Supply a Constant Load of 90 MW for One Week (168 h). The Unit Characteristics are as Follows:

Hydro plant:

$$q = 300 + 15P_H \text{ acre} - \text{ft} / \text{h}$$

$$0 \le P_s \le 100\,\text{MW}$$

Steam plant:

$$H_s = 53.25 + 11.27P_s + 0.0213P_s^2$$

$$12.5 \le P_s \le 50\,\text{MW}$$

Let the hydro plant be limited to 10000 MWh of energy. Solve for T_s^*, the run time of the steam unit.

Solution

The load is 90 x 168 = 15120 MWh, requiring 5120 MWh to generate the steam plant 5120 – 10000.
The steam plant's maximum efficiency is at

$$\sqrt{\frac{53.25}{0.0213}} = 50\,\text{MW}$$

Therefore, the steam plant will need to run for 5120/50 or 102.4 *h*. The resulting schedule will require the steam plant to run at 50 MW and the hydro plant at 40 MW for the first 102.4 h of the week and the hydro plant at 90 MW for the remainder.

Example 6.9

In The Previous Example, Instead of Specifying the Energy Limit on the Steam Plant, Let the Limit be on the Volume of Water that Can be Drawn from the Hydro Plant's Reservoir in One Week. Suppose the Maximum Drawdown is 250000 acre – ft. How Long Should the Steam Unit Run?

Solution

To solve this, we must account for the plant's q versus P characteristic. A different flow will occur when the hydro plant is operated at 40 MW than when it is operated at 90 MW. In this case,

$$q_1 = \left[300 + 15(40)\right] \times T_s \text{acre} - \text{ft}$$

$$q_2 = \left[300 + 15(90)\right] \times (168 - T_s)\text{acre} - \text{ft}$$

And,

$$q_1 + q_1 = 250000\,\text{acre} - \text{ft}$$

Solving for T_s, we get 36.27 h.

PROBLEMS

6.1 Write the condition for the optimal power dispatch in a lossless system.
6.2 Draw the incremental fuel cost curve.
6.3 What is meant by the spinning reserve?
6.4 Write the significance of Unit Commitment.
6.5 Draw the incremental cost curve of a thermal power plant.
6.6 Write the equality and inequality constraints considered in the economic dispatch problem.

6.7 Define spinning reserve constraint in unit commitment problem.
6.8 What is the incremental cost criterion?
6.9 What is the unit commitment problem's minimum up and minimum downtime?
6.10 Define the participation factor.
6.11 What is the participation factor concerning economic load dispatch?
6.12 Write the coordination equation taking losses into account.
6.13 What is meant by the incremental cost curve?
6.14 Compare with unit commitment and Economic load dispatch.
6.15 Define the penalty factor
6.16 List the few constraints that are accounted for the in-unit commitment problem.
6.17 What is meant by the priority list method?
6.18 Mention the assumption made in the formation of loss formula matrix B.
6.19 What are the advantages of using the forward dynamic programming method?
6.20 What is the unit commitment problem? Discuss the constraints that are to be accounted for the unit commitment problem.
6.21 Obtain the priority list of unit commitment using full-load average production cost for the given data:

$$\textit{Heat rate of unit1 } H_1 = 500 + 7.52P_{G1} + 0.0015\,P_{G1}^2\text{MW/hr}$$

$$\textit{Heat rate of unit2 } H_2 = 300 + 7.55P_{G2} + 0.002P_{G2}^2\text{MW/hr}$$

$$\textit{Heat rate of unit3 } H_3 = 80 + 8P_{G3} + 0.0039P_{G3}^2\text{MW/hr}$$

$$P_D = 500\text{MW}$$

Unit	Pmin(MW)	Pmax(MW)	Fuel Cost
1	150	500	1.1
2	100	450	1.0
3	50	250	1.2

6.22 The fuel inputs per hour of plants 1 and 2 are given as:

$$F_1 = 0.2P_1^2 + 40P_1 + 120\$/\text{hr}$$

$$F_2 = 0.25P_2^2 + 30P_2 + 150\$/\text{hr}$$

Determine the economic operating schedule and the corresponding generation cost if the maximum and minimum loading on each unit is 120MW and 25MW. Assume the transmission losses are ignored, and the total demand is 185 MW. Also, determine the saving obtained if both the units equally share the load.

6.23

i. Construct the priority list for the units given below.

$$H_1 = 510 + 7.20P_1 + 0.00142P_1^2.$$

$$\mathrm{P_{min}} = 150\,\mathrm{MW}, \mathrm{P_{max}} = 600\,\mathrm{MW}. \text{Fuel cost} = 1.1\$ / \mathrm{MBtu}.$$

$$H_2 = 310 + 7.85P_2 + 0.00194P_2^2$$

$$\mathrm{P_{min}} = 100\,\mathrm{MW}, \mathrm{P_{max}} = 400\,\mathrm{MW}. \text{Fuel cost} = 1.0\$ / \mathrm{MBtu}$$

$$H_3 = 78 + 7.97P_3 + 0.00482P_3^2$$

$$\mathrm{P_{min}} = 50\,\mathrm{MW}, \mathrm{P_{max}} = 200\,\mathrm{MW}. \text{Fuel cost} = 1.2\$ / \mathrm{MBtu}$$

ii. Derive the coordination equation with losses neglected.

6.24 The incremental fuel costs in \$ per MW.hr of the two units are given by

$$\frac{dF_1}{dP_1} = 0.0075\,P_1 + 9.2$$

and

$$\frac{dF_2}{dP_2} = 0.012\,P_2 + 7.2$$

Assume that both units are operating at all times, that the total load varies from 160 to 1300 MW, and each unit has a power range between 80 MW and 650 MW.

Find the incremental fuel cost and allocation of load between units for minimum cost.

6.25 A system consists of two plants connected by a transmission line. The only load is located at plant two. When 200 MW is transmitted from plant one to plant two, power losses in the line are 16 MW. Find the required generation for each plant and the power received by the load when λ for the system is \$11.25 per MW.hr.Assume that the incremental cost equation for each plant is given by:

$$\frac{dF_1}{dP_1} = 0.015\,P_1 + 7.5\frac{\$}{\mathrm{MW.hr}}$$

and

$$\frac{dF_2}{dP_2} = 0.0195\,P_2 + 8.2\frac{\$}{\mathrm{MW.hr}}$$

6.26 An area of the interconnected power system has two thermal units operating on economic dispatch. The fuel costs of these units are:

$$F_1 = 0.2P_1^2 + 40P_1 \$ / \text{hr}$$

$$F_2 = 0.25P_2^2 + 30P_2 \$ / \text{hr}$$

P_1 and P_2 are in MW. Total transmission loss for the area is

$$P_L = 2\times10^{-4}\,P_1^2 + 2.5\times10^{-5}P_1.P_2 + 4\times10^{-5}\,P_2^2\ \text{MW}$$

Determine the output of each unit, total transmission loss, total load demand, and total operating cost when the system $\lambda = 16\dfrac{\$}{\text{MW.hr}}$.

6.27 A plant has two generators supplying the plant bus, and neither is to operate below 100 MW or above 625 MW. Incremental costs with P_1 and P_2 in MW are

$$\frac{dF_1}{dP_1} = 0.015\,P_1 + 7.5\frac{\$}{\text{MW.hr}}$$

and

$$\frac{dF_2}{dP_2} = 0.018\,P_2 + 7.2\frac{\$}{\text{MW.hr}}$$

For economic dispatch, find the plant λ when $P_1 + P_2$ equal

i. 200 MW.
ii. 500 MW.
iii. 1150 MW.

6.28 An area of the interconnected power system has two thermal units operating on economic dispatch. The fuel costs of these units are:

$$F_1 = 1080\,P_1^2 + 850\,P_1\ \text{MBtu / hr}\ 0 \le P_1 \le 0.6\,\text{p.u}$$

$$F_2 = 150\,P_2^2 + 750\,P_2\ \text{MBtu / hr}\ 0 \le P_1 \le 1.4\,\text{p.u}$$

P_1 and P_2 are in MW. The total transmission loss for the area is

$$P_L = 0.0108\,P_1^2 + 0.015\,P_1.P_2 + 0.018\,P_2^2\ \text{p.u}$$

Determine the output of each unit, total transmission loss, total load demand, when total operating $P_{Load} = 1.5$ p. u.

7 Unit Commitment

UC is operational planning. The purpose of this planning is to determine a schedule called the UC schedule, which tells us ahead of time when and which units to start and shut down during the operation load curve over a prespecified time, such that the total operating cost for that period becomes minimum. This chapter focuses on the UC problem formulation and the Dynamic Programming (DP) method. The feasibility of load supply and generation and spinning reverse are discussed in this chapter. The time consideration in UC calculation is also discussed.

7.1 UC PROBLEM FORMULATION

In the UC problem, assume that there are N units available to us and that we have a forecast of the demand to be served. The purpose of solving the UC problem is to find what subset of units should be selected to turn on and supply the demand to minimize the cost.

There is a need to emphasize the essential difference between the UC and Economic Load Dispatch (ELD) problems. The ELD problem assumes that N units are already connected to the system. The purpose of the ELD problem is to find the optimum share for these N units. On the other hand, the UC problem selects proper N available units to connect to the system and supply forecasted demand. The UC problem is solved offline based on a forecast of demand in the future. At that moment in the future, the generation of different grid-connected units (selected in UC) is determined by solving online ELD problems based on actual measured demand.

The UC problem is solved over time (in the future), such as the 24 hours of a day or the 168 hours of a week. The UC problem is much more difficult to solve than the ELD problem. The selection procedures involve the ELD problem as a subproblem. Similar to ELD, the UC problem is solved for controllable load only.

Formulation of UC is similar to ELD except that, as mentioned before, the UC problem is solved over time, and the answer is in discrete form (on-grid or off-grid status of each unit at each hour of scheduling period).

Basic UC formulation:

$$\text{Min Cost} = \sum_{j=1}^{N_T}\sum_{i=1}^{N} F_i(P_i).U_{ij}$$

$$S.T. = \sum_{i=1}^{N} P_i^{\text{Max}}.U_{ij} \geq Demand_j \qquad j = 1, 2, \ldots, N_T$$

DOI: 10.1201/9781003293965-7

$$\sum_{i=1}^{N} P_i^{\text{Min}}.U_{ij} \le Demand_j \qquad j = 1, 2, \ldots, N_T$$

where:

N_T is the number of time intervals (usually 24 hours).
N is the number of available units.
$Demand_j$ is the estimated demand at hour j

U_{ij} is (state of unit i at hour j) $= \begin{cases} 1 & On \\ 0 & Off \end{cases}$.

Important Notes:
Decision variables in the UC problem are U_{ij} available, not P_is. UC determines what units are in the circuit at each hour and their generations (P_i) obtained approximately. Accurate value of P_i is calculated through ELD solving in real time using actual demand.

As shown in Table 7.1, the best combination is turning ON unit 1 and turning OFF units 2 and 3. Furthermore, the total generation cost is calculated by solving the ELD, and the minimum total generation cost is 3295. The UC problem solution is more complicated as compared to the ELD problem.

Example 7.1

A grid has three-generation units, the fuel cost equations, and the output power of each unit given:

$$F_1 = 320 + 5.2P_1 + 0.0015P_1^2 \ \frac{\$}{h} \qquad 250\text{MW} \le P_1 \le 550\text{MW}$$

$$F_2 = 140 + 6.2P_2 + 0.00265P_2^2 \ \frac{\$}{h} \qquad 200\text{MW} \le P_2 \le 480\text{MW}$$

$$F_3 = 80 + 7.1P_3 + 0.00402P_3^2 \ \frac{\$}{h} \qquad 30\text{MW} \le P_3 \le 200\text{MW}$$

The units generation supplies a load of 500 *MW*. What unit or combination of units should be used to supply this load most economically?

Solution

To solve this problem, we can simply try all combinations of the three units. Some combinations will be infeasible if the sum of all P^{Max} for the units committed is less than the load or if the sum of all P^{Min} for the units committed is greater than the load (demand). The units will be dispatched for each feasible combination by solving the ELD problem. The results are presented in Table 7.2. It is important to imply that because each unit has two states (ON and OFF), in this example with three units, totally $2^3 = 8$ states (units combinations) exist. These *eight* combinations are considered and, among them, feasible ones are compared based on ELD calculations.

TABLE 7.1
Operating Table of Example 7.1

Unit 1	Unit 2	Unit 3	Max. Generation	Min. Generation	P_1	P_2	P_3	F_1	F_2	F_3	Total Generation Cost (F_1 + F_2 + F_3)
OFF	OFF	OFF	0	0	Infeasible						
OFF	OFF	ON	200	30	Infeasible						
OFF	ON	OFF	480	200	Infeasible						
OFF	ON	ON	680	230	0	368.7	131	0	2786	1658	4444
ON	OFF	OFF	550	250	500	0	0	3295	0	0	3295
ON	OFF	ON	750	280	470	0	30	3095	0	297	3392
ON	ON	OFF	1030	450	300	200	0	2015	1486	0	3501
ON	ON	ON	1230	480	270	200	30	1833	1486	297	3616

Example 7.2

Three of the thermal units described in the following are running, and the output power of each unit is given:

$$H_1 = 510 + 7.2P_1 + 0.00142P_1^2 \ \frac{\text{MBtu}}{\text{h}} \qquad 150\text{MW} \le P_1 \le 600\text{MW}$$

$$H_2 = 310 + 7.85P_2 + 0.00194P_2^2 \ \frac{\text{MBtu}}{\text{h}} \qquad 100\text{MW} \le P_2 \le 400\text{MW}$$

$$H_3 = 78 + 7.79P_3 + 0.00482P_3^2 \ \frac{\text{MBtu}}{\text{h}} \qquad 50\text{MW} \le P_3 \le 200\text{MW}$$

The fuel cost of each unit in $/*MBtu* are given as:

$$C_1 = 1{,}1\frac{\$}{MBtu},\ C_2 = 1,\ 0\ \$/MBtu\ ,\ and\, C_3 = 1{,}2\ \$/MBtu$$

The units generation supplies a load of 550 MW. What unit or combination of units should be used to supply this load most economically?

Solution

To solve this problem, we can simply try all combinations of the three units. Some combinations will be infeasible if the sum of all P^{Max} for the units committed is less than the load or if the sum of all P^{Min} for the units committed is greater than the load (demand). The results are presented in Table 7.2. It is important to imply that because each unit have two states (ON and OFF), in this example with three units, totally $2^3 = 8$ states (units combinations) exist. These *eight* combinations are considered and, among them, feasible ones are compared based on ELD calculations.

The method of investing all feasible combinations is known as enumeration

$$F_1(P_1) = H_1(P_1) x C_1$$

$$F_2(P_2) = H_2(P_2) x C_2$$

$$F_3(P_3) = H_3(P_3) x C_3.$$

As shown in Table 7.2, the best combination is turning ON unit 1 and turning OFF units 2 and 3. Furthermore, the total generation cost is calculated by solving the ELD, and the minimum total generation cost is 5389. The UC problem solution is more complicated as compared to the ELD problem.

As can be seen, using the enumeration method to solve the UC problem is very difficult. For the simple system in the previous example with three units, it is necessary to solve ELD (with its complications) multiple times.

The UC problem is much more complicated than the above example because the scheduling of units for different hours cannot be considered independently.

TABLE 7.2
Operating Table of Example 7.2

Unit 1	Unit 2	Unit 3	Max. Generation	Min. Generation	P_1	P_2	P_3	F_1	F_2	F_3	Total Generation Cost ($F_1 + F_2 + F_3$)
OFF	OFF	OFF	0	0	Infeasible						
OFF	OFF	ON	200	50	Infeasible						
OFF	ON	OFF	400	100	Infeasible						
OFF	ON	ON	600	150	0	400	150	0	3760	1658	5418
ON	OFF	OFF	600	150	550	0	0	5389	0	0	5389
ON	OFF	ON	800	200	500	0	50	4911	0	586	5497
ON	ON	OFF	1000	250	295	255	0	3030	2440	0	5471
ON	ON	ON	1200	300	267	233	50	2787	2244	586	5617

Example 7.3

Assume the load curve shown in Figure 7.2 for the power system of the previous example and find a priority list for turning (ON/OFF) units in response to load changes. The ranges of the output power from each unit are as follows:

$$150\text{MW} \le P_1 \le 600\text{MW}$$

$$100\text{MW} \le P_2 \le 400\text{MW}$$

$$50\text{MW} \le P_3 \le 200\text{MW}$$

Solution

Based on the load curve of Figure 7.1, demand changes from 400 MW to 1000 MW. If considering different levels for demand in this range for such level, perform the operating schedule to find the best combination of units. The final results are listed in Table 7.3.

As can be seen, in the light loading (less than or equal to 400 *MW*), only unit 1 is needed. When demand increased greater than 800 *MW*, at least two units were required to be turned ON. Finally, if demand increases to 1100 *MW*, the third unit should be turned ON. In other words, the priority list for turning units ON is

1. Unit 1
2. Unit 2
3. Unit 3.

The result is also shown graphically in Figure 7.2 on the load curve.

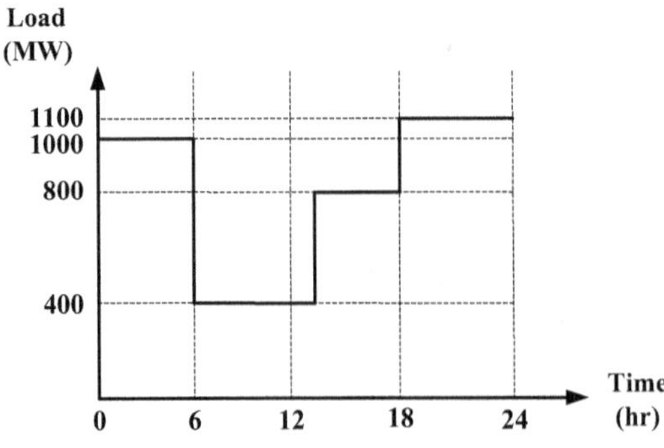

FIGURE 7.1 Load curve demand.

TABLE 7.3
The Best Combination of Units

Time	Load	Optimum Combination		
		Unit 1	Unit 2	Unit 3
0–6	1000	ON	ON	ON
		ON	ON	OFF
6–14	400	ON	OFF	OFF
		OFF	ON	OFF
		ON	ON	OFF
		ON	OFF	ON
		OFF	ON	ON
		ON	ON	ON
14–18	800	ON	ON	OFF
		ON	OFF	ON
		ON	ON	ON
18.24	1100	ON	ON	ON

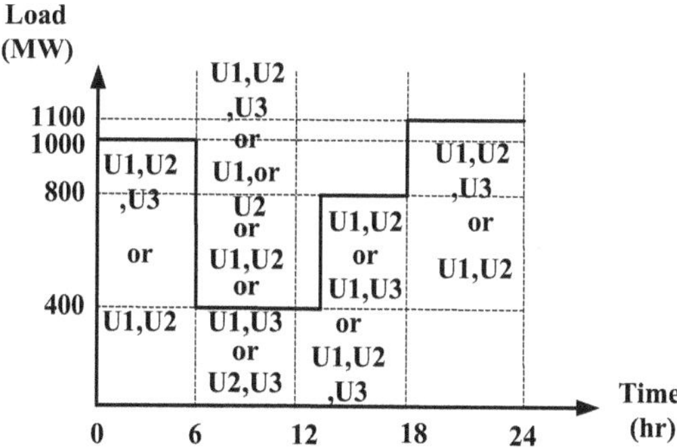

FIGURE 7.2 Results of load curve demand.

For example, if one specific unit needs to be turned OFF at specific hours based on the obtained priority list and load curve changes, it cannot be turned ON again at the next hour. Because starting a thermal unit is a complicated, costly, and time-consuming process. Some units need to remain turned ON (or OFF) for certain hours. Also, some constraints about units crew and changes in different fuel costs in different hours are other reasons that prevent us from solving the UC problem independently. Because of dependency between hours in UC, the total number of states considered in the enumeration method becomes very large. If we have (N) available generation units, the total number of possible combinations in each hour ($2^N - 1$) (one combination, which is all units turned OFF, is always infeasible). Therefore, for a 24 hours schedule, the total number of states, considering hours dependency, is $(2^N - 1)^{24}$. This is a very large number, even for the low number of units (N). Consider a system with ($N = 5, 10, 20, 40$) units. The value of $(2^N - 1)^{24}$ becomes the following in Table 7.4.

These huge numbers show that the enumeration method is not applicable even using modern high-speed computers. Therefore, it is necessary to find alternative solution methods which search the possible solution space smartly and find the answer faster. The most talk-about techniques for the solution of UC problem are as follows.

TABLE 7.4
The Total Number of Possible Combinations

N	$(2^N - 1)^{24}$
5	$6.2 x 10^{35}$
10	$1.73 x 10^{72}$
20	$3.12 x 10^{144}$
40	*(Too big)*

1. Priority-list scheme.
2. Dynamic programming (DP).
3. Intelligent search methods (genetic algorithm, particle swarm optimization, etc.).

Before introducing these methods, it is necessary to be familiar with UC problem constraints. As mentioned before, UC has some additional constraints compared to the ELD problem, which causes UC solutions for different hours to depend on each other. These constraints include the following:

1. Feasibility of load supply.
2. Minimum and maximum generation limits for units.
3. Spinning reserve.
4. Minimum uptime.
5. Minimum downtime.
6. Crew constraints.
7. Starting cost.
8. Must run.
9. Fuel constraints.
10. Hydro constraints.

Detailed explanations about these techniques will be presented in the next part.

7.2 DYNAMIC PROGRAMMING METHOD

The DP method helps solve various problems and significantly reduces the computational effort to find optimal trajectories. The main idea behind this method is to divide the main problem into some smaller sub-problems with a more straightforward solution. At first, we can see the power of the DP method through a simple example to solve a famous mathematic problem.

The algorithm for solving UC problems using the DP method is presented in Figure 7.9. Note that X is the number 0 for the feasible state (combination) in hour K in Figure 7.10. N is the number of all feasible states (combinations) in hour $K - 1$ number of paths from the state in hour $K - 1$ to each state in hour K.

M = total hours is scheduling period (e.g., 24 for one-day scheduling).

Example 7.4

Figure 7.3 represents the cost of transporting a unit shipment from node *A* to node *N*. The values on the arcs are the costs or values of shipping the unit from the originating node to the terminating node of the arc. The problem is to find the route from (*A*) to (*N*) with minimum cost.

Solution

One way to derive a solution is by evaluating all possible routes (enumeration method). However, using the DP method, most non-optimal routes will be ignored. The DP method includes dividing the main problem into proper stages. The value of variables calculated and non-optimal sub-routes is neglected in each stage. The main idea behind DP is the theorem of optimality. **An optimal route must contain only optimal sub-routes.**

In this example, transfer between two consecutive regions from source (*A*) and target (*N*) is considered one stage. In each stage, it is necessary to calculate the total cost of all sub-routes to that repeat except neglected sub-routes in previous stages. Then we can neglect sub-routes that are certainly not a part of the final complete route, for example, in the first stage. Figure 7.4 shows the first step of the cost of transporting.

For each node, the previous node in the subroute and the total cost of the sub-route are written. In the first stage, we cannot neglect any sub-route. Therefore, we can go forward to the next stage.

For example, consider node (*E*). There are two routes to this node, one from (*B*) and the other from (*C*). The total cost of sub-route passing from (*B*) is (18) (12 from *B* to *E* plus a total of 6 for the previous sub-route until *B*). Similarly, the total cost of sub-route passing from (*C*) is 12 (9 + 3). Figure 7.5 shows the second step of the cost of transporting.

Now we can select optimal sub-route among evaluated sub-routes based on the theorem of optimality. **If the final optimal route needs to pass from the node (*E*), it is better to come from (*C*) than (*B*)** because the optimal route is established from optimal sub-routes. In other words, neglecting remained route, passing through (*E*) has less cost if it comes from (*C*) as compared to (*B*). Therefore, all routes (in enumeration method) which include sub-route (*ABE*) can be certainly neglected and not evaluated in the next stages. Similarly, all sub-routes to nodes (*F*) and (*G*) can be evaluated and the best sub-routes selected.

Please note that we do not know which node at this stage (*E, F or G*) is present in the final optimum route, but we can determine the best sub-route terminated to each node and neglect non-optimal sub-routes.

A similar algorithm is implemented in the next stage. It should be noted that only the best answers (min *cost*) of the previous stage should be used to evaluate sub-routes in this stage. For example, evaluating the arc from (*E*) to (*H*), the total cost is 16 (4 plus 12, the best cost to reaching *E*).

We can continue this process until reaching the final node, (*N*) then we can find the optimal route from the results.

Based on Figure 7.6, the minimum cost from (*A*) to (*N*) is 24. The optimal route can be obtained by moving backward in the solution process. Based on the results (see Figure 7.6), the optimal sub-route to node (*N*) is from (*L*). The optimal sub-route to node (*L*) is from (*I*), and to reach (*I*), the best way is from (*E*). Similarly, the node (*C*) is the best way to reach (*E*). Therefore, the optimal route (with min. cost) from (*A*) to (*N*), which is highlighted in Figure 7.6, is (*ACEILN*). As can be seen from this example, the DP method can find specific answers to the problem significantly faster than the enumeration method.

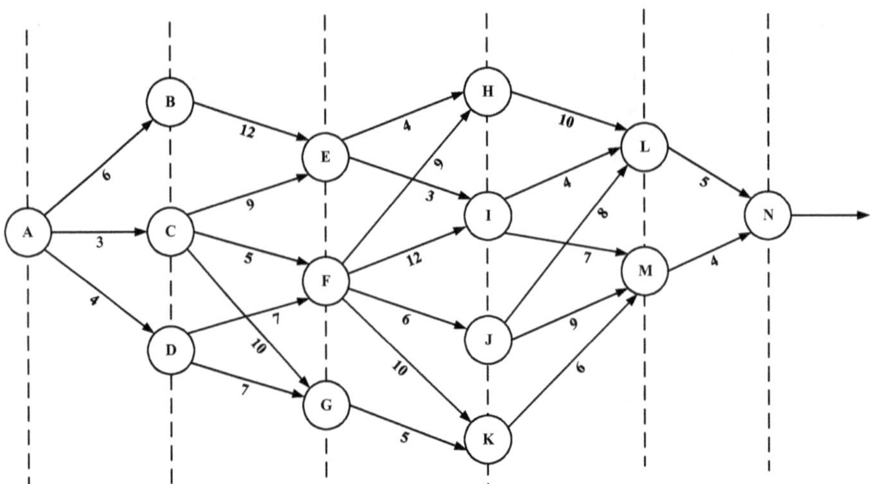

FIGURE 7.3 The cost of transporting of Example 7.4.

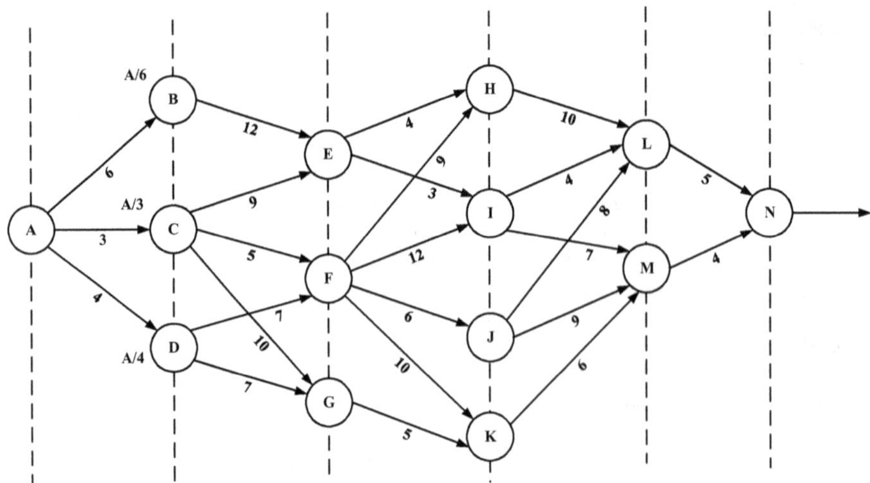

FIGURE 7.4 The first step of the cost of transporting of Example 7.4.

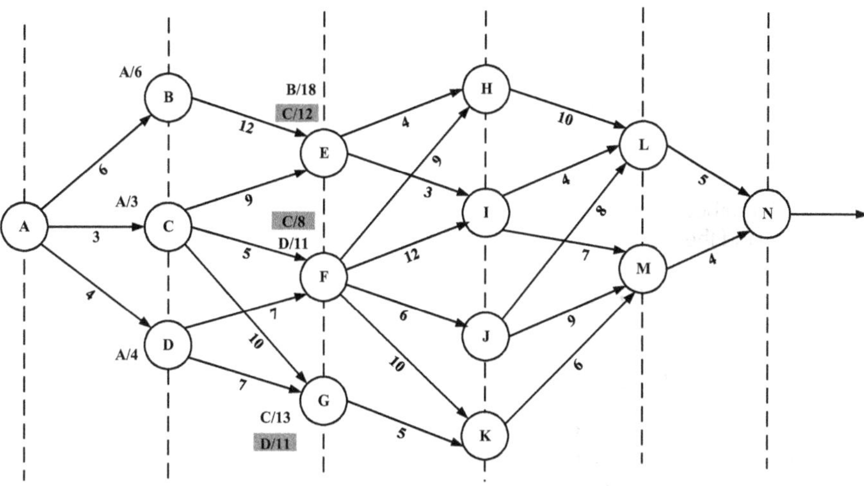

FIGURE 7.5 The second step of the cost of transporting of Example 7.4.

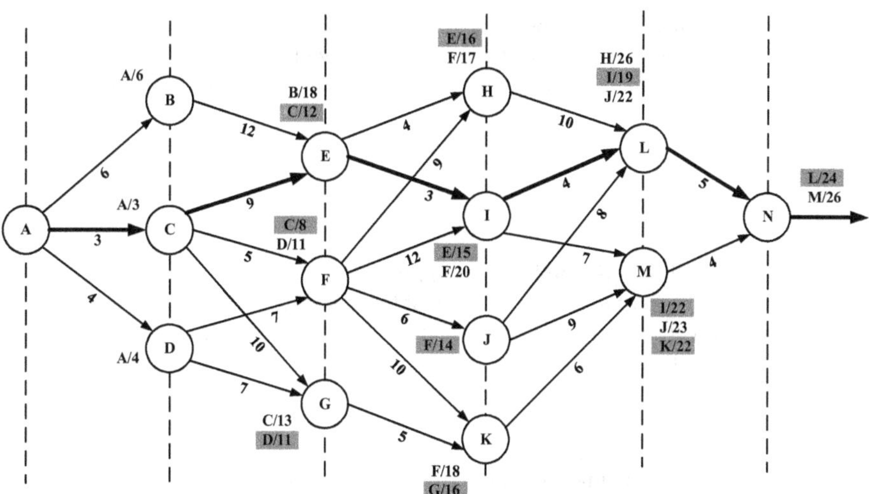

FIGURE 7.6 The last step of the cost of transporting.

Example 7.5

Now, assume that in the previous example, passing through each node also has cost shown in Figure 7.7.

Solution

The DP method can be used in the same way as the previous example. You only need to add the cost of each node to the total cost of each sub-route. For example, cost of sub-route (AB) is 11 (5 is the cost of the route from (A) to (B) plus 6, which is the cost of node B). Complete this example using the DP method presented in Figure 7.8. The optimal route is ($ADGKMN$) *and the* minmum cost = 54.

Application of DP Method to UC Problem

The DP method can similarly solve the UC problem as in two previous examples. Each hour of scheduling can be considered one stage in the DP method. Each node in each state is the feasible combination of turned-on units, and the cost of each node is obtained from the ELD solution considering that combination and forecasted load. The cost of the way from one node in a given hour (stage) to any node in the next hour is the summation of start-up cost for units that must be turned on in moving between these two nodes. If moving from one combination at a given hour to another in the next hour does not meet any constraint for any one of the units (e.g. min. up/downtime crew constraints, etc.), start-up cost between these two nodes should be considered infinitely or of a large value to present a selection of this sub-route in the final answer.

The equation to complete the minimum cost in an hour (K) with a combination (I) is

$$F_{cost}(K,I) = \min_{\cdot(L)}\left[P_{cost}(K,I) + S_{cost}(K-1,L:K,I) + F_{cost}(K-1,L)\right]$$

where:

State (K, I) is the $I - th$ combination in hour K.
$F_{cost}(K, I)$ is the least total cost to arrive at the state (K, I)
$P_{cost}(K, I)$ is generation cost for the state (K, I) obtained from ELD
(L) is set of all combinations in hour $K - 1$.
$S_{cost}(K - 1, L \rightarrow K, I)$ is transition (start-up) cost from state $S_{cost}(K - 1, L)$ to state $S_{cost}(K, I)$.

The algorithm for solving UC problems using the DP method is presented in Figure 7.9. Note that X is the number 0 for the feasible state (combination) in hour K in Figure 7.10. N is the number of all feasible states (combinations) in hour K – 1 number of paths from the state in hour K – 1 to each state in hour K.
M = total hours is scheduling period (e.g., 24 for one-day scheduling).
For example, if N = 3 and X = 5.

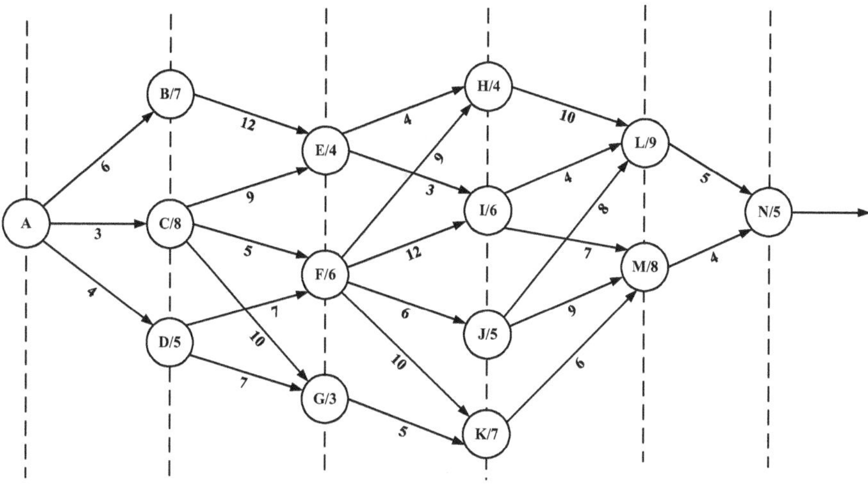

FIGURE 7.7 The second step of the cost of transporting.

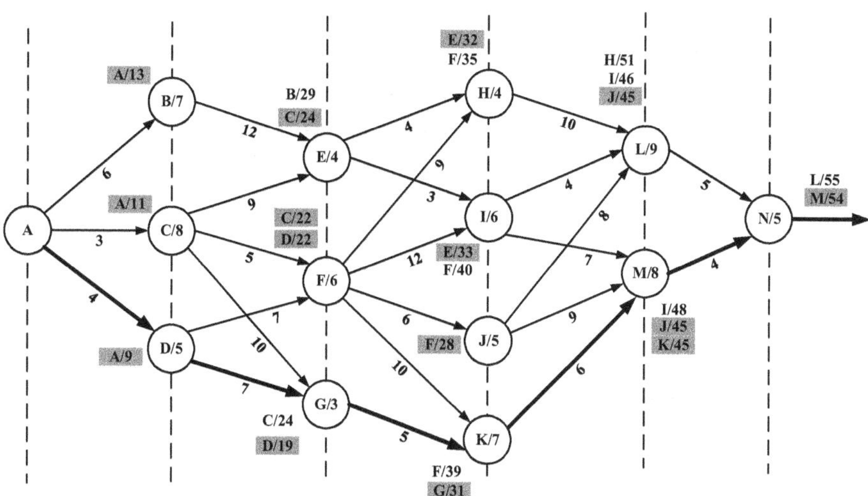

FIGURE 7.8 The last step of the cost of transporting.

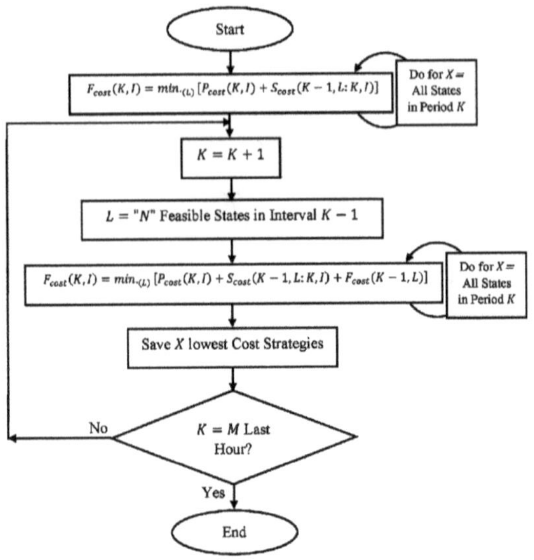

FIGURE 7.9 The algorithm for solving UC problems using the DP method.

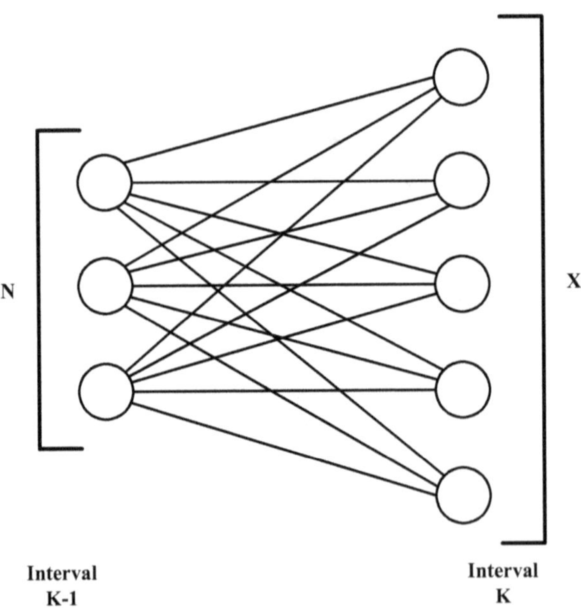

FIGURE 7.10 The network is solving UC using the DP method.

7.3 UNIT COMMITMENT PROBLEM METHOD

The problem methods of the UC are follows.

7.3.1 Feasibility of Load Supply and Generation Limits

Like the ELD problem, equality of generation and demand (now approximately) and generation limits of units (P_i^{min} and P_i^{max}) should be met at the optimum point.

7.3.2 Spinning Reserve

Total grid-connected capacity should be greater than forecasted demand for a response to unplanned events and errors of demand forecast. This reserve is usually considered as a certain percentage of the load. Some important parameters which affect the reserve amount are unit response speed, maximum changeable load, maximum available unit (which may be failed), and limitations on transmission lines.

The above example shows that proper selection of spinning reserve could be so complicated.

Beyond spinning reserves, the UC problem may involve various classes of the scheduled reserve or offline reserves. These include quick-start not connected units such as gas-turbine or most of the hydro units and pumped-storage hydro units.

Example 7.6

Suppose a power system consists of two isolated regions: A and B. As shown in the Figure 7.11, five units have been committed to supplying 3090 MW. The two regions are separated by transmission tie lines that can transfer a maximum of 500 MW in either direction. What can we say about the allocation of spinning reserves in this system?

Solution

The data for this system is given in Table 7.5. Except for unit 4, the loss of any unit on this system can be covered by the spinning server on the remaining units. However, unit 4 presents a problem. If unit 4 were to be lost and unit 5 were to be run to its maximum of 600 MW, the B region would still need 590 MW to cover the load in that region. The 590 MW would have to be transmitted over the tie lines from the A region, which can easily supply 590 MW from its reserve. However, the tie capacity of only 550 MW limits the transfer. Therefore, the loss of unit 4 can't be covered even though the entire system has enough reserve. The only solution to this problem is to commit more units to operate in the B region.

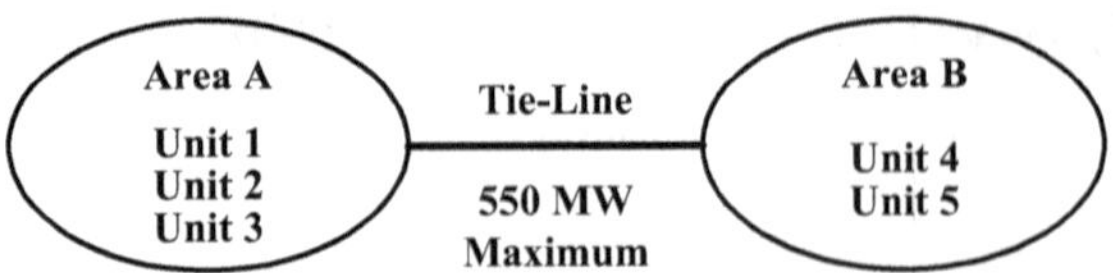

FIGURE 7.11 Two area system of Example 7.5.

TABLE 7.5
Data of Example

Region	Unit	Unit Capacity (MW)	Unit Output (MW)	Regional Generation (MW)	Spinning Reserve	Regional Load (MW)	Interchange (MW)
A	1	1000	900	1740	100	1900	160 in
	2	800	420		380		
	3	800	420		380		
B	4	1200	1040	1350	160		160 out
	5	600	310		290		
Total	1-5	4400	3090	3090	1310	3090	

7.4 UNIT COMMITMENT TIME CONSIDERATION

These units that are timed to come up to full capacity should be considered.

7.4.1 Minimum Up Time

Once a thermal unit is de-committed, there is a minimum time before it can be re-committed.

7.4.2 Crew Constraints

If a plant consists of two or more units, they can't both be turned on at the same time since there are not enough crew members to attend both units while starting up.

7.4.3 Starting Cost

Figure 7.12 shows a start-up cost within several hours. Because the temperature and pressure of the thermal unit must be moved slowly (because of thermodynamic constraints), a certain amount of energy must be expended to bring the unit online. This energy is only consumed to warm up the boiler and other thermodynamic systems and does not result in any MW generation from the unit. The cost of this energy is

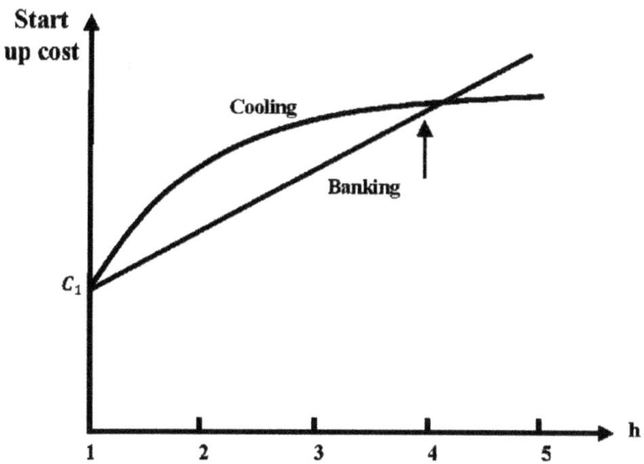

FIGURE 7.12 A start-up cost/hr curve.

brought into the UC problem as a start-up cost. This cost is an essential constraint in the UC problem and the main reason for the dependence of UC solutions at different hours. The start-up cost can vary from a maximum cold-start value to a much smaller one if the unit was turned off recently and is still relatively close to operating temperature. There are two approaches to treating a thermal unit during its down period. The first allows the unit's boiler to cool down and then heat back up to operating temperature in the time for a scheduled turn on. The second (called banking or warm-start) requires sufficient energy to be input to the boiler to maintain operating temperature. The costs for the two can be compared so that, if possible, the best approach (cooling or boiling) can be chosen.

7.5 UNIT COMMITMENT SOLUTION METHODS

The simplest UC solution method creates a priority list of units. As we saw in the previous part, a simple shut-down rule or priority-list scheme could be obtained after an exhaustive enumeration of all unit combinations at each load level. The priority list could be obtained in a much simpler manner by noting each unit's full-load average production cost, where the full-load average production cost simplifies the net heat rate at full load multiplied by the cost.

$$Full-load\ average\ prodcution\ cost = \frac{F_i\left(P_{if}\right)}{P_{if}}$$

$$P_{if} = Fullload\ of\ i-th unit.$$

Example 7.7

Construct a priority list for Example 7.6, based on the full-load average production cost.

Solution

At first, the full-load average production cost will be calculated in Table 7.6. A strict order for these units, based on the average production cost, would order them as follows:

1. Unit 2
2. Unit 1
3. Unit 3

Note that such a scheme would not completely parallel the list obtained in the previous example because this is an approximate method. The commitment scheme would (ignoring min up/downtime, start-up costs, etc.) simply use only the following combinations in Table 7.7.

Most priority list schemes are built around a simple shut-down algorithm that might operate as follows:

- i. At each hour when the load is dropping, determine whether dropping the next unit on the priority list will leave sufficient generation to supply the load plus spinning reserve requirements. If not, continue operating as is; if yes, go on to the next step.
- ii. Determine the number of hours, H, before the unit will be needed again. Assuming that the load is dropping and will go back up some hours later.
- iii. If H is less than the minimum shut-down time for the unit, keep commitment and go to the last step; if not, go to the next step.
- iv. Calculate two costs. The first is the sum of the hourly generation costs for the next H hours with the unit up. Then recalculate the same sum for the unit down and add the start-up cost for either cooling the unit or banking it, whichever is less expensive. If there is sufficient saving from shutting down the unit, it should be shut down; otherwise, keep it on.
- v. Repeat this entire procedure for the next unit on the priority list. If it is also dropped, go to the next and so forth.

TABLE 7.6
The Full-Load Average Production Cost

Unit	Full Load Average Production Cost (R/MWh)
1	9.79
2	9.48
3	11.188

TABLE 7.7
The Commitment Scheme

Combination	Min MW from Combination	Min MW from Combination
2+1+3	300	1200
2+1	250	1000
2	100	400

7.6 ECONOMIC DISPATCH vs. UNIT COMMITMENT

UC and economic dispatch have the same objective of cost minimization. UC has limited access and can minimize the generation/fuel cost it deals with the system operation up to the utility end. More precisely, UC provides the best possible (in the sense of economy and power balance) pre-planned schedule to turn on/off units and their power generation level based on the following:

1. Units' cost characteristics (shut down, start-up, emission, fuel, and maintenance costs).
2. Their availability/non-availability conditions at upcoming time/interval, and
3. Percentage of reserve assigned to each unit. Some constraints at the generation level should also be satisfied, like max/min power capacity of units, ramp rates, minimum up/downtime of units, etc.

Economic dispatch next checks for transmission network constraints like transmission loss, the transmission capacity of lines, and voltage limits at buses, etc., and again optimizes the problem subject to these added constraints in such a way to find the modified schedule at minimum cost. Power system operators then finalize this modified schedule in load dispatch centers for real-time operations.

Most of the time, UC and ED problem is not solved back-to-back rather combined in a single formulation and optimized to achieve the objectives decided.

PROBLEMS

7.1 A grid had three units given here:

$$H_1 = 400 + 7.7P_1 + 0.0015P_1^2 \ \frac{\text{MBtu}}{\text{h}} \qquad 250\text{MW} \le P_1 \le 600\text{MW}$$

$$H_2 = 300 + 7.9P_2 + 0.002P_2^2 \ \frac{\text{MBtu}}{\text{h}} \qquad 200\text{MW} \le P_1 \le 400\text{MW}$$

$$H_3 = 100 + 8.2P_3 + 0.005P_3^2 \ \frac{\text{MBtu}}{\text{h}} \qquad 80\text{MW} \le P_1 \le 200\text{MW}$$

We want to supply a load of 600 MW. What unit or combination of units should be used to supply this load most economically?

7.2 A grid had three units given here:

$$F_1 = 380 + 5P_1 + 0.00156P_1^2 \ \frac{\$}{h} \qquad 230\text{MW} \le P_1 \le 550\text{MW}$$

$$F_2 = 240 + 6P_2 + 0.0027P_2^2 \ \frac{\$}{h} \qquad 180\text{MW} \le P_1 \le 480\text{MW}$$

$$F_3 = 70 + 7P_3 + 0.0046P_3^2 \ \frac{\$}{h} \qquad 40\text{MW} \le P_1 \le 220\text{MW}$$

We want to supply a load of 600 MW. What unit or combination of units should be used to supply this load most economically?

7.3 Figure 7.13 represents the cost of transporting a unit shipment from node *A* to node *N*. The values on the arcs are the costs, or values, of shipping the unit from the originating node to the terminating node of the arc. The problem is to find the route from (*A*) to (*N*) with minimum cost.

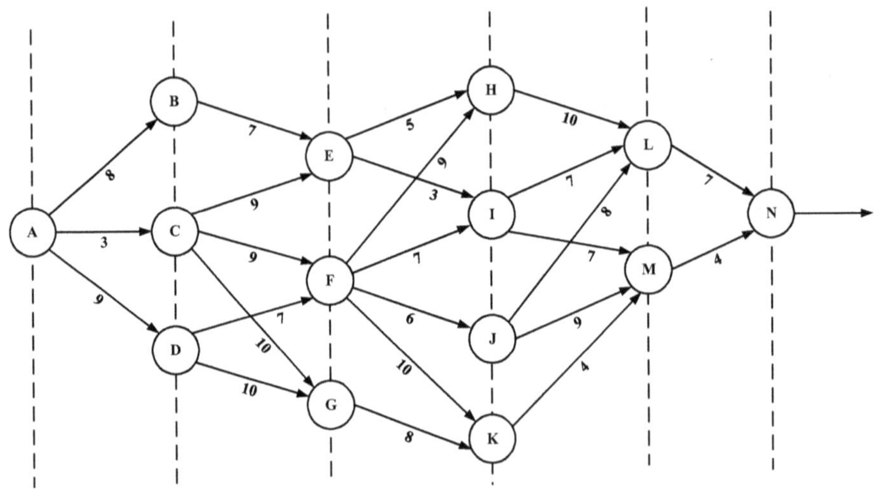

FIGURE 7.13 The cost of transporting of Problem 7.3.

7.4 Now, assume that in the previous problem, passing through each node also has cost like this Figure 7.14.

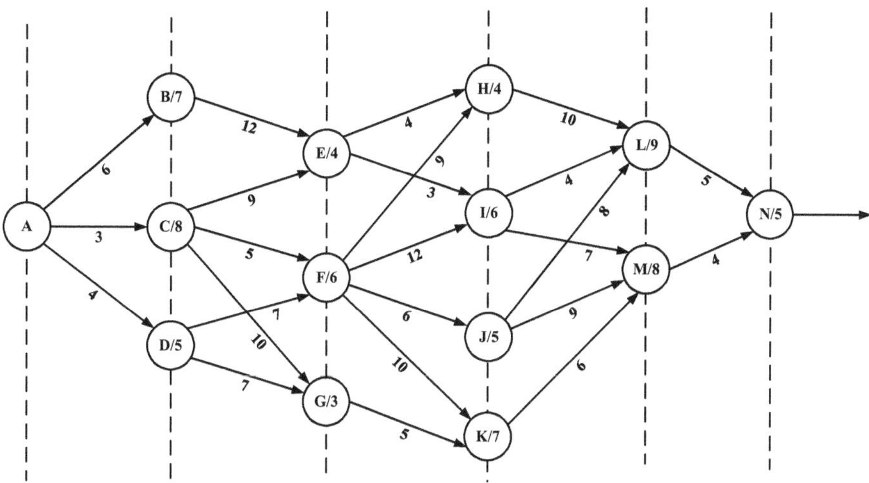

FIGURE 7.14 The second step of the cost of transporting of Problem 7.4.

8 Power Systems State Estimation

With the rise in energy consumption, the need for structured security in the power system network has become increasingly complicated. As a result, there is a real-time model for monitoring and controlling power system networks. State estimation tools are necessary to develop realistic models, especially for non-linear systems like the power grid. The general framework of state estimation in power systems is discussed in this chapter. It also covers the most prevalent linear and non-linear systems strategies and numerical examples to demonstrate the various algorithms.

8.1 GENERAL STATE ESTIMATION DEFINITION AND FUNCTIONS

State estimation (S.E.) is a technique for determining unknown state variable values based on erroneous data. The real value of unknown variables is estimated using statistical criteria. One criterion to use is the sum of the squares of the mistake and is one of the most popular criteria. Because the estimated value may be bigger or smaller than the true value, and thus cancelation of any of the errors is forbidden, the sum of squares of the error is used instead of merely the sum of the errors. Since the early nineteenth century, the previously indicated criterion has been used.

However, it was only later that engineers began incorporating it into computer programs to estimate a system's state variable. It was first employed in the aerospace industry to tackle difficulties involving estimating the location of an aerospace vehicle, which may be a missile, an airplane, or even a spacecraft.

State estimate has been applied in various sectors during the twentieth century, ranging from NASA space applications to control engineering and power systems. This chapter will go over the application of state estimation in the realm of power systems.

The voltage magnitude in Volts, active and reactive power in Watts and VARs, and even ampere flow measurements are all poor measures of the power system or inputs for state estimators. As a result, voltage magnitudes, reactive phase angles, and voltage phase angles are sometimes regarded unknown state variables at the system's nodes or buses. Even if the measures aren't always correct, exact, or even near to be precise, they are necessary for the system's security, control, and economic dispatch limitations.

8.2 ENERGY MANAGEMENT SYSTEM

Electrical power systems are extremely important in our modern culture. The population's demand for electric power has risen drastically throughout the years, demonstrating the importance of energy in today's world. Electricity is now required in homes, hospitals, military bases, industries, commercial usage, and even transportation. People believe in electric power's near-perfect reliability to suit their daily needs.

DOI: 10.1201/9781003293965-8

Regularly, people turn on the light switch, plug in their phones to charge, or turn on an X-ray machine, for example, and expect them to work automatically. However, people do not consider the efforts, studies, and research conducted by electrical engineers, universities, electric utilities, and organizations to maintain the high reliability of the power system.

Normal, emergency, and restorative states are the three conceivable states of power systems. When the system is in a normal state, all state variables operate within their operating boundaries. When system parameters exceed their operational range, the system is deemed at risk. When such contingencies occur, the power system becomes unstable and vulnerable. This is regarded as an emergency status of the electricity system. Some control procedures and activities must be carried out to return the system to a normal operational state after it has been in an emergency status. The system, control devices, and other monitoring and regulatory applications are utilized. At this point, the system is in a restorative condition, which means it must be "restored" to its original state of balance and operation with no system violations.

Many electric power systems throughout the world are currently undergoing critical operational upgrades. Many electric power business actors emphasize the commercial side over the technical. The electric grid is experiencing new challenges for which it was not intended. Backflow of power is one of these issues, as are congestions, various contingencies, new and unanticipated states, and others. As a result, electric power companies would like to have dynamic data regarding the statuses of the system. The key need for any simulation model is to be as close to the real-time scheme as possible.

This is where the Energy Management System (EMS) comes in, by giving measured data and employing computer applications to identify other variables that can help govern the network. State estimators and contingency constrained Optimal Power Flow are two of these applications.

Electric power grids are built for huge power systems, and one of the primary ways to ensure their reliability is to provide adequate information and feedback to the control center that governs them. Over the years, one of the primary goals for electrical engineers has been to build grid monitoring. Obtaining the measurements for every value and comparing it to its limit is important in keeping the power system secured and stable. If power system operators know where the system stands in terms of control and stability, they will know exactly what to upgrade, improve, add power generation, or solve any fault in the system. Corrective actions can be made rapidly if the system models have up-to-date information about system variables beyond their bounds.

If the control center is given exact measurements and information about the status of the power system, engineering choices concerning operations will be easier. Meters and electrical power equipment, on the other hand, may not always offer correct information and measurements. This can happen for various causes in electronics, including when the communication line between the device and the control center is broken or out of service, confusing. As a result, the control center will get a zero value, which is erroneous. To sum it up, measurement instruments are prone to mistakes and malfunction. Because mistakes compound and are difficult to detect, even the tiniest faults may significantly impact the system. Some equipment might be wired backward, resulting in a negative value for that parameter.

8.3 IMPORTANCE OF STATE ESTIMATORS IN POWER SYSTEMS

State estimators are critical engineering tools in current EMS systems. State estimate is a critical tool for power system monitoring. It also has a crucial function as a "dry cleaner," detecting and identifying undesirable measurement mistakes and eliminating them. It's used to tackle difficulties including "smoothing out" tiny random errors in meter readings, detecting and identifying massive measurement inaccuracies, and "filling in meter readings that have failed owing to communications failures," among others. The EMS also uses state estimators for various objectives, including refining algorithms for Locational Marginal Pricing (LMP), which involves pricing power by location and the flow of electricity from and to the grid to alleviate power system congestion. The type of data required for state estimate is highly dependent on the approach used. The maximum likelihood criterion, the weighted least-squares criterion, and the lowest variance criterion all employ input data, such as active and reactive power measurements, as well as voltage and current measurements.

Power system status estimation is a critical function of EMS applications. As a result, the most crucial factor in making this logic function is to have the monitoring station always connected to the field. Sending measurements from a field or station to the control room over the phone is uncommon with today's technology. It is common knowledge that technology is employed for communication and data transfer. Some exceptions may exist where power plant equipment statuses are manually maintained. However, it is difficult to get accurate measurements for an urgent decision during transients in some emergencies. Therefore, state estimation is one way to overcome and reliably approximate the unknown values and filter the errors. This chapter focuses on this concept and how to overcome the challenges using state estimation techniques.

Example 8.1

The following Bus-3 system has two generators. One is located at the swing bus, bus 1, generating 70 M.W. into the system. The second generator generates 30 M.W. at bus 2, as shown in Figure 8.1. A load at bus three consumes 100 MW of power: the bus's data voltage, type, and initial phase angle are given in Table 8.1. Line impedances and power flows between buses are shown in Table 8.2. All values are stated based on a 100 MVA base. Calculate the phase angle at each bus.

Solution

Since bus 1 is the reference bus, we can take two-line power flows with corresponding reactance to calculate the unknown state variables, θ_2 and θ_3. Therefore, based on the following power flow equations, θ_2 and θ_3 can be calculated from the "True" measurements:

$$P_{12} = \frac{1}{x_{12}}(\theta_1 - \theta_2)$$

$$P_{13} = \frac{1}{x_{13}}(\theta_1 - \theta_3)$$

$$P_{23} = \frac{1}{x_{23}}(\theta_2 - \theta_3)$$

Thus, knowing that bus 1 is the reference, then $\theta_1 = 0$ rad.

$$P_{12} = 0.1\text{pu} = -5\theta_2$$

Then

$$\theta_2 = -0.02\,\text{rad}$$

$$P_{13} = 0.45\,\text{pu} = -\frac{1}{0.3}\theta_3$$

Giving

$$\theta_3 = -0.135\,\text{rad}$$

The state variable computations and values are based on "True" line power flow measurements. However, measurements provided to the control center from substations or measuring equipment are not always precise in practice. Consider the following parameters, which have been obtained from the measuring devices, and need the engineers at the control center to determine the values of the state variables in the same manner as previously explained.

Imperfect measurements: $P_{12} = 0.12$ pu = 12 MW; $P_{13} = 0.48$ pu = 48 MW

$$P_{23} = 0.5\,pu = 50\,\text{MW}$$

If two of these measurements are taken as before, then using Equation 8.1, θ_2 and θ_3 can be calculated. With the same steps and procedure as performed earlier,

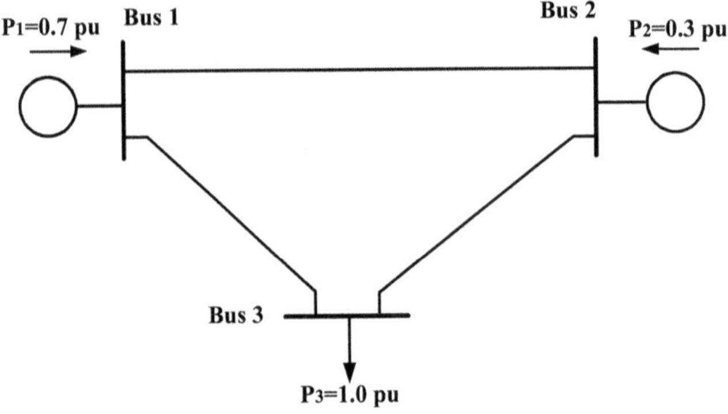

FIGURE 8.1 Bus-3 network configuration.

TABLE 8.1
Bus Data of Example 8.1

Bus #	Type	P(pu)	θ_i
1	Slack	0.7	0
2	Voltage controller	0.3	-
3	Load	1.0	-

TABLE 8.2
Transmission Lines Data of Example 8.2

Busbar		Line Reactance X(pu)	Line Flows	
From Bus	To Bus		P(MW)	P(pu)
1	2	0.2	10	0.1
1	3	0.3	45	0.45
2	3	0.4	55	0.55

substituting P_{12} and P_{13} would result in −0.024 and −0.144 radians for θ_2 and θ_3, respectively. To see the effect of these two imperfect measurements, in calculating the state variables, on the third power flow between bus two and bus 3, we refer to the P_{23} equation. Using the estimated values for the state variables, P_{23} is predicted as follows:

$$P_{23} = \frac{1}{0.4}(\theta_2 - \theta_3) = 2.5\theta_2 - 2.5\theta_3 = 2.5 - 0.024 - (-0.144) = 0.3\,\text{pu}.$$

It is noticed that 0.3 per unit or 30 MW for the real power flowing from bus 2 to bus 3 is relatively lower than the original value of 55 MW or 0.55 pu.Similarly, if other values were considered to calculate the same state variables, it would be impossible to predict the system's actual behavior based on the badly analyzed estimations. Therefore, this shows the importance of developing a method or an algorithm to estimate and find a close estimate to the "true" values of the state variables. This is where state estimation techniques are used to build the tool needed to model a close to real power system model for system security, protection, regulation, and stability.

8.4 SUPERVISORY CONTROL AND DATA ACQUISITION, AND PHASOR MEASUREMENT UNITS

Engineers used to believe that the real-time data produced by the Supervisory Control and Data Acquisition (SCADA) system was accurate. However, they noticed that the measurements are vulnerable to inaccuracies even when employing SCADA and are occasionally unavailable. SCADA takes data from Remote Terminal Units such as activeand reactive powers, voltage and current magnitudes, and phase angles (RTUs). For a long time, obtaining the voltage phase angle was challenging since a time reference was required to synchronize the measurements. Engineers began using the global positioning system (GPS) to synchronize measurements to tackle this difficulty. GPS receivers were installed in Phasor Measurement Units (PMUs). By doing so, the readings are supplied to the control center with a GPS time stamp, which aids in determining the voltage phase angles. As a result, when employed correctly in state estimators, PMUs enhance voltage angle accuracy. Because of the relationships between phasor measurements and state variables, PMUs are much faster than traditional approaches. They also help to improve the reliability and robustness of state estimates. As a result, the notion of replacing the SCADA with PMUs was proposed, but it was discarded due to the high costs of substitution. PMUs, on the other hand, can be used to help improve the traditional methods employed in state estimators based on SCADA data.

A renewable generation has been increasing in recent years, producing green energy. However, with the rapid development in renewable energy, increased power penetration has also increased. Due to such an occurrence, lower voltage systems throughout the power system network. Because of the continual unbalance produced by demand variability and renewable energy penetration, validating the power system models is critical at this time. As a result, state estimators are crucial in monitoring the system and its states.

State estimators can be upgraded and enhanced in various ways to ensure that they function effectively and efficiently. One method for accomplishing this is to convert single-phase measurements to full three-phase measurements. However, this means that the entire system will develop in complexity and, as a result, will be difficult to fix and costly. An easier and simpler method to this improvement problem, on the other hand, is to employ the synchro-phasor measurements provided from the PMU, as previously stated.

There is no need to work with non-linear equations or algorithms if PMU measures are used instead of SCADA measurements. The use of PMU data has the advantage of collecting an adequate number of measurements, allowing the system to be completely visible. The measurement equations become linear in this situation, resulting in a simple non-iterative solution for the system state variables. Unlike the traditional process of transforming and then feeding data into the model, measurements from PMUs are in the form of either voltage or current phasors, making them easier to import and transform into the model than the normal power flows and injection data from SCADA observations.

Several factors contributed to the development of synchro-phasors to where they are now. One key challenge when dealing with power network parameters is dealing with three-phase measurements such as V_{ab}, V_{bc}, and V_{ac}. If it hadn't been for the Symmetrical Component Discrete Fourier Transform (SCDFT) concept or technique, it would have been difficult for synchro-phasors to advance. This recursive approach made it easier to determine the symmetrical components of the power system's phase voltage and phase current.

The GPS is another significant development that has significantly improved and promoted the use of PMU. The GPS technology enabled the time stamping of the synchronized measurements taken by the PMUs. Due to this, it became easier to synchronize those measurements with network research processes such as state estimation approaches. Furthermore, with this design, monitoring dynamic behavior, such as that found in electric power systems, will be less complicated and more manageable.

8.5 ESTIMATORS OF STATE IN PRACTICAL IMPLEMENTATION

Previously, integrated utilities oversaw regulating and stabilizing the electric power system. However, control and monitoring capability have lately been delegated to a distinct body known as Independent System Operator (ISO). This organization is an autonomous, non-profit organization tasked with taking complete control of the electricity grid and ensuring its security. Each area of the United States has its ISO. California ISO, Midcontinent ISO, New York ISO, New England ISO, Alberta Electric System Operator, PJM Interconnection, Southwest Power Pool, Electric Reliability Council of Texas, New Brunswick System Operator, and finally Ontario Independent Electricity System Operator are among the ISOs and Regional Transmission Operators (RTOs).

With today's massively complicated power system, the adoption of computer programs is the most crucial step in grid control. Several algorithms have been developed to regulate and monitor the electricity network; however, none of these systems effectively controlled and secured the grid. As a result, ISOs began incorporating state estimators into their control techniques and monitoring applications. State estimators assisted in estimating unknown parameters and enhanced many other applications such as contingency analysis. Furthermore, state estimators have aided in developing contingency management systems such as the LMP.

As a result, these applications utilize state estimators to compute and anticipate the system's state, which aids the market-pricing technique, LMP, in controlling congestions on the bulk power grid. Similarly, state estimators are used by contingency analysis and dispatch tools to avoid contingencies from happening in the system because state estimators give updated data for system models.

8.6 METHODS OF STATE ESTIMATION

State estimate is utilized to find an approximation for the unknown state variables in the system based on poor observations. The SCADA system sends erroneous measurements to the control center. State estimation tools are used to process these faulty measurements, filter them, discover faults, eliminate inaccurate data, and determine the best estimate for each substation or bus. These state variables are complex voltages, which are the magnitude and angle of the voltage.

The following are the three most utilized methodologies or criteria in state estimation:

Maximum Likelihood: This strategy maximizes the likelihood that the estimated value for the state variable is close to the real value. Section 8.6.1 will provide a quick overview of this strategy.

Weighted Least Squares (WLS): This is the most often used approach in the business. Its primary goal is to minimize the sum of the squares of the difference between the estimated and actual values of the state variable, often known as the error. Section 8.6.2 will go into further depth about this method.

Minimum Variance: Like the WLS, the minimum variance minimizes the anticipated value of the total of the mistakes as indicated in the WLS approach. Section 8.6.3 will provide an overview of this strategy.

8.6.1 Maximum Likelihood Method

Maximum Likelihood estimation is a statistical technique. To begin, a measurement vector is constructed from all the measurements. After that, a probability function for the measurement vector is computed. This is accomplished by multiplying each measurement's probability density functions (pdf) in the vector by each other. The density function is depicted in Equation 8.1 below, where the standard deviation is the anticipated value or mean, and z is the pdf's random variable. $f(z)$ is the function that will output the likelihood of the measurement we receive from SCADA, for example, being the accurate value has a probability closer to 1 or 100%. The probability density function for the mistakes is assumed to have a normal distribution.

$$f(z) = \frac{1}{\sigma\sqrt{2\pi}} e^{-0.5\left(\frac{z-\mu}{\sigma}\right)^2} \tag{8.1}$$

The likelihood function, which was previously explained, is shown in Equation 8.2 below. The maximum likelihood method's goal is to maximize this function to enhance the likelihood of predicting the "actual" value of the state variable. The number of measurements is subscript m in the following equation, and the subscript denotes the ith measurement i.

$$f_m(z) = f(z_1) \times f(z_2) \times \ldots f(z_n) = \prod_{i=1}^{m} f(z_i) \tag{8.2}$$

The likelihood function, $f_m(z)$ Measures the likelihood of identifying certain measurements in the measurement vector. Functions $f(z)$ are independent of each other, as observed and assumed in Equation 8.1. However, $f(z)$ is affected by the mean and standard deviation. As a result, the likelihood function may be represented in logarithmic form by alternating and modifying those parameters to optimal values. With m measurements, this logarithmic function technique alters the function to become as follows:

$$L = \log f_m(z) = \sum_{i=1}^{m} \log f(z_i) \tag{8.3}$$

$$L = -\frac{1}{2}\sum_{i=1}^{m}\left(\frac{z-\mu}{\sigma_i}\right)^2 - \frac{m}{2}\log 2\pi - \sum_{i=1}^{m}\log\sigma_i \tag{8.4}$$

By doing so, the maximum likelihood approach may be solved by either reducing the sum of the squares of the standardized normal variable, as illustrated below, or by maximizing the likelihood function's logarithm, L.

$$Maximize \Rightarrow \log f_m(z)$$

$$Minimize \Rightarrow \sum_{i=1}^{m}\left(\frac{z-\mu}{\sigma_i}\right)^2 \tag{8.5}$$

Assuming that the minimization component of the preceding equation is true and is the residual of the vector z measurements, a simple adjustment is made to simplify the process. Furthermore, the squares of the standard deviations are stated as matrices, and the entire equation is written as follows:

$$\sum_{i=1}^{m}\left(\frac{z-\mu}{\sigma_i}\right)^2=\sum_{i=1}^{m}\left(\frac{r_i}{\sigma_i}\right)^2=\sum_{i=1}^{m}\left(W_{ii}\,.\,r_i\right)^2 \tag{8.6}$$

where:

r_i: the ith measurement's residual
Wii: the residuals' weight.

The WLS approach, which minimizes the sum of the squares of a specified weight on the residuals, is detailed in-depth in Section 8.6.2.

8.6.2 Weighted Least Squares Method

One of the most often used power system state estimators approaches is the WLS method. This chapter describes how state estimate methods attempt to forecast the closest feasible approximation to the state variable. The maximum likelihood principle seeks to improve the likelihood of arriving at a value near the "real" value; whereas, another technique seeks to reduce the gap between the estimated value and the actual state variable value. This is precisely what the WLS approach does. Its primary goal is to reduce the squares of measurement mistakes. This section discusses the WLS algorithm, the measurement function, the Jacobian matrix, and an example of how the WLS approach may be applied to a network.

Assume that the SCADA system measurements are of size m and are represented as a vector z as follows:

$$Z^{meas}=\begin{bmatrix} Z_1 \\ \vdots \\ Z_m \end{bmatrix}$$

Following that, a state vector represented by x is examined. Because the power system is a complex network, state estimation is governed by a non-linear function h(x), as seen in Equation 8.7:

$$h(x)=\begin{bmatrix} h_1(x_1\ldots x_n) \\ \vdots \\ h_m(x_1\ldots x_n) \end{bmatrix} \tag{8.7}$$

where n denotes the total number of system state variables.

True measurement values exist for these state variables, of course. As a result, the derived estimate from $h(x)$ will differ from the real value by an unknown amount represented by e. This mistake is presented in Equation 8.8 as a vector of identical size to the measurement vector z. It is worth noting that each mistake is independent of the others, having a zero mean and independent covariance.

Thus, the "real" measurements, z, are identical to the calculated measurements of the unknown state variables, x, using the non-linear function, h(x), as well as the error vector to adjust for the discrepancy between the actual and computed value. The state equation after derivation is shown in Equation 8.9 below.

$$e = \begin{bmatrix} e_1 \\ \vdots \\ e_m \end{bmatrix} \tag{8.8}$$

$$Z^{meas} = h(x) + e \tag{8.9}$$

An objective function must be reduced to solve the state estimate issue utilizing the WLS state estimating approach. The Jacobian matrix is made up of the sum of the squares of the measurement errors, as well as a weighted matrix known as the covariance matrix, which is detailed further below:

$$\min J(x) = \sum_{i=1}^{m} \left(\frac{Z_i - h_i(x)}{R_{ii}} \right)^2 \tag{8.10}$$

where:

$J(x)$: Jacobian matrix, also known as the residual measurement function.
Z_i: i[th] measurement.
m: total number of measurements.
σ_i^2: i[th] measurement variance.
R_{ii}: is called "the covariance matrix of measurement errors." It is a diagonal matrix, which means that the off-diagonal values are equal to zero since the measurements are independent of one another, as are their mistakes. The covariance matrix is shown in Equation 8.11:

$$\operatorname{cov} e = R = \begin{bmatrix} R_{11} & 0 & 0 & 0 \\ 0 & R_{22} & \dots & 0 \\ 0 & \vdots & \ddots & \vdots \\ 0 & 0 & \dots & R_{nm} \end{bmatrix} = \begin{bmatrix} \sigma_1^2 & 0 & 0 & 0 \\ 0 & \sigma_2^2 & \dots & 0 \\ 0 & \vdots & \ddots & \vdots \\ 0 & 0 & \dots & \sigma_{nm}^2 \end{bmatrix} \tag{8.11}$$

The partial derivatives of $J(x)$ concerning the state variables x must be calculated to minimize the objective function. As a result, the objective function is given as

$$g(x) = \frac{\partial J(x)}{\partial x} = -\frac{\partial h(x)^T}{\partial x} \left[R(z) \right]^{-1} - h(x) = 0 \tag{8.12}$$

The objective function is indicated as $g(x)$, as most power system state estimate literature does. Furthermore, because it is part of the equation, the measurement function $h(x)$ undergoes a partial derivative, as indicated in Equation 8.13. Matrix $H(x)$ is defined and referred to as the measurement Jacobian matrix in terms of $g(x)$ and $H(x)$is equal to

$$H(x)^T = \frac{\partial h(x)^T}{\partial x} \tag{8.12a}$$

$$g(x) = \frac{\partial J(x)}{\partial x} = -H(x)^T\left[R(z)\right]^{-1} - h(x) = 0 \tag{8.12b}$$

where T: The matrix's transpose.

One of the most critical tasks in the WLS algorithm is calculating the measurement Jacobian matrix, H. As a result, understanding how to calculate this matrix is critical. For example, in electric power systems, the connection between various metrics collected via the SCADA system is non-linear. As a result, determining the measurement Jacobian matrix is slightly more involved than that of linear systems. The creation of matrix H is detailed below, illustrating the partial derivatives of real powers, reactive powers, and current magnitudes in terms of the state variables, which in power system analysis are the voltage magnitude and phase angle, as covered earlier in this chapter.

$$H = \begin{bmatrix} \frac{\partial P_{ii}}{\partial \theta} & \frac{\partial P_{ii}}{\partial |V|} \\ \frac{\partial P_{ij}}{\partial \theta} & \frac{\partial P_{ij}}{\partial |V|} \\ \frac{\partial Q_{ii}}{\partial \theta} & \frac{\partial Q_{ii}}{\partial |V|} \\ \frac{\partial Q_{ij}}{\partial \theta} & \frac{\partial Q_{ij}}{\partial |V|} \\ \frac{\partial I_{mag}}{\partial \theta} & \frac{\partial I_{mag}}{\partial |V|} \\ \frac{\partial V_{mag}}{\partial \theta} & \frac{\partial V_{mag}}{\partial |V|} \end{bmatrix} \tag{8.13}$$

where:

$\frac{\partial P_{ii}}{\partial \theta}$: Real power injection measurements at bus i concerning the phase angles.

$\frac{\partial P_{ii}}{\partial |V|}$: Real power injection measurements at bus i concerning the voltage magnitudes.

$\frac{\partial Q_{ii}}{\partial \theta}$: Reactive power injection measurements at bus i concerning the phase angles.

$\frac{\partial Q_{ii}}{\partial |V|}$: Reactive power injection measurements at bus i concerning the voltage magnitudes.

$\frac{\partial P_{ij}}{\partial \theta}$: Real power flow measurements from bus i to bus j concerning the phase angles.

$\frac{\partial P_{ij}}{\partial |V|}$: Real power flow measurements from bus i to bus j concerning the voltage magnitudes.

$\frac{\partial Q_{ij}}{\partial \theta}$: Reactive power flow measurements from bus i to bus j concerning the phase angles.

$\frac{\partial Q_{ij}}{\partial |V|}$: Reactive power flow measurements from bus i to bus j concerning the voltage magnitudes.

$\frac{\partial I_{mag}}{\partial \theta}$: Current magnitude measurements from bus i to bus j concerning the phase angles.

$\frac{\partial I_{mag}}{\partial |V|}$: Current magnitude measurements from bus i to bus j concerning the voltage magnitudes.

$\frac{\partial V_{mag}}{\partial \theta}$: Voltage magnitude measurements concerning the phase angles.

$\frac{\partial V_{mag}}{\partial |V|}$: Voltage magnitude measurements concerning their corresponding voltage magnitude at bus i or bus j.

Such that:

$$\frac{\partial |V_i|}{\partial |V_i|} = 1$$

$$\frac{\partial |V_i|}{\partial |V_j|} = 0$$

$$\frac{\partial |V_i|}{\partial \theta_i} = 0$$

$$\frac{\partial |V_i|}{\partial \theta_j} = 0$$

The higher-order of Equation (8.12) may be ignored using Taylor series expansion, allowing the problem to be solved repeatedly using Gauss-Seidel.

Alternatively, the Newton-Raphson approach can also be used to solve this. The formula below depicts the iterative process used to solve the state variable, x, in the previously described measurement. The Jacobian matrix, covariance matrix, measurement vector, and measurement function are denoted in Equation (8.14) as follows:

$$[x]^{k+1} = [x]^k + \left[H(x^k)^T.[R]^{-1}.H(x^k)\right]^{-1}.H(x^k)^T.[R]^{-1}.\left[z - H(x^k)\right]$$

$$[x]^{est} = \left[H(x^k)^T.[R]^{-1}.H(x^k)\right]^{-1}.H(x^k)^T.[R]^{-1}.\left[z - H(x^k)\right] \quad (8.14)$$

Equation 8.12 may be rewritten as follows, as seen in Equation 8.15:

$$[x]^{k+1} = [x]^k + G(x^k)^{-1}.g(x^k) \quad (8.15)$$

where:

$$G(x^k)^{-1} = \left[H(x^k)^T.R^{-1}.H(x^k)\right]^{-1}$$

$G(x^k)$: known as the gain matrix.

$$g(x^k) = H(x^k)^T.[R]^{-1}.\left[z - H(x^k)\right]$$

$g(x^k)$: the nonlinear function.
k: iteration key.
x^k: solution vector for the kth iteration.

"$G(x^k)$ is sparse, positive, definite, and symmetric provided that the system is fully observable." Notice that both $g(x^k)$ and $G(x^k)$ contain the measurement function, $h(x)$, the Jacobian measurement matrix, $H(x)$, and the covariance matrix, R. The required data can be calculated using the power system model, branch parameters, measurement location, and type. However, calculating $G(x^k)$ is a bit trickier than expected. Since $G(x^k)$ is a sparse matrix. Therefore, according to Equation 8.15, the inverse is a full matrix. To avoid the full matrix on the right-hand side, some adjustments and manipulations are made to Equation 8.15. First, x^k is moved to the left-hand side of the equation, which yields the difference between the two consecutive iterations, $\Delta x = (x^{k+1}) - (x^k)$. Then the whole equation is multiplied by $G(x^k)$ to get rid of it from the right-hand side and move it to the left part. By doing this, inverting it was avoided, as discussed. The modified equation now becomes Equation 8.16:

$$G(x^k)\cdot\Delta x = g(x^k) \quad (8.16)$$

$$G(x^k)\cdot\Delta x = H(x^k)^T\cdot[R]^{-1}\cdot\left[z - H(x^k)\right]$$

To begin, set $k = 0$ (initial iteration) and the initial state value, x^0 as 1.0 per unit (pu) for the voltage magnitude and 0° degree for the voltage phase or angle, then

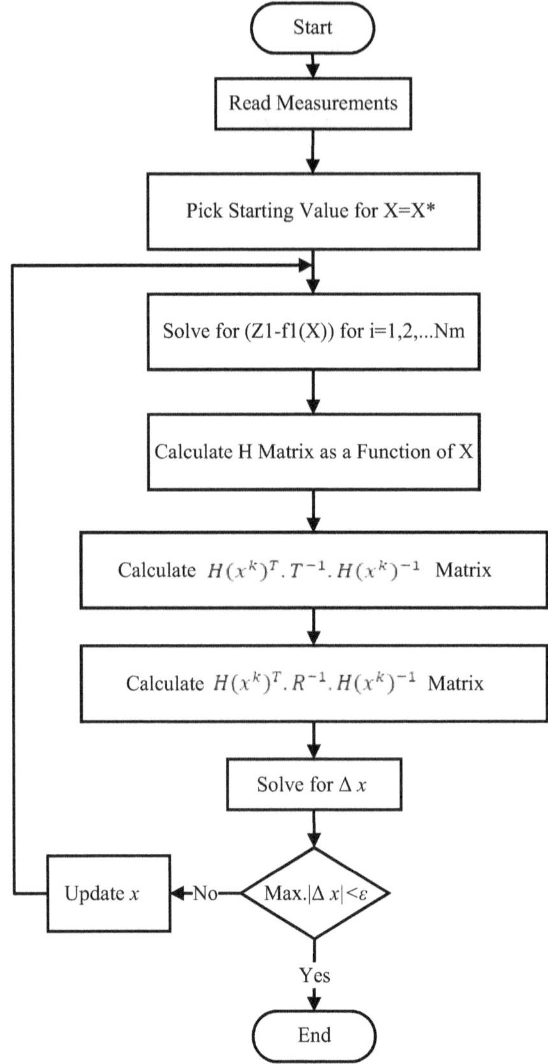

FIGURE 8.2 Weighted Least Square algorithm.

compute $g(x^k)$ and $G(x^k)$. To determine the next iteration state variable, x^1, use Equation 8.15. Repeat the technique for the remaining iterations until the difference between $x^{(k+1)}$ and $x^{(k)}$ is less than a pre-determined threshold. The WLS technique or algorithm is depicted in Figure 8.2.

Overdetermined Case: Note that the original Equation 8.14 was derived and formed based on the assumption that the total number of measurements, m, is greater than the total number of parameters, n, calculated. The previous case, where $m > n$, is considered an "overdetermined" state estimation problem.

$$[x]^{est} = \left[H(x^k)^T.[R]^{-1}.H(x^k)\right]^{-1}.H(x^k)^T.[R]^{-1}.z^{meas} \quad (8.17)$$

Completely Determined Case: When the total number of measurements equals the estimated parameters, $m = n$, the equations can be adjusted such that the estimated state variables, $[x]^{est}$, fit the number of measurements in the z vector precisely. As a result, the following equation is used to obtain the predicted state variables for this "fully specified" state estimation problem:

$$[x]^{est} = \left[H(x^k)\right]^{-1}.z^{meas} \quad (8.18)$$

Underdetermined Case: Finally, for an "underdetermined" state estimation problem, the total number of measurements is less than the estimated parameters, $m < n$. This case frequently happens in the power system network when measurements are lost or cannot be sent through the SCADA system. Therefore, the underdetermined problem is unsolvable since $G(x^k)$ is singular and has no inverse. As seen from Figure 8.3, if only M_{12} is known, then the only relationship that can be concluded is

$$[x]^{est} = H(x^k)^T.\left[H(x^k)^T.H(x^k)^T\right]^{-1}.z^{meas} \quad (8.19)$$

The one between bus one and bus 2. Hence, there is not enough information or data between bus one or bus 2 with bus 3. So neither the voltage magnitude nor the

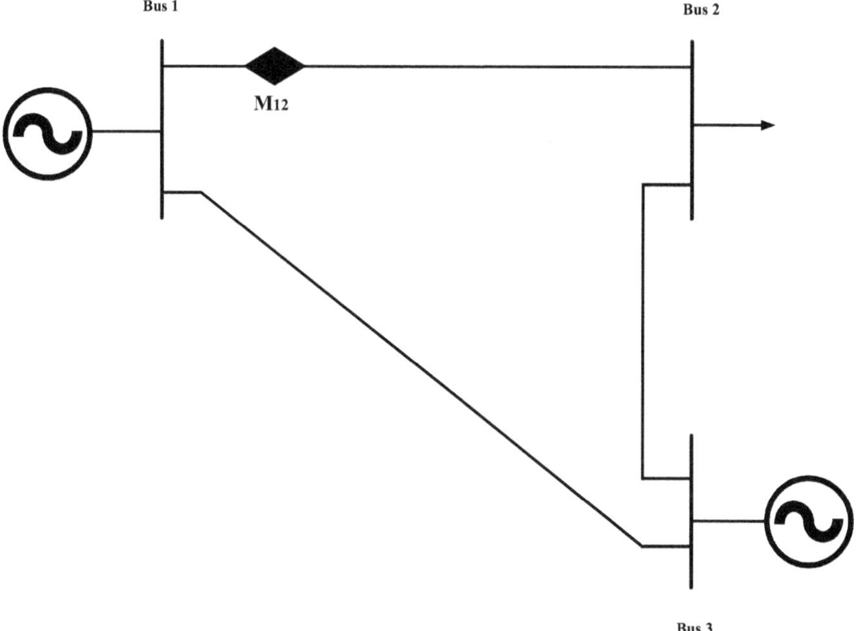

FIGURE 8.3 Case measurement network with uncertainty.

voltage phase at bus three can be calculated. The network is considered an unobservable network, meaning that the state variables cannot be observed or

The missing measurements, such as the actual or reactive powers, must be estimated using methods other than a direct measurement to remedy this issue. The measurements obtained by an indirect method are referred to as faux measurements. As a result, such measurements can replace the missing real data to continue the state estimate procedure. The challenge is, how do you determine which numbers to use to replace the missing measurements? One possible option must be to contact the power plant and manually enter the measurements into the status estimation tool. Another alternative is to utilize historical data that shows the link between the buses and their loads concerning the total system load. As a result, in the underdetermined scenario, Equation 8.18 would be utilized to estimate the state variables.

$$\left[x\right]^{est} = \left[H\left(x^k\right)^T .R^{-1}.H\left(x^k\right)\right]^{-1} z^{meas} \tag{8.20}$$

Example 8.2

Perform the state estimation using the WLS method for the power system shown in Figure 8.4, the reference angle equal to zero; calculate unknown angles at 2 and 3. The $\sigma_i^2 = 0.0001$.

Solution

$H(x)$ in this case becomes $H(\theta_2,\theta_3)$, since θ_2 and θ_3 are the state variables. Therefore, it is essential to express the measurements P_{12}, P_{13}, and P_{23} in terms of the state variables θ_2 and θ_3.

Thus, knowing that bus 1 is the reference, then $\theta_1 = 0$ rad.

$$P_{12} = \frac{1}{X_{12}}\left(\theta_1 - \theta_2\right) = \frac{1}{0.2}\left(\theta_1 - \theta_2\right) = -5\theta_2$$

$$P_{13} = \frac{1}{X_{13}}\left(\theta_1 - \theta_3\right) = \frac{1}{0.3}\left(\theta_1 - \theta_3\right) = -\frac{1}{0.3}\theta_3$$

$$P_{23} = \frac{1}{X_{23}}\left(\theta_2 - \theta_3\right) = \frac{1}{0.4}\left(\theta_2 - \theta_3\right) = 2.5\theta_2 - 2.5\theta_3$$

$$P_{12} = 0.1\text{pu} = -5\theta_2$$

Then

$$\theta_2 = -0.02\,\text{rad}$$

$$P_{13} = 0.45\,\text{pu} = -\frac{1}{0.3}\theta_3$$

Giving

$$\theta_3 = -0.135\,\text{rad}$$

Yielding:

$$H(\theta_2,\theta_3) = \begin{bmatrix} -5 & 0 \\ 0 & -3.33 \\ 2.5 & 2.5 \end{bmatrix}$$

Since σ_{P12}, σ_{P13}, and σ_{P32} is the same value for the whole system, then the covariance matrix can be calculated as follows:

$$R = \begin{bmatrix} 0.01^2 & 0 & 0 \\ 0 & 0.01^2 & 0 \\ 0 & 0 & 0.01^2 \end{bmatrix} = \begin{bmatrix} 0.0001 & 0 & 0 \\ 0 & 0.0001 & 0 \\ 0 & 0 & 0.0001 \end{bmatrix}$$

The measurement matrix is calculated as follows:

$$Z^{meas} = \begin{bmatrix} P_{12} \\ P_{13} \\ P_{23} \end{bmatrix} = \begin{bmatrix} 0.12 \\ 0.48 \\ 0.50 \end{bmatrix}$$

$$x^{est} = \begin{bmatrix} \theta_2^{est} \\ \theta_3^{est} \end{bmatrix} = H(x)^T.R^{-1}.H(x).H(x)^T.R^{-1}.z^{meas}$$

$$\begin{bmatrix} \theta_2^{est} \\ \theta_3^{est} \end{bmatrix} = \left[\begin{bmatrix} -5 & 0 \\ 0 & -3.33 \\ 2.5 & 2.5 \end{bmatrix}^T . \right.$$

$$\begin{bmatrix} 0.0001 & 0 & 0 \\ 0 & 0.0001 & 0 \\ 0 & 0 & 0.0001 \end{bmatrix}^{-1} .$$

$$\left.\begin{bmatrix} -5 & 0 \\ 0 & -3.33 \\ 2.5 & 2.5 \end{bmatrix}\right]^{-1} .$$

$$\begin{bmatrix} -5 & 0 \\ 0 & -3.33 \\ 2.5 & 2.5 \end{bmatrix}^T .$$

$$\begin{bmatrix} 0.0001 & 0 & 0 \\ 0 & 0.0001 & 0 \\ 0 & 0 & 0.0001 \end{bmatrix}^{-1} . \begin{bmatrix} 0.12 \\ 0.48 \\ 0.50 \end{bmatrix}$$

$$\begin{bmatrix} \theta_2^{est} \\ \theta_3^{est} \end{bmatrix} = \begin{bmatrix} -0.012992 \\ -0.1668961 \end{bmatrix} \text{rad}$$

By substituting the estimated values, θ_2 and θ_3, in the real power equations earlier and based on the assumption that all of the measuring devices have metering error variance of 0.01, resulting in the following P_{12}, P_{13}, and P_{32}.

$$P_{12} = -5\theta_2 = (-5).(-0.012992) = 0.0649\,\text{pu} = 6.496\,\text{MW}$$

$$P_{13} = -\frac{1}{0.3}\theta_3 = -\frac{1}{0.3}.(-0.1668961) = 0.55632\,\text{pu} = 55.632\,\text{MW}$$

$$P_{23} = \frac{1}{X_{23}}(\theta_2 - \theta_3) = 2.5\theta_2 - 2.5\theta_3 = 2.5(-0.012992) - 2.5(-0.1668961)$$

$$= 0.3847\,\text{pu} = 38.476\,\text{MW}$$

When these values are compared to the original power flow numbers, it is evident that they are incorrect. However, there are various reasons why the figures change so greatly from their original values. The measurement is one of the explanations. Taking the real power flowing between bus two and bus three as an example, the measurements differ from the "True" figure by about 10%. The error variance of the measurement instruments is the second cause. In this situation, it was set to 0.01.

The objective function behind the WLS method is to minimize the Jacobian matrix:

$$J(\theta_2, \theta_3) = \sum_{i=1}^{m} \left(\frac{z_i - h_i(x)}{R_{ii}} \right)^2 = \left(\frac{z_i - h_i(x)}{R_{ii}} \right)^2 + \left(\frac{z_i - h_i(x)}{R_{ii}} \right)^2 + \left(\frac{z_i - h_i(x)}{R_{ii}} \right)^2$$

$$= \left(\frac{0.12 - 0.06496}{0.01} \right)^2 + \left(\frac{0.45 - 0.55632}{0.01} \right)^2 + \left(\frac{0.55 - 0.38476}{0.01} \right)^2 = 416.376$$

Engineers have labored for decades because a residual is large to develop efficient power electronic measuring systems that provide more precise readings. In this example, we used 3MW as the accuracy measure; however, the accuracy level is considerably higher currently. So, if the precision of one of the measuring devices is lowered to one-tenth of its original value, the covariance matrix will alter to reflect the new value, which is 0.001 rather than 0.01. The value of the residual, state variables and their accompanying computed power flows will vary dramatically with the following revised covariance matrix and fresh data.

The new covariance matrix, R, based on the more accurate measurements is as follows:

$$R = \begin{bmatrix} 0.001^2 & 0 & 0 \\ 0 & 0.01^2 & 0 \\ 0 & 0 & 0.01^2 \end{bmatrix} = \begin{bmatrix} 0.000001 & 0 & 0 \\ 0 & 0.0001 & 0 \\ 0 & 0 & 0.0001 \end{bmatrix}$$

New measurements:

$$Z^{meas} = \begin{bmatrix} P_{12} \\ P_{13} \\ P_{23} \end{bmatrix} = \begin{bmatrix} 0.09 \\ 0.44 \\ 0.57 \end{bmatrix}$$

The estimated state variables become:

$$\begin{bmatrix} \theta_2^{est} \\ \theta_3^{est} \end{bmatrix} = \left[\begin{bmatrix} -5 & 0 \\ 0 & -3.33 \\ 2.5 & 2.5 \end{bmatrix}^T \cdot \begin{bmatrix} 0.000001 & 0 & 0 \\ 0 & 0.0001 & 0 \\ 0 & 0 & 0.0001 \end{bmatrix} \cdot \begin{bmatrix} -5 & 0 \\ 0 & -3.33 \\ 2.5 & 2.5 \end{bmatrix} \right]^{-1} \cdot \begin{bmatrix} -5 & 0 \\ 0 & -3.33 \\ 2.5 & 2.5 \end{bmatrix}^T \cdot$$

$$\begin{bmatrix} 0.000001 & 0 & 0 \\ 0 & 0.0001 & 0 \\ 0 & 0 & 0.0001 \end{bmatrix}^{-1} \cdot \begin{bmatrix} 0.12 \\ 0.48 \\ 0.50 \end{bmatrix} = \begin{bmatrix} -0.017818 \\ -0.173112 \end{bmatrix}$$

The corresponding power flows are:

$$P_{12} = 8.9\,\text{MW}$$

$$P_{13} = 57.504\,\text{MW}$$

$$P_{23} = 38.8235\,\text{MW}$$

It is observed that with the new weight that was put on the measuring device on the line between bus 1 and bus 2, the value of P_{12} is close to the original value as given in Example 8.1. However, with the new weight being put on P_{12}, it is witnessed that P_{13} and P_{32} values were affected as well. P_{12} value got closer to the "True" measure and improved the estimated value from 6.5MW to 8.9MW, closer to the 10MW. However, this negatively affects the estimated values P_{13} and P_{23}. Both values moved further from the previous example, from 55MW to 57MW for P_{13} and 38.4MW to 38.8MW for P_{23}.Tables 8.3 and 8.4 shows a bus data and transmisstion line data of Example 8.2.

TABLE 8.3
Bus Data of Example 8.2

Bus #	Type	P(pu)	θ_i
1	Slack	0.7	0
2	Voltage controller	0.3	-
3	Load	1.0	-

TABLE 8.4
Transmission Lines Data of Example 8.2

Busbar		Line Reactance	Line Flows	
From Bus	To Bus	X(pu)	P(MW)	P(pu)
1	2	0.2	10	0.1
1	3	0.3	45	0.45
2	3	0.4	55	0.55

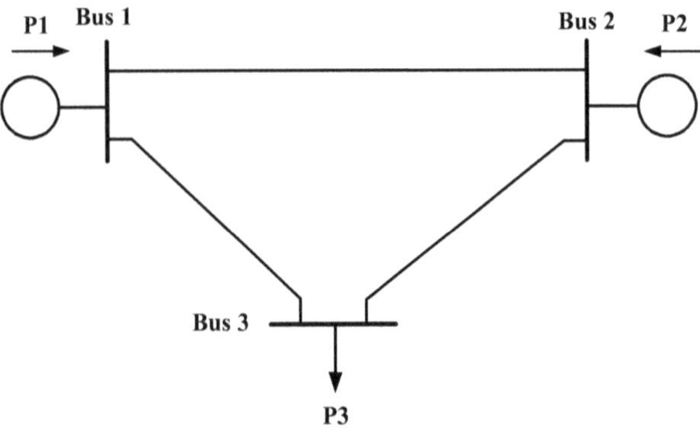

FIGURE 8.4 Bus-3 power system of Example 8.2.

8.6.3 Minimum Variation

As previously stated, the minimal variance criterion is very similar to the WLS approach; however, the least variance criterion minimizes the predicted value of the sum of the errors specified in the preceding section. In other words, it is a statistical strategy for individual grouping components based on the residual errors of those individual points from the clustered ones.

Joe H. Ward Junior was the first to propose the minimal variance requirement. This approach is extensively used since it is simple and does not need a comprehensive statistical explanation of the random mistakes. In this procedure, the only two variables required are the first and second orders of the measurement errors.

8.7 DETECTION AND IDENTIFICATION OF ERRONEOUS DATA

Errors in technological instruments or meters are common due to faults, life cycles, faulty wiring, or other factors. As a result, detecting incorrect measurement data is critical at this stage of power system modeling. If erroneous data is constantly fed into the state estimators, the system's state variables will deviate from the desired estimations. An algorithm must be integrated with the state estimation technique to identify and delete incorrect measurements from the system. Detecting faulty data may be accomplished using various strategies, one of which is the Chi-Square test. The identity is discovered using the "biggest normalized residual test," one of the most often used methods for poor data testing. These two strategies are briefly explained in Section 8.7.1.

8.7.1 IDENTIFYING BAD DATA

The amplitude of the residual, $J(x)$, which is a result of state estimation as mentioned in this chapter, is one signal for the presence of corrupted data in the set of measurements. A relatively modest $J(x)$ value suggests that the measurement set is free of distorted data. However, if the residual converges to a bigger value, there is a chance that faulty data is present.

To identify the border or range within which the residual magnitude may be regarded as a good or bad number, the residual, like the measurement function, must follow a specified distribution. As a result, the residual $J(x)$ is thought to have a probability distribution function of the Chi-Square distribution, 2 (K). The degree of freedom parameter of this distribution is known as K, and it is the difference between the number of measurements and the number of states, as illustrated below:

$$K = N_m - N_s = N_m - (2n - 1) \tag{8.21}$$

where:

N_m: the number of measurements;
N_s: the number of states;
m: number of buses.

For the residual, the user must choose a threshold. It has been noticed that faulty data results in a standard deviation greater than the error bound of 3. Assume the threshold is defined as L_J, the "limit" or "border" for the residual. A hypothesis test would be run on the Chi-Square distribution to check whether the residual falls within the "acceptable range." As a result, selecting the appropriate threshold is critical at this moment. If the limit is set too low, the test may fail so that the residual is detected to be greater than the threshold, causing the test to fail even though the data is not necessarily poor. On the other hand, the boundary limit might be set excessively high, allowing all data to be considered acceptable; nevertheless, the set of measurements could contain a significant number of incorrect parameters.

Because a hypothesis test will be run on the residual value, there is no need to define or select a threshold value as long as the significant number is known. The significant figure denotes the likelihood that the residual, J(x), will be greater than a certain threshold, L.J. An example of this procedure is provided below for more clarification and illustration.

Example 8.3

Consider six independent measurements for the quantities listed in the table are obtained from a substation for that single bus, understanding that they were taken from a sample with a normal distribution. Detect Bad Data Using Chi-Square Distribution method.

Solution

The Chi-Square distribution is used with a 95% confidence level for poor data identification. Based on the issue description, the significant figure is 100-95 = 5, and the degree of freedom is determined as follows:

$$K = N_m - (2n-1) = 6 - (2\times 1 - 1) = 6 - 1 = 5$$

where n is 1, since a single bus data was considered, N_m is 6.

Therefore, to obtain the probability of getting this value using the Chi-Square method, the sum of the squares of those six measurements has to be calculated.

$$J(x) = \sum_{i=1}^{6} x_i^2$$

$$= 1.7^2 + 0.8^2 + (-2.3)^2 + 0.3^2 + (-1.5)^2 + 2.7^2 = 18.45$$

Using the Chi-Square distribution table and the derived values, the degree of freedom, and the significant Figure 8.5, a threshold value of $L_j = 11.070$ was obtained. As a result, as shown in Figure 8.5, the total of the squares occurred to be in the rejected region. As $J(x) = 18.45 > L_j = 11.070$, the value L was more than the threshold, L_j. This signifies that the obtained data failed the test, indicating "bad" data.

TABLE 8.5
Quantity Measurement Data at a Random Bus Stop

Variable	x_1	x_2	x_3	x_4	x_5	x_6
Measured Value	1.7	0.8	−2.3	0.3	−1.5	2.7

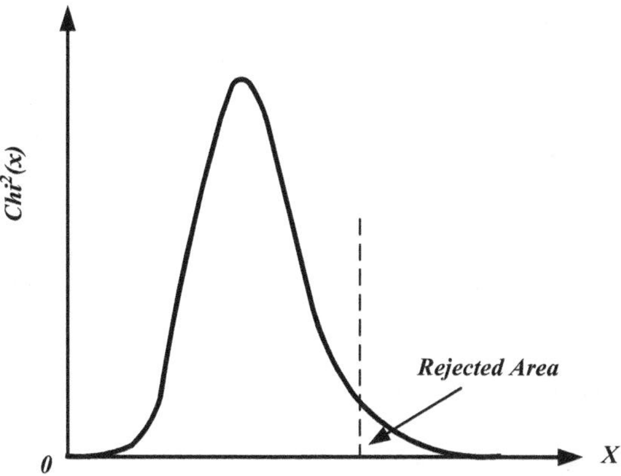

FIGURE 8.5 Chi-Square Distribution Threshold Probability Testing Function.

8.7.2 Bad Data Detection in the Weighted Least Square Approach

The WLS technique may compute the residual measurement function, $J(x)$. As a result, this objective function may detect incorrect data in a system. Using the Chi-Square test with data from $J(x)$, the following procedures may be used to assess if a batch of data contains faulty data or not.

i. As explained earlier in Section 8.3, the residual measurement function or objective function, $J(x)$, is first solved using Equation 8.10. As a reaffirmation, the following equation is used to calculate $J(x)$:

$$J(x) = \sum_{i=1}^{m} \left(\frac{z_i - h_i(x)}{h_{ii}} \right)^2$$

ii. A number from the Chi-Square distribution table is then entered for that probability as a threshold, L_J Based on the pre-stated confidence percentage and degree of freedom in the issue description.
iii. To assess faulty data in the system, J(x) and L_J are compared.

Therefore, if $J(x) \geq L_J$, then bad data is present. Otherwise, the measurement set is presumed to be acceptable.

8.7.3 Identification and Removal of Bad Data

The greatest normalized residual test is one of the approaches used to detect and eliminate damaged data. Depending on the type of incorrect data, the approach is implemented differently. A single poor data point represents the first instance with a huge error. The second situation corresponds to many errors in the set of measurements.

Detection of one versus several faulty data sets:

Because only one bad data point must be discovered and retrieved, the single bad data detection situation is easy to handle. In this case, the incorrect data measurement corresponds to the greatest normalized residual mentioned shortly after. The other scenario, in which many corrupted data sets are identified, may be classified into three categories:

1. Non-interacting: Because the residuals in this set of faulty data have a weak link, other neighboring ones are not considerably influenced.
2. Interaction: If many faulty data points interact, it might imply that excellent measurements include erroneous mistakes. This occurs when there is a substantial correlation between good and poor data.
3. Interacting but non-conforming: this group is explained in the same way as the "Interacting group," except that when the mistakes are matching or compatible, they will look consistent with each other, making it difficult to distinguish the corrupted from the clean data.

The amount of the residual can indicate the presence of erroneous data; however, J(x) did not indicate which measurement was tainted. Following the Chi-Square test, the biggest normalized residual test is run based on the findings to identify the faulty values. This technique simply determines the difference between the estimated and "actual" measurements, then divides the result by the appropriate element in the residual covariance matrix., and outlines the steps for carrying out this test:

1. The first step is to calculate the residual, which, as previously stated, is an output of the state estimation procedure. The residual or error, e, as defined previously in Equation 8.9, may be rewritten as follows:

$$e = r = z^{meas} - h(x)$$

$$r_i = z_i - h_i(x), i = 1, 2, \ldots, m.$$

2. The residuals are then normalized in the second stage. As previously explained, the residual is split by its corresponding member in the residual covariance matrix. However, one minor change is made to the covariance element to highlight its sensitivity. As a result, instead of dividing by the square root of the covariance, the revised modification, with the inclusion of the sensitivity factor, is as follows:

$$r_i^N = \frac{|r_i|}{R_{ii}S_{ii}} = \frac{|r_i|}{\Omega_{ii}} i = 1, 2, \ldots, N_m \tag{8.22}$$

3. The normalized residual is then compared to a pre-determined threshold, t. This value is normally set to 3.0. The residual is detected as faulty data and is deleted from the set of measurements if and only if the following conditions are met:

$$r_i^N \geq t$$

4. Following the removal and elimination of the corrupted measurements, the state estimation procedure is restarted after the faulty data is extracted. Steps 1 through 3 are then performed to eliminate any remaining distorted data readings.

8.8 TECHNIQUES OF STATE ESTIMATION FOR NON-LINEAR SYSTEMS

Even though the WLS technique may be used to properly answer unknown state variables in a power system, it is still a "static approach." This means that the WLS technique takes measurements from the power network at a certain point in time and calculates or guesses values for the unknown state variable near the "True" value. WLS, on the other hand, cannot be utilized to estimate dynamic states.

Dynamic state estimate is a technique for predicting future state variables before collecting observations. In other words, without knowing the measurements, dynamic state estimators give a solution for the current snapshot at time t and the next forthcoming state vector at time t+1. Dynamic approaches may detect faulty data and identify problems like the static approach. Furthermore, it performs well under regular and emergency settings, making it ideal for power system status estimate improvement. As a result, engineers can anticipate what kinds of difficulties the system expects, which is a significant benefit. This gives you more time to act and respond to key situations and unforeseen events.

Although dynamic state estimation was investigated concurrently with static state estimate, the technology was insufficient to develop and enhance it.

Currently, state estimate is on the rise. PMUs and synchrophasor measurements give engineers rapid and precise data for improving estimators and, as a result, the power system model.

One of the most used approaches for estimating dynamic states is the Kalman Filter. The standard Kalman Filter approach is utilized for linear dynamic state estimation, which might be useful in some electric power systems. However, as previously noted, power system problems are non-linear. Thus various tweaks and enhancements to the traditional Kalman Filter approach were made to satisfy those demands. The "Extended Kalman Filter" (EKF) and "Unscented Kalman Filter" are two refinements to the conventional Kalman Filter approach that will be covered in this section.

8.8.1 Classical Kalman Filter

The classic Kalman Filter is an algorithm utilized in many applications, including spacecraft, aircraft, cars, power systems, and other technical domains. It is utilized for various purposes, including guiding, prediction, control, navigation, signal processing, and many more. The Kalman Filter concept determines accurate values for unknown variables in a system based on faulty measurements or inputs. These erroneous numbers are known to be inaccurate, with noise and mistakes. Because the Kalman Filter is a dynamic state estimator, it works overtime by watching the input data and determining the best estimate for the original system state variables.

Some assumptions have been made for this approach to perform properly. The first assumption is that the system under consideration must be linear. The second assumption is that the measurement error has a Gaussian distribution. Figure 8.6 depicts the basic processes of prediction and updating.

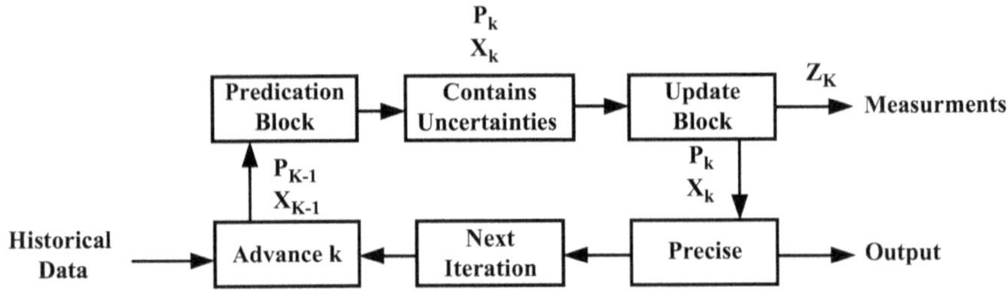

FIGURE 8.6 Kalman Filter algorithm.

where:

P_k: The error covariance matrix at time k.
X_k: The state estimate at time k.

The graphic above can be used to make some observations. First, the mean and state covariance change with time. Second, the mean and covariance of the state are the mean and covariance of the Kalman Filter, respectively. The third and final point to mention is that once the measurement data is uploaded into the system, both the mean and the covariance are rapidly updated.

To follow the process of the classical Kalman Filter and calculate or estimate the state variable of the system, one should create the following matrices:

F_k: the state-transition model applied to the previous state variable (k-1).
x_{k-1}: previous state variable.
X_k: actual state variable.
U_k: control vector.
B_k: the control-input model applied to the control vector for every timeframe.
$C_{.K.}$: system error or noise, following zero mean (normal distribution) and Q_k covariance below.
e_k: measurement error or noise, following zero mean (normal distribution) and R_k covariance

$$R = \begin{bmatrix} e_1^2 & 0 & 0 \\ 0 & \ddots & 0 \\ 0 & 0 & e_k^2 \end{bmatrix} \quad (8.23)$$

$$Q = \begin{bmatrix} \left(\frac{q_1}{L}\right)^2 & 0 & \cdots & 0 \\ 0 & \left(\frac{q_1}{L}\right)^2 & & 0 \\ \vdots & & \ddots & \vdots \\ 0 & 0 & \cdots & \left(\frac{q_1}{L}\right)^2 \end{bmatrix} \quad (8.24)$$

H_k: uses measurement data to translate the observed state of the system to the real one, similar to the H matrix in the WLS approach.
z_k: the measurements matrix at time k.

As a result, two formulae for state prediction and state updating are generated based on the set mentioned above of matrices. The equation for predicting state is given by

$$x_k = F_k x_{k-1} + B_k F u_{k-1} + C_{k-1} \quad (8.25)$$

This equation has been modified further and split into the predicted state estimation equation and the estimated covariance.

$$x_{k|k-1} = F_k x_{k-1|k-1} + B_{k-1} U_{k-1} \quad (8.26)$$

$$P_{k-1|k-1} = F_k P_{k-1|k-1} F_k^T + Q_k \quad (8.27)$$

State update equation:

$$Z_k = H_k x_k + e_k \quad (8.28)$$

The state update equation is further enlarged to represent the error residual y_k, residual covariance S_k, Kalman Gain K_k, and, lastly, the state estimate and covariance estimate for the update stage, x_k and $P_{k|k}$, respectively, as in the prediction phase.

$$y_k = Z_k - H_k x_{k|k-1} \quad (8.29)$$

$$S_k = H_k P_{k|k-1} H_k^T + P_k \quad (8.30)$$

$$K_k = P_{k|k-1} H_k^T S_k^{-1} \quad (8.31)$$

$$x_k = x_{k|k-1} + K_k S_k \quad (8.32)$$

$$P_{k|k} = \left(I - K_k H_k\right) P_{k|k-1} \quad (8.33)$$

Concentrating on Equation 8.32 will explain the entire concept behind the Kalman Filter approach. The error residual, which is the difference between the actual measurement and the estimated or forecast measurement supplied by the SCADA or synchro-phasors, is used to assess the state estimation value. As a result, if the error residual is zero, indicating that the measurements were nearly identical, the projected value of the state variable at time k, is regarded as the same as the state at $k-1$, so it does not require an update. If the error residual is not equal to zero, the Kalman Gain K_k determines how much to adjust to get to the actual estimate at time k. As a result, the equation section is referred to as a "correction term."

The classical method is unsuitable for non-linear power system network problems. Thus, the following sections explain the different kinds of non-linear Kalman Filter methods developed for such problems.

8.8.2 Non-Linear Kalman Filter Methods

Some improvements and new techniques were presented to handle non-linear systems since the classical methods are not very suitable for non-linear systems. Therefore, the next sections will discuss two extensions of the classical methods, namely the EKF method and the unscented Kalman Filter method.

8.8.3 The Extended Kalman Filter Method

Based on linearization approaches such as the Taylor Series expansion method, the EKF methodology was reformed. The system models do not have to be linear for the Kalman Filter to work in this technique. The following procedure describes how the EKF linearizes the non-linear function at the current timestamp estimation.

A similar notion is used to the classical process, as shown in Equations 8.34 and 8.35. However, only minor changes were made to guarantee that the equations were consistent with and worked effectively with non-linear systems.

$$x_k = f\left(x_{k-1}, u_{k-1}\right) + C_{k-1} \tag{8.34}$$

$$Z_k = h\left(x_k\right) + e_k \tag{8.35}$$

where:

$f(x_{k-1}, u_{k-1})$: is any linear or non-linear function used to compute the predicted state from the previous estimate;

$h(x_k)$: is a similar function to f(x, u). However, it is used for the predicted measurement calculation;

e and c: system and estimation error at a specified time, just as defined earlier, following a Gaussian distribution, with zero mean and discrete covariance matrices of R_k and Q_{k-1}, respectively.

The Jacobian matrix, a partial derivative matrix, must be created and calculated for the covariance estimate. This technique employs Equation 8.25 for prediction and the

Equations from 8.28 to 8.29 for updating. However, partial derivatives provide a new meaning for the H and F matrices.

A state transition matrix is provided.

$$F_{k-1} = \frac{\partial f}{\partial x}|_{x_{k-1\#k-1},U_{k-1}} \tag{8.36}$$

State observation matrix is given as:

$$H_k = \frac{\partial h}{\partial x}|_{x_{k|k-1}} \tag{8.37}$$

Of course, the measurement residual must be adjusted accordingly, and therefore, Equation 8.34 is rewritten as:

$$y_k = Z_k - h\left(x_{k|k-1}\right) \tag{8.38}$$

The EKF approach is useful for estimating state variables in our power system network challenges. However, one of the EKF method's drawbacks is that it does not produce an ideal answer or estimate. The estimate is based on two factors. First, consider the precision of the collection of measurements. Another critical factor is the power system's transition model and how near its transitions are. The closer these two keys are to being linear, the more similar the EKF approach is to the standard Kalman Filter method. The EKF is the most often utilized estimate technique for non-linear systems. However, more than 45 years of estimating community experience has demonstrated that it is difficult to implement, complex to adjust, and only reliable for virtually linear systems on the time scale of the updates. "Many of these challenges derive from the use of linearization."

8.8.4 The Unscented Kalman Filter Method

Certain limitations were found following the EKF technique. To mention a few, the approximation it produces is not an accurate estimate since the method linearizes the function while ignoring the non-linear components. In terms of technology, it would be costly to enhance that process by iterative methods or other comparable means to get closer to the "True" value. As a result, a new technique was devised that does not linearize the function of the power system network as the EKF does.

The UKF approach is comparable to the EKF method. It is, nevertheless, recognized as the "derivative-free" approach. In contrast to the linearization of the EKF, the UKF employs a non-linear function. It then obtains the state estimation by approximating the non-linear system's probability distribution. The Unscented Transform is a powerful mathematical function that has been created to augment the Kalman Filter (U.T.). U.T. is a mathematical or statistical procedure used to transform non-linear functions into a probability distribution matching a finite set. Those altered probability points are referred to as sigma points. Following the transformation, an estimate of the outcomes is generated in the mean vector and associated error covariance matrix by applying the known non-linear function h(x) to each vector.

The Unscented Kalman Filter approach relies heavily on the unscented transform. The non-linear function is employed in this strategy to assist obtain the estimate as close to the "True" value as feasible. Because the function is not linearized, and the high-order components are used rather than ignored, the UKF estimate is more accurate than the EKF estimate. Another advantage of employing this approach is that the derivation step is eliminated. Because the power system model is so large and complicated, building a Jacobian matrix like the one used in EKF would be both expensive and time-consuming. As a result, the Jacobian matrix will not be required when employing the UKF approach. Another significant advantage of employing the unscented transformation technique in the Kalman Filter is that it contains the third order, which improves the accuracy of the mean transformation.

A simple method to demonstrate this is to use the non-linear function to propagate a collection of samples around the system's last known state. As discussed later in this section, weighted matrices calculate an estimated mean and covariance from those sampling points.

Example 8.4

A power system with two voltmeters and one wattmeter is located, as shown in Figure 8.7 to measure the following quantities:

Z_1 The voltage magnitude of Bus number 1.
Z_2 The voltage magnitude of Bus number 2.
Z_3 The reactive power is injected into Bus number 1.
Z_4 The magnitude of active power flow from Bus number 1 to Bus number 2,
Z_5 The magnitude of reactive power flow from Bus number 2 to Bus number 1.

Using WLS, estimate the system states.

Solution

Z_1 The voltage magnitude of Bus number 1.

$$Z_1 = |V_1|$$

Z_2 The voltage magnitude of Bus number 2.

$$Z_2 = |V_2|$$

Z_3 The reactive power is injected into Bus number 1.

$$Z_3 = Q_1$$

Z_4 The magnitude of active power flow from Bus number 1 to Bus number 2,

$$Z_4 = P_{12}$$

Z_5 The magnitude of reactive power flow from Bus number 2 to Bus number 1.

$$Z_5 = Q_{12}$$

Select bus number 1 reference so $\delta_1 = 0$
Suppose the state variables are $x_1 = \delta_2$

$$x_2 = |V_1|$$

$$x_3 = |V_2|$$

So,

$$P_{12} = P_L = -P_2 = -\frac{1}{X_{12}}|V_1||V_2|\sin(\delta_2 - \delta_1) = -2x_2x_3\sin x_1$$

$$Q_{12} = Q_L = -Q_2 = \frac{1}{X_{12}}\left[|V_2|^2 - |V_1||V_2|\cos(\delta_2 - \delta_1)\right] = 2\left(x_3^2 - x_2x_3\cos x_1\right)$$

$$Q_1 = \frac{1}{X_{12} + X_{10}}|V_1|^2 - \frac{1}{X_{12}}|V_1||V_2|\cos(\delta_2 - \delta_1) = 2.125x_2^2 - 2x_2x_3\cos x_1$$

The measurement errors in the ith iteration as

$$e_1^i = Z_1 - h_1\left(x_1^i + x_2^i + x_3^i\right) = Z_1 - x_2^i$$

$$e_2^i = Z_2 - h_2\left(x_1^i + x_2^i + x_3^i\right) = Z_2 - x_3^i$$

$$e_3^i = Z_3 - h_3\left(x_1^i + x_2^i + x_3^i\right) = Z_3 - \left[2.125\left(x_2^i\right)^2 - 2x_2^i\, x_3^i\cos x_1^i\right]$$

$$e_4^i = Z_4 - h_4\left(x_1^i + x_2^i + x_3^i\right) = Z_4 - \left(-2x_2^i\, x_3^i\cos x_1^i\right)$$

$$e_5^i = Z_5 - h_5\left(x_1^i + x_2^i + x_3^i\right) = Z_5 - 2\left[\left(x_3^i\right)^2 - x_2^i\, x_3^i\sin x_1^i\right]$$

$$H_x^i = \begin{bmatrix} \frac{\partial h_1}{\partial x_1}\Big|^i & \frac{\partial h_1}{\partial x_2}\Big|^i & \frac{\partial h_1}{\partial x_3}\Big|^i \\ \frac{\partial h_2}{\partial x_1}\Big|^i & \frac{\partial h_2}{\partial x_2}\Big|^i & \frac{\partial h_2}{\partial x_3}\Big|^i \\ \frac{\partial h_3}{\partial x_1}\Big|^i & \frac{\partial h_3}{\partial x_2}\Big|^i & \frac{\partial h_3}{\partial x_3}\Big|^i \\ \frac{\partial h_4}{\partial x_1}\Big|^i & \frac{\partial h_4}{\partial x_2}\Big|^i & \frac{\partial h_4}{\partial x_3}\Big|^i \\ \frac{\partial h_5}{\partial x_1}\Big|^i & \frac{\partial h_5}{\partial x_2}\Big|^i & \frac{\partial h_5}{\partial x_3}\Big|^i \end{bmatrix}$$

$$h_1\left(x_1^i + x_2^i + x_3^i\right) = x_2^i$$

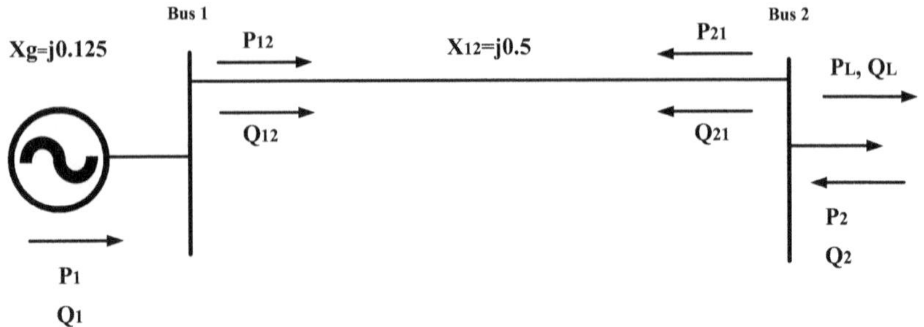

FIGURE 8.7 Power system configuration of Example 8.4.

$$h_2\left(x_1^i + x_2^i + x_3^i\right) = x_3^i$$

$$h_3\left(x_1^i + x_2^i + x_3^i\right) = 2.125\left(x_2^i\right)^2 - 2x_2^i\, x_3^i \cos x_1^i$$

$$h_4\left(x_1^i + x_2^i + x_3^i\right) = -2x_2^i\, x_3^i \cos x_1^i$$

$$h_5\left(x_1^i + x_2^i + x_3^i\right) = 2\left[\left(x_3^i\right)^2 - x_2^i\, x_3^i \sin x_1^i\right]$$

$$H_x^i = \begin{bmatrix} 0 & 1 & 0 \\ 0 & 0 & 1 \\ 2x_2^i\, x_3^i \sin x_1^i & -4.25\, x_2^i - 2\, x_3^i \cos x_1^i & -2x_2^i \cos x_1^i \\ 2x_2^i\, x_3^i \sin x_1^i & -2\, x_3^i \cos x_1^i & -2x_2^i \cos x_1^i \\ -2\, x_2^i\, x_3^i \cos x_1^i & -2\, x_3^i \sin x_1^i & 4\, x_3^i - 2x_2^i \sin x_1^i \end{bmatrix}$$

Example 8.5

A power system with two voltmeters and one wattmeter is located, as shown in Figure 8.7, to measure the following quantities:

Z_1 The voltage magnitude of Bus number 1.

$$Z_1 = |V_1| = 0.95\,\text{pu}$$

Z_2 The voltage magnitude of Bus number 2.
$Z_2 = |V_2| = 1.03$ pu
Z_3 The reactive power is injected into Bus number 1.

$$Z_3 = Q_1 = 0.75\,\text{pu}$$

Z_4 The magnitude of active power flow from Bus number 1 to Bus number 2,

$$Z_4 = P_{12} = 0.9\,\text{pu}$$

Z_5 The magnitude of reactive power flow from Bus number 2 to Bus number 1.

$$Z_5 = Q_{12} = 0.5\,\text{pu}$$

The variance of the measurement errors is given in per unit:

$$\sigma_1^2 = \sigma_2^2 = \sigma_3^2 = 0.0004$$

$$\sigma_4^2 = 0.0009$$

$$\sigma_5^2 = 0.0016$$

Calculate the weighted sum of squares of the errors after the first iteration.

Solution

$$x_1^0 = \delta_2^0 = 0^\circ$$

$$x_2^0 = |V_1|^0 = 1.0\,\text{pu}$$

$$x_3^0 = |V_2|^0 = 1.0\,\text{pu}$$

$$e_1^0 = Z_1 - h_1\left(x_1^0 + x_2^0 + x_3^0\right) = Z_1 - x_2^0 = 0.95 - 1 = -0.05$$

$$e_2^0 = Z_2 - h_2\left(x_1^0 + x_2^0 + x_3^0\right) = Z_2 - x_3^0 = 1.03 - 1 = 0.03$$

$$\begin{aligned} e_3^0 &= Z_3 - h_3\left(x_1^0 + x_2^0 + x_3^0\right) = Z_3 - \left[2.125\left(x_2^0\right)^2 - 2x_2^0\,x_3^0\cos x_1^0\right] \\ &= 0.75 - [2.125\times(1.0)^2 - 2\times1\times1\times\cos 0^\circ] = 0.625 \end{aligned}$$

$$e_4^0 = Z_4 - h_4\left(x_1^0 + x_2^0 + x_3^0\right) = Z_4 - \left(-2x_2^0\,x_3^0\cos x_1^0\right) = 0.9 - \left(-2\times1\times1\times\cos 0^\circ\right) = 2.9$$

$$\begin{aligned} e_5^0 &= Z_5 - h_5\left(x_1^0 + x_2^0 + x_3^0\right) = Z_5 - 2\left[\left(x_3^0\right)^2 - x_2^0\,x_3^0\sin x_1^0\right] \\ &= 0.5 - 2\left[1.0^2 - 1\times1\times\sin 0^\circ\right] = -1.5 \end{aligned}$$

$$H_x^0 = \begin{bmatrix} 0 & 1 & 0 \\ 0 & 0 & 1 \\ 2x_2^0\,x_3^0\sin x_1^0 & -4.25\,x_2^0 - 2\,x_3^0\cos x_1^0 & -2x_2^0\cos x_1^0 \\ 2x_2^0\,x_3^0\sin x_1^0 & -2\,x_3^0\cos x_1^0 & -2x_2^0\cos x_1^0 \\ -2x_2^0\,x_3^0\cos x_1^0 & -2\,x_3^0\sin x_1^0 & 4\,x_3^0 - 2x_2^0\sin x_1^0 \end{bmatrix}$$

$$=\begin{bmatrix} 0 & 1 & 0 \\ 0 & 0 & 1 \\ 0 & -6.25 & -2 \\ 0 & -2 & -2 \\ -2 & 0 & 4 \end{bmatrix}$$

The gain matrix $G_x^0 = H_x^{0T} R^{-1} H_x^0$

$$= \begin{bmatrix} 0 & 1 & 0 \\ 0 & 0 & 1 \\ 0 & -6.25 & -2 \\ 0 & -2 & -2 \\ -2 & 0 & 4 \end{bmatrix}^T \times \begin{bmatrix} \frac{1}{0.0001} & 0 & 0 & 0 & 0 \\ 0 & \frac{1}{0.0001} & 0 & 0 & 0 \\ 0 & 0 & \frac{1}{0.0001} & 0 & 0 \\ 0 & 0 & 0 & \frac{1}{0.0009} & 0 \\ 0 & 0 & 0 & 0 & \frac{1}{0.0016} \end{bmatrix}$$

$$\times \begin{bmatrix} 0 & 1 & 0 \\ 0 & 0 & 1 \\ 0 & -6.25 & -2 \\ 0 & -2 & -2 \\ -2 & 0 & 4 \end{bmatrix}$$

$$= \begin{bmatrix} 0.0250 & 0 & -0.0500 \\ 0 & 4.0507 & 1.2944 \\ -0.0500 & 1.2944 & 0.6444 \end{bmatrix} \times 10^4$$

$$x_1^0 = \delta_2^0 = 0.0^\circ$$

$$x_2^0 = |V_1|^0 = 1.0\,\text{pu}$$

$$x_3^0 = |V_2|^0 = 1.0\,\text{pu}$$

$$\begin{bmatrix} x_1^1 \\ x_2^1 \\ x_3^1 \end{bmatrix} = \begin{bmatrix} x_1^0 \\ x_2^0 \\ x_3^0 \end{bmatrix} + G_x^{0-1} H_x^{0T} R^{-1} \begin{bmatrix} e_1^0 \\ e_2^0 \\ e_3^0 \\ e_4^0 \\ e_5^0 \end{bmatrix}$$

$$\begin{bmatrix} 0.0 \\ 1.0 \\ 1.0 \end{bmatrix} + \left[\begin{bmatrix} 0.0250 & 0 & -0.0500 \\ 0 & 4.0507 & 1.2944 \\ -0.0500 & 1.2944 & 0.6444 \end{bmatrix} \times 10^4\right]^{-1}$$

$$\times \begin{bmatrix} 0 & 1 & 0 \\ 0 & 0 & 1 \\ 0 & -6.25 & -2 \\ 0 & -2 & -2 \\ -2 & 0 & 4 \end{bmatrix}^T$$

$$= \times \begin{bmatrix} \frac{1}{0.0001} & 0 & 0 & 0 & 0 \\ 0 & \frac{1}{0.0001} & 0 & 0 & 0 \\ 0 & 0 & \frac{1}{0.0001} & 0 & 0 \\ 0 & 0 & 0 & \frac{1}{0.0009} & 0 \\ 0 & 0 & 0 & 0 & \frac{1}{0.0016} \end{bmatrix}$$

$$\times \begin{bmatrix} -0.05 \\ 0.03 \\ 0.625 \\ 2.9 \\ -1.5 \end{bmatrix}$$

$$\begin{bmatrix} x_1^1 \\ x_2^1 \\ x_3^1 \end{bmatrix} = \begin{bmatrix} 0.1471 \\ 0.9827 \\ 0.6986 \end{bmatrix}$$

$$x_1^1 = |V_1|^1 = 1.0\,\text{pu}$$

$$x_1^1 = \delta_2^1 = 0.1471\,\text{rad} = 8.428°$$

$$x_2^1 = |V_1|^1 = 0.9827\,\text{pu}$$

$$x_3^1 = |V_2|^1 = 0.6986\,\text{pu}$$

$$e_1^1 = Z_1 - h_1\left(x_1^1 + x_2^1 + x_3^1\right) = Z_1 - x_2^1 = 0.95 - 0.9827 = -0.0327$$

$$e_2^1 = Z_2 - h_2\left(x_1^1 + x_2^1 + x_3^1\right) = Z_2 - x_3^1 = 1.03 - 0.6986 = 0.3314$$

$$e_3^1 = Z_3 - h_3\left(x_1^i + x_2^i + x_3^i\right) = Z_3 - \left[2.125\left(x_2^1\right)^2 - 2x_2^1\, x_3^1 \cos x_1^1\right]$$
$$= 0.75 - [2.125(0.9827)^2 - 2\times 0.9827 \times 0.6986 \times \cos 0.1471] = 0.05667$$

$$e_4^1 = Z_4 - h_4\left(x_1^i + x_2^i + x_3^i\right) = Z_4 - \left(-2x_2^1\, x_3^1 \cos x_1^1\right)$$
$$= 0.9 - (-2\times 0.9827 \times 0.6986 \times \cos 0.1471) = 2.2582$$

$$e_5^1 = Z_5 - h_5\left(x_1^i + x_2^i + x_3^i\right) = Z_5 - 2\left[\left(x_3^1\right)^2 - x_2^1\, x_3^1 \sin x_1^1\right]$$
$$0.5 - 2\,[0.6986^2 - 0.9827 \times 0.6986 = \times \sin 0.1471] = -0.2748$$

The weighted sum of squares of the errors is

$$\hat{f} = \sum_{j=1}^{5}\left(\frac{\hat{e}_j}{\sigma_j}\right)^2 = \left(\frac{\hat{e}_1}{\sigma_1}\right)^2 + \left(\frac{\hat{e}_2}{\sigma_2}\right)^2 + \left(\frac{\hat{e}_3}{\sigma_3}\right)^2 + \left(\frac{\hat{e}_4}{\sigma_4}\right)^2 + \left(\frac{\hat{e}_5}{\sigma_5}\right)^2$$

$$= \left(\frac{-0.0327}{0.02}\right)^2 + \left(\frac{0.3314}{0.02}\right)^2 + \left(\frac{0.05667}{0.02}\right)^2 + \left(\frac{2.2582}{0.03}\right)^2 + \left(\frac{-0.2748}{0.04}\right)^2 = 4998.53$$

PROBLEMS

8.1 The following Bus-3 system has two generators. One is located at the swing bus, bus 2, generating 50 M.W. into the system. The second generator generates 90 M.W. at bus 1, as shown in Figure 8.8. A load at bus 3 consumes 140 MW of power: the bus's data voltage, type, and initial phase angle are given in Table 8.1. Line impedances and power flows between buses are given in Table 8.2. All values are stated based on a 100 MVA base. Calculate the phase angle at each bus.

8.2 Perform the state estimation using the WLS method for the power system shown in Figure 8.9, the reference angle equal to zero. Calculate unknown angles at 2 and 3. The $\sigma_i^2 = 0.0005$.

8.3 Consider six independent measurements for the quantities listed in the table are obtained from a substation for that single bus, understanding that they were taken from a sample with a normal distribution. Using Chi-Square Distribution to Detect Bad Data.

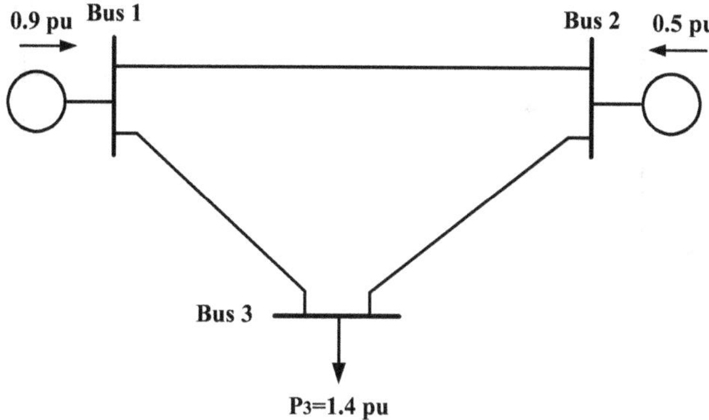

FIGURE 8.8 Bus-3 network configuration of Problem 8.1.

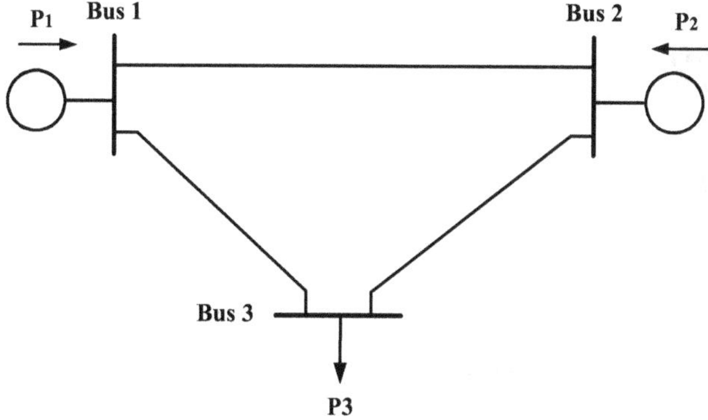

FIGURE 8.9 Bus-3 power system of Example 8.2.

8.4 A power system with two voltmeters and one wattmeter are located as shown in Figure 8.10 to measure the following quantities:

Z_1 The voltage magnitude of Bus number 2.
Z_2 The voltage magnitude of Bus number 1.
Z_3 The reactive power is injected into Bus number 1.
Z_4 The magnitude of active power flow from Bus number 1 to Bus number 2,
Z_5 The magnitude of reactive power flow from Bus number 2 to Bus number 1.

Using WLS, estimate the system states.

8.5 A power system with two voltmeters and one wattmeter are located as shown in Figure 8.11 to measure the following quantities:

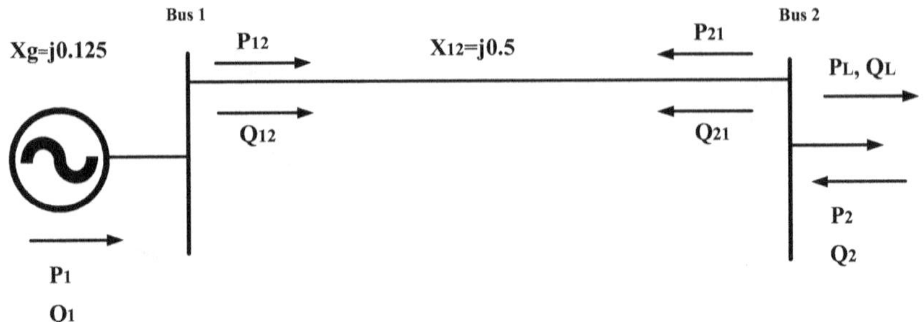

FIGURE 8.10 Power system configuration of Problem 8.4.

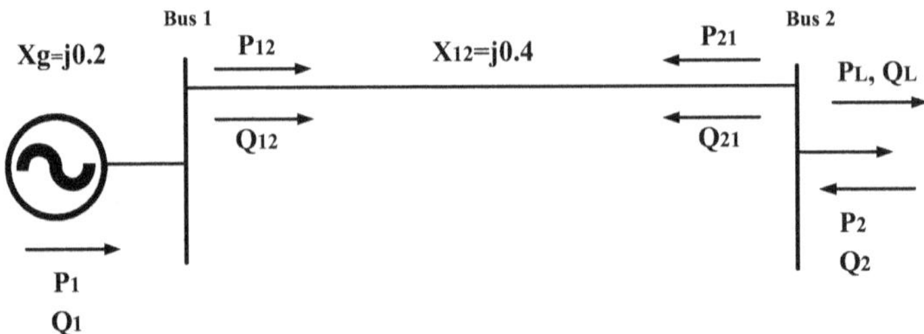

FIGURE 8.11 Power system configuration of Problem 8.5.

Z_1 The voltage magnitude of Bus number 1.

$$Z_1 = |V_1| = 1.035\,\text{pu}$$

Z_2 The voltage magnitude of Bus number 2.

$$Z_2 = |V_2| = 0.935\,\text{pu}$$

Z_3 The reactive power is injected into Bus number 1.

$$Z_3 = Q_1 = 0.152\,\text{pu}$$

Z_4 The magnitude of active power flow from Bus number 1 to Bus number 2,

$$Z_4 = P_{12} = 0.584\,\text{pu}$$

Z_5 The magnitude of reactive power flow from Bus number 2 to Bus number 1.

$$Z_5 = Q_{12} = 0.52\,\text{pu}$$

The variance of the measurement errors are given in per unit:

$$\sigma_1^2 = \sigma_2^2 = \sigma_3^2 = 0.0005$$

$$\sigma_4^2 = 0.0008$$

$$\sigma_5^2 = 0.0067$$

Calculate the weighted sum of squares of the errors after the first iteration.

9 Load Forecasting

Load forecasting is one of the necessary parameters in power system analysis. Therefore, several statistical forecasting techniques are used for medium-term load forecasting to predict the monthly peak load of a specific network over a period of time.

To supply high-quality electric energy to the customer securely and economically, an electric company faces many economic and technical problems in operating, planning, and controlling electric energy systems. For optimal planning and operation of the system, modern system theory and optimization techniques are applied to exact considerable cost savings. In achieving this goal, the knowledge of future power system load is the first prerequisite. Therefore long-, medium-, and short-term load forecasting are very important.

9.1 LOAD FORECASTING SOLUTION TECHNIQUES

The techniques used in load forecasting solutions are as follows:

1. Additive seasonal.
2. Regression model.
3. Brown's smoothing method.
4. Seasonal box – Jeninns.
5. Artificial neural method.

There are three types of load forecasting:

1. Long-Term Forecasts: Annual, monthly, or weekly load data with the scope of 10 to 50 years for the problem of capacity expansion planning, generation, transmission, or distribution system additions to be resolved.
2. Medium-Term Forecasts: Monthly or weekly load data with a horizon of 1 week to 10 years, for the activities of operation planning: maintenance scheduling, station refueling (determine the future fuel requirements for the station and especially for the nuclear stations), system security and reliability, energy interchange contracts, estimation of revenues from sales, etc., to be determined.
3. Short-Term Forecasts: Hourly data with a horizon of 1 day to 1 week, for the activities of control (operation condition), unit commitment (cost-saving), and scheduling of power system, and also as input to the load flow study or contingency analysis.

The load forecast is the primary input data to most computerized applications dealing with capacity expansion planning, like reliability, etc.

DOI: 10.1201/9781003293965-9

Forecasting techniques can be categorized into three groups. The first is called qualitative, where all information relating to an item is used to forecast the demands. This technique is often used when little demand history is available. An example of this technique is the regression model (the quadratic least square method), where the estimation value of y(t) will be as follows:

$$Y(t) = a + bt + ct \wedge 2 \tag{9.1}$$

The second group is called causal, where a cause-and-effect relation is sought. Here the forecaster seeks a relationship between demands and other factors, such as business, industrial plants, national indices, temperature, number of consumers, etc.

The relationship is used to forecast the future demands of the entity. An example of this technique is the economic forecasting model, where the estimation value of $y(t)$ is:

$$Y(t) = aA + bB + cC \tag{9.2}$$

where:

A is the number of consumers
B is the effect of weather
C is the economic factor, etc.
a, b, c are constants.

The third group is called time-series analysis, where a statistical analysis of past demand is used to generate forecasts. A basic assumption is that the underlying trend of the past will continue, which means that the estimation value of y(t) will function from past demands.

$$y(t) = f\left(y(t-i)\right),\ i = 1,\ 2,\ 3\ldots \tag{9.3}$$

where $y(t - i)$ is the estimation value of $y(t)$. As an example, *the additive seasonal model.*

Generated electrical energy cannot be stored. It has to be generated whenever there is demand for it. Therefore, it is imperative for the electrical power utilities that the load on their system can be estimated in advance. This is commonly known as load forecasting, "short-term" prediction can provide a forecast every few minutes depending upon which load change in the power system can be made. However, "medium- and long-term" predictions allow time to plan the generation to meet the varied situation in system expansion and load growth.

9.2 LOAD CURVES AND FACTORS

When choosing the generation type (thermal, hydroelectric, and nuclear), several points must be considered for selecting the size and number of generating units.

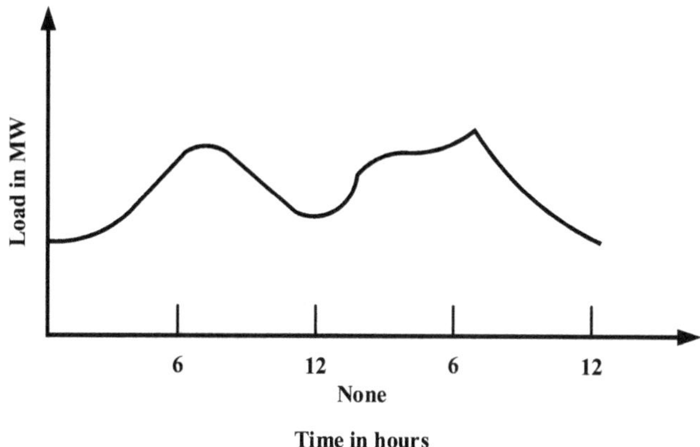

FIGURE 9.1 Daily load curve.

Some of these considerations are the fuel available and its cost, availability of suitable sites for a hydro station, and the nature of the load supplied.

The load that a power system must supply is never constant because of variable demands at different times. The variations can be seen from the prediction *load curve.*

Load curve: It is a graphic record showing the demand for power for every instant during the hour, the day, the month of the year. Figure 9.1 represents the daily load curve.

9.2.1 Important Terms and Factors

Variable load problem have introduced the following terms and factors in power plant generation:

1. **Connected load:** The sum of the continuous rating of all the equipment connected in the supply system.
2. **Demand:** Demand for installation or system for the load drawn from the supply at a specified time interval. It is expressed in K.W.s, K.V.A.s, or Amperes.
3. **Maximum demand:** The greatest demand for the load on the power station during a given period. Maximum demand is very important to determine the installed capacity of the station. The station must be capable of meeting the maximum demand.
4. **Demand factor:** It is the ratio of connected load to the maximum demand.

$$\textbf{\textit{Demand factor}} = \frac{Connected\ load}{\max .demand} \tag{9.4}$$

Demand factor < 1 is used to determine the capacity of the plant equipment. And may indicate the degree to which the total connected load is operated simultaneously.

5. **Average load:** The average load occurring on the power station in a given period (day or month, or year) is known as average load or average demand.

$$\textbf{\textit{Daily average load}} = \frac{Total\ energy\ generated\ in\ day\ KWh}{No.of\ hours\ in\ a\ day\ (24\ hours)} \tag{9.5}$$

or

$$\textbf{\textit{Daily average load}} = \frac{\textbf{\textit{No.of units KWh generated in day}}}{No.of\ hours\ in\ a\ day\ (24\ hours)} \tag{9.6}$$

Also

$$\textbf{\textit{Monthly average load}} = \frac{Total\ energy\ generated\ during\ a\ month\ KWh}{No.of\ hours\ in\ a\ month} \tag{9.7}$$

$$\textbf{\textit{Yearly average load}} = \frac{Total\ energy\ generated\ during\ a\ month\ KWh}{No.of\ hours\ in\ a\ year\ (8760\ h)} \tag{9.8}$$

In general, the average load

$$\textbf{\textit{average load}} = \frac{Total\ energy\ generated\ during\ T\ period\ KWh}{No.of\ hours\ in\ T\ period} \tag{9.9}$$

6. **Load factor:** It is the ratio of the average load to the maximum demand.

$$\textbf{\textit{Load factor}}\ (\textbf{\textit{L.F.}}) = \frac{\textbf{\textit{Average demand}}}{max\ demand} \tag{9.10}$$

If T = 24 hours, the load factor is called the daily load factor.
L.F < 1;

$$\textbf{L.F}\ \alpha\ \frac{\mathbf{1}}{max.demand}$$

Cost Plant α Capacity of Station α Max. Demand

$$\textbf{Cost of plant}\ \alpha\ \frac{\mathbf{1}}{\textbf{\textit{L.F of power station}}}$$

The load factor is very important to determine the plant's overall cost, and it indicates the degree to which the peak load is sustained during the period.

7. **Diversity factor**: It is defined as the ratio of the sum of the maximum demands of the various part of a system to the maximum coincident demand

of the whole system. The maximum demands of the individual consumers of a group do not occur simultaneously. Thus, there is a diversity in the occurrence of the load. Due to this diverse nature of the load, a full load power supply to all the consumers at the same time is not required.

$$\text{Diversity factor} = \frac{Sum\ of\ indiviual\ max.demands}{max.demand\ on\ power\ station} \tag{9.11}$$

$$D.F \geq 1; \text{and } D.F\,\alpha\,\frac{1}{max.demand}$$

$$\therefore cost\ of\ plant\,\alpha\,\frac{1}{D.F}$$

8. **The capacity factor:** It indicates the extent of the use of the generating station. If the power generation unit is always running at its rated capacity, its capacity factor is 100% or 1. It is also expressed regarding peak load and load factor.

$$Capacity\ factor = \frac{Actual\ energy\ produced}{max.energy\ that\ could\ have\ been\ produce} \tag{9.12}$$

or

$$Capacity\ factor = \frac{Average\ load(demand) \times T}{max.energy\ that\ could\ have\ been\ produce} \tag{9.13}$$

$$Capacity\ factor = \frac{Average\ load(demand) \times T}{max.demand \times T} \tag{9.14}$$

and

$$Capacity\ factor = \frac{Average\ load(demand)}{Plant\ capacity} \tag{9.15}$$

9.

$$Reserve\ Capacity = Plant\ Capacity - max.Demand \tag{9.16}$$

10.

$$Plant\ use\ factor = \frac{Station\ output\ in\ KWh}{Plant\ capacity\ hours\ of\ use} \tag{9.17}$$

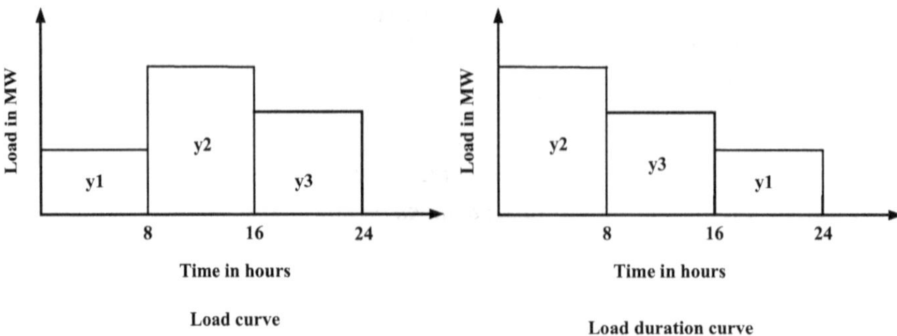

FIGURE 9.2 Load curve and load duration.

Example 9.1

A 20 MW power station produces an annual output of $6.5X10^6$ KWh and remains in operation for 2100 hours in a year, find the plant use factor.

Solution

$$Plant\, use\, factor = \frac{Station\ output\ in\, KWh}{Plant\ capacity\ hours\ of\ use}$$

$$Plant\, use\, factor = \frac{6.5\times 10^6 \times 10^3}{20\times 10^6 \times 2100} = 15.4\%$$

9.3 LOAD DURATION CURVE

When the load elements of a load curve are arranged in descending magnitudes, the curve thus obtained is called a load duration curve. It gives the data in a more presentable form. Figure 9.2 represents the daily load curve and daily load duration curve.

The area under the daily load curve = Area under the daily load duration curve.

$$= Total\ energy\ generated\left(KWh \right) on\, the\, day$$

9.4 LOAD CURVES AND SELECTION OF THE NUMBER AND SIZES OF THE GENERATION UNITS

The number 3 and size of 200 MW each generating unit are selected so that they correctly fit the station load curve as shown in Figure 9.3.

Table 9.1 gives the operating load table; the periods 0–7 and –12 –14 the load requires 150 MW while the rating of each unit 200 MW, so need to only operate one unit. The periods 7–12, 14–17, and 22–24 the load requires 300 MW and needs two

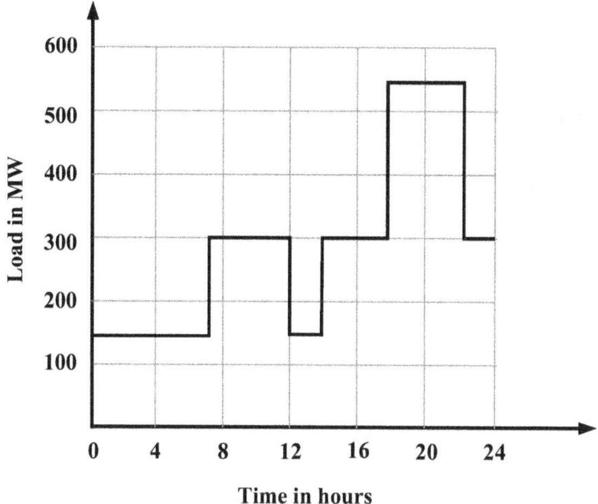

FIGURE 9.3 Load curve.

TABLE 9.1
Generation Schedule

Time	Units in Operation
0–7	1
7–12	1 + 2
12–14	1
14–17	1 + 2
17–22	1 + 2 + 3
22–24	1 + 2

units to operate to cover the demand. The periods 17–22, the load requires 550 MW, so three units will have to cover the power demand.

The important points in the selection of generator units are as follows:

a. The selection of units should approximately fit the station's annual (yearly) load curve.
b. The plant's capacity should be made 15 to 20% more than the maximum demand.
c. One unit should be kept as a spare generating unit (stand-by unit).
d. Using identical units (having the same capacity) saves the station's cost.
e. The load curve can be very accurate if a large number and small capacity of units are selected; this is one side. On the other side, the investment cost per kW of capacity increases as the size of the units decreases.

Example 9.2

A power station is to supply three consumers. The daily demands of three consumers are given in Table 9.2.

Plot the load curve of the power station and find:

1. Load factor of the individual consumer.
2. Diversity factor of the power station.
3. Load factor of the power station.

Solution

1. load factor

$$\text{Load factor}(\text{LF.}) = \frac{\text{Average demand}}{\text{max demand}}$$

$$\text{Load factor 1} = \frac{200\times 6 + 600\times 8 + 0\times 4 + 800\times 6}{800\times 24} = 56.25\%$$

$$\text{Load factor 2} = \frac{100\times 6 + 800\times 8 + 600\times 4 + 0\times 6}{800\times 24} = 61.45\%$$

$$\text{Load factor 3} = \frac{0\times 6 + 400\times 8 + 400\times 4 + 600\times 6}{600\times 24} = 58.3\%$$

2. Figure 9.4 shows the load curve of the power station.

$$\text{Diversity factor} = \frac{\text{Sum of maximum individual demands}}{\text{maximum demand on the power station}}$$

From the load curve, the max. demand on the power station is 1800 KW.

$$\text{Diversity factor} = \frac{800 + 800 + 600}{1800} = 1.22$$

3.

$$\text{The load factor of power station} = \frac{300\times 6 + 1800\times 8 + 1000\times 4 + 1400\times 6}{1800\times 24} = 66.2\%$$

TABLE 9.2
Loads Schedule of Example 9.2

Time (hours)	Consumer (1)	Consumer (2)	Consumer (3)
0–6	200 KW	100 KW	No-load
6–14	600 KW	800 KW	400 KW
14–18	No - load	600 KW	400 KW
18–24	800 KW	No - load	600 KW

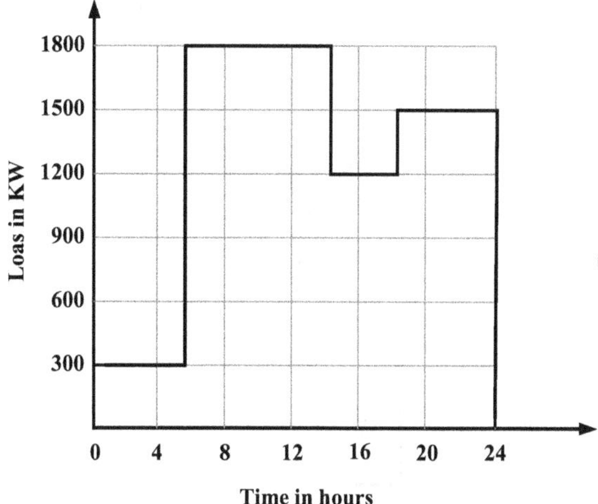

FIGURE 9.4 Load curve.

Example 9.3

A power station is to supply two loads. The daily load curve of the station is shown in Figure 9.5. If the load factor of the two loads are 0.416 and 0.458, respectively, and the diversity factor of the power station is 1.142, find

1. The load factor of the power station.
2. The energy consumed per day and the maximum demand of each load.

Solution

1. load factor (*L.F*) = 100

$$\text{Load factor}\left(\text{LF.}\right)=\frac{\text{Average demand}}{\text{max demand}}$$

$$=\frac{18\times400+6\times400+10\times400+0.5\times6\times400+0.5\times6\times600}{1400\times24}=0.49$$

2. Let
   E1 – energy consumed by load 1
   E2 – energy consumed by load 2
   M1 – Max. the demand for load 1
   M2 – Max. the demand for load 2

$$L.F1=\frac{E_1}{24\times M_1}$$

$$0.416=\frac{E_1}{24\times M_1}$$

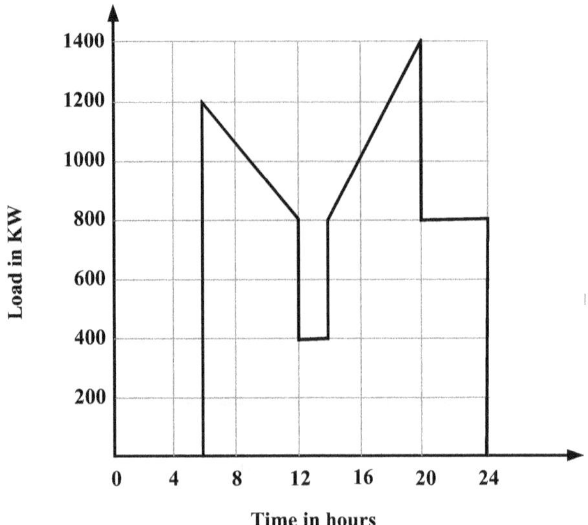

FIGURE 9.5 Load curve.

$$10M_1 = E_1$$

$$L.F2 = \frac{E_2}{24 \times M_2}$$

$$0.448 = \frac{E_2}{24 \times M_2}$$

$$11M_2 = E_2$$

$$\text{Diversity factor} = \frac{Sum\,of\ indiviual\,\max.demands}{max.demand\,on\,power\ station}$$

$$1600 = M_1 + M_2$$

$$\text{Energy generated from power station} = E_1 + E_1$$

$$16600 = E_1 + E_1$$

Solving to find

M1 = 1000 KW
M2 = 600 KW
E1 =10000 KWh
E2 = 6600 KWh.

9.5 PREDICTION OF LOAD AND ENERGY REQUIREMENTS

The nature of the load to be supplied affects the choice of plant to a considerable extent. The load that a power system must supply is never constant because of variable demands at different times. The variations can be seen from the predicated LOAD CURVE.

The minimum capacity of generating stations must be such as to meet the maximum demand. At the same time, the power system always needs to maintain reliability and continuity of power supply.

Therefore, to determine load and energy requirement prediction, the load must meet the demand curve. Some methods for this purpose are as follows:

1. Load survey.
2. Methods of extrapolation.
3. Mathematical methods.
4. Mathematical methods using economic parameters.

9.6 ADDITIVE SEASONAL

One of the load forecasting models applicable to time series with seasonal demand patterns is *additive seasonal*. This method includes both the horizontal and trend seasonal demand patterns, where the seasonal influences are additive. The demand swings due to seasonal influences that have the same magnitude every year. The expected demand at (t) is:

$$y(t) = a + bt + d(t) \tag{9.18}$$

where (a+bt) gives the trend influence and d(t) represents the seasonal influence and is called the seasonal increment at time t.

When d(t)=0, month t has no seasonal influences. When d(t)>0, month t has a higher expected demand than an average month, (i.e.) y(t) > a+bt, and with d(t) < 0, the expected demand is less, (i.e.)...y(t) < a+bt. In this pattern, seasonal influence is the same at every cycle regardless of the magnitude of the level.

Since the seasonal increments must balance out over a year, their sum over a calendar year is zero, as explained in Figure 9.6, i.e.,

$$\Sigma d(t) = 0,\ \mathrm{t} = 1\,\mathrm{to}\,12 \tag{9.19}$$

Input to the additive seasonal method are past loads, and its output is the load forecasted for a given month

$$y(t) = (a + bt) + dt. \tag{9.20}$$

9.7 THE ADDITIVE SEASONAL ARCHITECTURE

This model uses three smoothing parameters in the process. These are "L, U, K," where each lies in the interval (0,1). The smoothing parameters "L, U, K" are used to find a,b,d. Two phases are carried out in the implementation of this model. The first

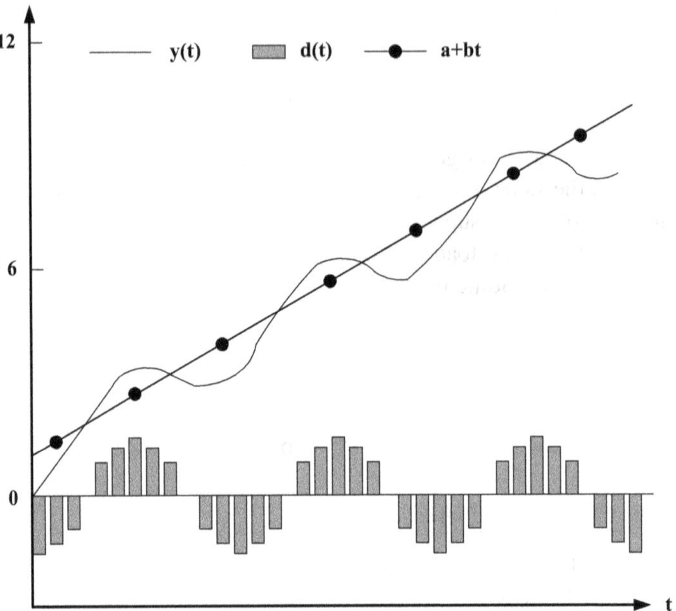

FIGURE 9.6 The component of the additive seasonal demand pattern.

is an initialization phase, where past demands ($X1, X2, X3.........XT$) are used to seek the estimates needed to get started. The second is an updating phase whereby the coefficients are modified as each new demand entry becomes available.

Initialization: To get started, the forecaster assembles the prior demand entries (X1, X2, X3......... XT). These are conveniently listed into yearly clusters:

$$X(1), X(2).......................X(12)$$
$$X(13), X14,.....................X(24)$$
$$\cdot$$
$$\cdot$$
$$X(T-1)...........................X(T).$$

Let j = T/12 represent the number of full years of demand entries available, where T is the amount of data. With this data, the initialization is carried out using the following ten steps:

1. The average demand per month is found for years one through j. These are:

$$Y(1) = \frac{\left[X(1) + ... + X(12)\right]}{12} \tag{9.21}$$

$$Y(2)=\frac{\left[X(T-11)+........+X(T)\right]}{12} \tag{9.22}$$

2. The slope is estimated from:

$$b=\left(Y(j)\right)-Y(1)/(T-12) \tag{9.23}$$

3. Now, the level at $t = 0$ is estimated using:

$$a=Y(1)-\left(\frac{12+1}{2}\right)\times b \tag{9.24}$$

4. The level for each month from t = 1 to t = T is found from

$$A(t)=a+bt \tag{9.25}$$

5. The seasonal increments for months $t = l$ to $t = T$ are estimated from

$$d(t)=X(t)-at \tag{9.26}$$

6. An average value of the seasonal increments for every 12 months is obtained. These are:

$$D(1)=\left(d(1)+d(13)+........+d(T-11)\right) \tag{9.27}$$

$$D(12)=\left(d(12)+d(24)+........+d(T)\right) \tag{9.28}$$

7. The 12 seasonal increments are normalized so that their sum becomes zero. This step is carried out according to the following procedure:

$$dd=\frac{\left[D(1)+D(2)+.......+D(12)\right]}{12} \tag{9.29}$$

$$C(t)=d(t)-dd.....for\, t=1,\,2,\,3,.........12 \tag{9.30}$$

8. Using A(0) = a and B(0) = b, the following three estimates are calculated: ' at each period from t = 1 to t = T

$$N(t)=U*\left(X(t)-C(t)\right)+(1-U)\left(A(t-1)+B(t-1)\right) \tag{9.31}$$

$$\begin{aligned}P(t)&=K*\left(N(t)-N(t-1)\right)+(1-k)\times *p(T-1)C(t+12)\\&=L*\left(X(t)-N(t)\right)+(1-k)*C(t)\end{aligned} \tag{9.32}$$

9. Using the following equation to estimate the value of $X(t)$

$$R(t) = N(t) + P(t) + C(t+12) \tag{9.33}$$

10. Finally, the most current seasonal increments(C(T+1) ……,C(T+12)) are normalized so that their sum becomes zero. The following procedure carries this out:

$$ad = \frac{\left[C(T+1)+...........+C(T+12)\right]}{12} \tag{9.34}$$

$$d(60+t = C(T+t) - ad \tag{9.35}$$

Having completed the ten steps above, the initialization process is completed. It's now possible to generate forecasts for time T for the Tth future period. The forecasts demand is

$$Xt(t) = N(t) + P(t) * t + d(t+T) \tag{9.36}$$

If $t > 12$ is needed, the seasonal increments that are used for this purpose are:

$$d(T+13) = d(T+1) \tag{9.37}$$

$$d(T+14) = d(T+2). \tag{9.38}$$

9.8 FORECASTING MODELING

This section describes two forecasting models applicable for the data with a trend demand pattern. The first is *the regression model*, in which the method of least square is used to generate the forecasts. This method allows the forecaster to give higher weights to the more current demand pattern entries. A linear load model is proposed, and the parameters of the linear load models are estimated using the backpropagation with the moment algorithm. The second method is *brown's smoothing method.*

9.8.1 The Regression Models

In the regression model, the forecasted future demand is not a function of the last demand pattern, where the relationship can define the expected demand

$$y_t = a_1 f_1(t) + a_2 f_2(t) + \ldots + a_k f_k(t) \tag{9.39}$$

In Equation 9.39 the expected demand includes k terms with k unknown coefficients $(a_1, a_2, \ldots, a_k)$ and k is a known function of time $(f_1(t), f_2(t), +. \ldots, f_k(t))\cdot$ Each function $f_i(t)$, $i = 1, 2,\ldots k$ is defined using a relationship with t. Some common examples of the functions are 1, t, t^2, t^3, $sin\ (ct)$, $cos\ (ct)$ where c is constant.

Or the functions may be another demand pattern. The horizontal demand pattern is defined as:

$$y_1(t) = a_1 = a_1 f_1(t)\, where\ f_1(t) = 1 \tag{9.40}$$

In the trend demand pattern

$$y_1(t) = a_1 + a_2.t\ where\ f_1(t) = 1\, and\ f_2(t) = t$$

The quadratic demand pattern becomes

$$y_t = a_1 + a_2.t + a_3.t^2 \tag{9.41}$$

where $f_1(t) = 1, f_2(t) = t,\ and\ f_3(t) = t^2$

Some other examples are

$$y_t = a_1 + a_2 \sin(ct) + a_2 \cos(ct)) = a_1 f_1(t) + a_2 f_2(t) + a_3 f_3(t) \tag{9.42}$$

where $f_1(t) = 1, f_2(t) = \sin\ (ct)$, and $f_3(t) = \cos\ (ct)$

$$y_t = a_1 + a_2\, e^{ct} = a_1 f_1(t) + a_2 f_2(t) \tag{9.43}$$

where $f_1(t) = 1,\ and\ f_2(t) = e^{ct}$

Another example of a different demand pattern is

$$y_t = a_1 + a_2.t + a_3.t^2 + a_4.TM(t) = a_1 f_1(t) + a_2 f_2(t) + a_3 f_3(t) + a_4 f_4(t) \tag{9.44}$$

where $f_1(t) = 1, f_2(t) = t, f_3(t) = t^2$, and $f_4(t) = TM(t)$.

$TM(t)$ is like temperature, number of consumers, etc.

The role of the forecasting model is to find a coalesce with past demands (X_1, X_2,……… XT) that correspond to the demand pattern assumed by the forecaster. This entails finding estimates for a_1, a_2, a_3, …, a_k respectively. The first T demand entries (X_1, X_2,…… XT) are used in the fitting process for an assumed demand pattern of the type

$$y_t = a_1 f_1(t) + a_2 f_2(t) + \cdots + a_k f_k(t) \tag{9.45}$$

The following data is required in carrying out the calculations.

Using the data given in Table 9.3, the k equations listed below are generated

TABLE 9.3
The Role of the Forecasting Model

X(t)	$f_1(t)$	$f_2(t)...f_k(t)$
X_1	$f_1(t)$	$f_2(t)..f_k(t)$
X_2	$f_1(t)$	$f_2(t)..f_k(t)$
.	.	.
.	.	.
$X_{T.}$	$f_1(t)$	$f_2(t)..f_k(t)$

$$
\begin{aligned}
g_1(T) &= a_1F_{11}(T)+a_2F_{12}(T)+\ldots+a_kF_{1k}(T) \\
g_2(T) &= a_1F_{21}(T)+a_2F_{22}(T)+\ldots+a_kF_{2k}(T) \\
&\ldots.. \\
&\ldots.. \\
g_k(T) &= a_1F_{k1}(T)+a_2F_{k2}(T)+\ldots+a_kF_{kk}(T)
\end{aligned} \tag{9.46}
$$

t: The number of demands i=1, 2, ….,12

$$
\begin{aligned}
F_{in}(T) &= f_i(1)f_n(1)+f_i(2)f_n(2)+\ldots+f_i(T)f_n(T) \\
F_{in}(T) &= \Sigma\left(f_i(t)f_n(t)\right)\text{i \& n} = 1,2,\ldots\ldots\ldots
\end{aligned} \tag{9.47}
$$

The relations above are now used to seek a_1, ………, a_k. In this endeavor, the relations are conveniently listed in their equivalent matrix form, i.e.

$$
\begin{bmatrix} g_1(T) \\ g_2(T) \\ .. \\ g_k(T) \end{bmatrix} = \begin{bmatrix} F_{11}(T) & F_{12}(T) & \ldots & F_{1k}(T) \\ F_{21}(T) & F_{22}(T) & \ldots & F_{2k}(T) \\ \ldots & \ldots & & \ldots \\ F_{k1}(T) & F_{k1}(T) & \ldots & F_{kk}(T) \end{bmatrix} \begin{bmatrix} a_1 \\ a_2 \\ .. \\ a_k \end{bmatrix} \tag{9.48}
$$

$$
\begin{bmatrix} a_1 \\ a_2 \\ .. \\ a_k \end{bmatrix} = \begin{bmatrix} F_{11}(T) & F_{12}(T) & \ldots & F_{1k}(T) \\ F_{21}(T) & F_{22}(T) & \ldots & F_{2k}(T) \\ \ldots & \ldots & & \ldots \\ F_{k1}(T) & F_{k1}(T) & \ldots & F_{kk}(T) \end{bmatrix}^{-1} \begin{bmatrix} g_1(T) \\ g_2(T) \\ .. \\ g_k(T) \end{bmatrix} \tag{9.49}
$$

Now to explain the estimation and prediction using the regression model in the following steps:

1. Find the [F] matrix using equation (9.47).
2. Find the [G] matrix using equation (9.48).
3. Find the [a] matrix using equation (9.49).
4. Using the following equation to estimate the value of X

$$x_t(T) = a_1 f_1(t) + a_2 f_2(t) + \ldots + a_k f_k(t) \tag{9.50}$$

5. The prediction for the Tth future time is

$$x_t(T) = a_1 f_1(T+t) + a_2 f_2(T+t) + \ldots + a_k f_k(T+t) \tag{9.51}$$

As each new demand entry becomes available, the summations $g_i(T)$ and $F_{in}(T)$ are updated, and new estimates of the coefficients are found. These are applied as before to generate data forecasts.

9.8.2 Brown's Smoothing Method

The brown exponential smoothing computation is one elementary form of a learning model. The "rules of learning" can be formalized to give the basis for revaluating the average estimate. To go from linear to quadratic smoothing when the basic underlying pattern of the data is quadratic or higher-order on could go from first, double, and then quadratic or so higher order of smoothing. This can be illustrated in the following equations:

$$S'_t = S'_{t-1} + a\left[X_{t-1} - S'_{t-1}\right] \tag{9.52}$$

By rearranging terms in Equation 9.51, we get

$$S'_t = a\,X_{t-1}\,S'_{t-1} + (1-a)S'_{t-1} \text{ first smoothing} \tag{9.53}$$

$$S''_t = a\,S'_t + (1-a)S''_{t-1} \text{ second smoothing} \tag{9.54}$$

$$S'''_t = a\,S''_{t-1} + (1-a)S'''_{t-1} \text{ quadratic smoothing} \tag{9.55}$$

$$a_t = 3S'_t - 3S''_t + S'''_t \tag{9.56}$$

$$b_t = \frac{a\left[(6-5a)S'_t - (10-8a)3S''_t + (4-3a)S'''_t\right]}{\left[2(1-a)^2\right]} \tag{9.57}$$

$$c_t = \frac{a^2\left(s'_t - s''_t + s'''_t\right)}{(1-a)^2} \tag{9.58}$$

$$F(t+m) = a_t + b_t\,m + \left(c_t m^2\right)/2 \tag{9.59}$$

where X_{t-1} is the actual value at time $t-1$; S'_t, S'_{t-1} are the first smoothing values at time $t-1$; S''_t , S''_{t-1} are the second smoothing values at time $t-1$; S'''_t , S'''_{t-1} are the quadratic smoothing values at time $t-1$; a: is the smoothing constant ($0.1 < a < 0.9$); Ft+m: is the forecasted values at time t+m; m: is the forecasting period. The initialization processor for this method can be a very simple setting

$$S'_1 = S''_2 = S'''_3 = X \tag{9.60}$$

where X_1 is the first actual value of the time-series data.

9.8.3 Load Forecasting Using the Additive Seasonal Model

Using the ten steps described in Section 9.7 to forecast the monthly peak load given in Table 9.4, the parameters (U, L, K) can be changed incrementally (trial and error) to find the set which gives the minimum square error (MSE).

Table 9.5 gives the forecasted results from applying the seasonal additive model to forecast the monthly peak load for (96 months: 8 years), optimizing the number of historical data input to the computing program.

Figures 9.7 and 9.8 depict the forecasted annual load curves previous eight years using this method.

TABLE 9.4
The Actual Demand for Years 2002–2021

	Jan	Feb	Mar	Apr	May	Jun	Jul	Aug	Sep	Oct	Nov	Dec
2002	2794	2848	2887	2746	3359	3273	3488	3706	3555	3252	2761	2992
2003	3087	2905	2777	2914	3209	3435	3882	3888	3748	3278	2915	3888
2004	3010	2814	2782	2966	3440	3741	4063	4264	3798	3569	3128	3398
2005	3452	3320	3288	3151	3968	4158	4334	4305	4022	3815	3683	3719
2006	4050	3993	3556	3906	4275	4504	4720	4761	4262	4153	3855	4028
2007	4271	4089	3925	3926	4596	4807	5162	4883	4306	3815	3512	4075
2008	3854	3705	3650	3540	4362	4550	4732	4702	4320	3915	3266	4115
2009	4005	3800	3802	3650	3522	4665	3941	4832	4511	4512	3444	4300
2010	4108	3910	3965	3758	3735	4870	4132	5001	4600	4580	3500	4410
2011	4333	3970	4160	3990	4900	5088	5400	5277	4640	4644	3564	4555
2012	4434	4083	4275	4095	5003	5192	5544	5379	4672	4742	3677	4662
2013	4657	4516	4525	4439	5298	5476	5804	5532	4838	4462	3944	4609
2014	4997	5148	4743	4896	5501	5656	6034	5830	4990	4601	4044	4967
2015	5202	5474	4985	1980	5799	5943	6421	5888	5256	4853	4255	5025
2016	5690	5445	5273	5301	6100	6226	6809	6382	5535	5118	4477	5331
2017	5914	5649	5554	5745	6399	6510	7169	6541	5820	5399	4711	5545
2018	6195	6315	5816	6064	6700	6793	7580	6934	6140	5695	4957	5718
2019	6539	6780	6103	6309	6999	7077	7850	7049	6469	6007	5215	6024
2020	6880	6879	6374	6744	7299	7300	8230	7485	6811	6336	5487	6152
2021	7180	7034	6648	7118	7599	7644	8563	7656	7172	6683	5774	6461

TABLE 9.5
The Forecasting Demands in MW of Eight Years 2022–2029 by Using the Additive Seasonal Technique

	2022	2023	2024	2025	2026	2027	2028	2029
Jan	7953	8210	8466	8724	8980	9237	9494	9750
Feb	7813	8070	8326	8583	8840	9097	9353	9610
Mar	7429	7685	7942	8199	8456	8713	8969	9226
Apr	7901	8158	8415	8672	8929	9185	9442	9699
May	8394	8650	8907	9164	9421	9678	9934	10191
Jun	8451	8707	8964	9221	9478	9734	9991	10248
Jul	9386	9643	9899	10150	10413	10670	10926	11183
Aug	8490	8747	9004	9261	9518	9774	10031	10288
Sep	8014	8270	8527	8784	9041	9289	9554	9811
Oct	7528	7784	8041	8298	8555	8812	9068	9325
Nov	6616	6873	7129	7387	7643	7900	8157	8413
Dec	7302	7559	7816	8073	8329	8586	8843	9152

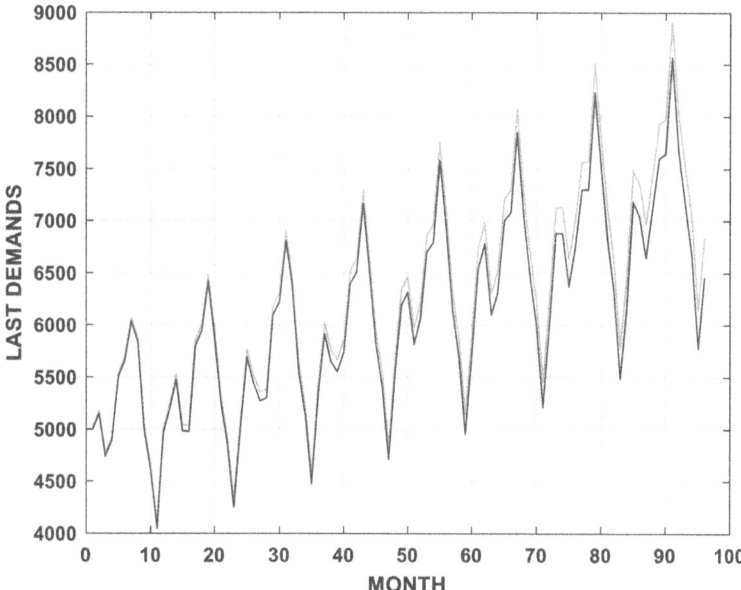

FIGURE 9.7 Compression between actual and forecasted demands of years 2022–2029 using additive seasonal model.

9.8.4 Trend Model

Load Forecasting with trend model plots the historical data for monthly peak loads as shown in Figure 9.9. These curves are nearly identical to the trend curves. Therefore, we take the data for each month separately and then perform the estimation and prediction using the trend forecasting techniques (the regression model).

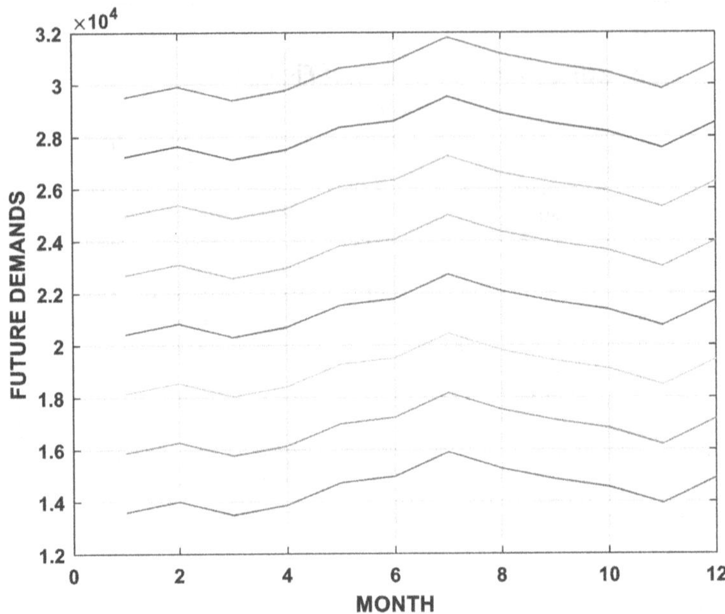

FIGURE 9.8 The forecasting demands of years 2022–2029 using the seasonal additive model.

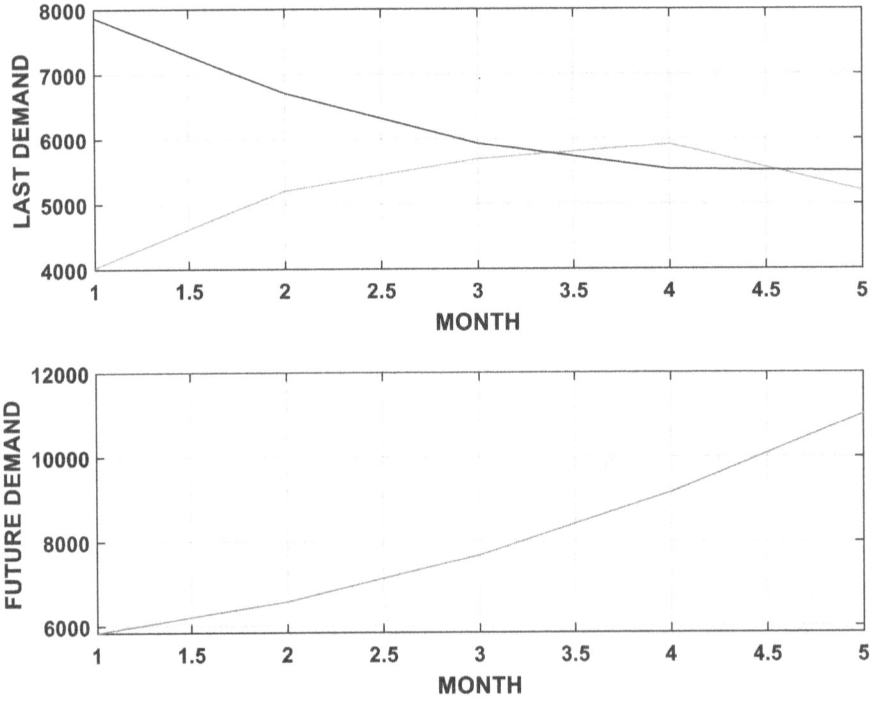

FIGURE 9.9 Real and last forecasted demands.

TABLE 9.6
The Forecasting Demands in MW of Eight Years 2022–2029 by Using the Regression Model Technique

	2022	2023	2024	2025	2026	2027	2028	2029
Jan	6546	6863	7182	7502	7822	8144	8467	8792
Feb	6514	6382	7175	7525	7889	8266	8657	9061
Mar	6096	6373	6651	6931	7212	7495	7779	8065
Apr	6354	6731	7130	7553	7999	8468	8960	9476
May	6999	7299	7599	7899	8199	8499	8799	9100
Jun	7065	7343	7621	7897	8172	8447	8720	8993
Jul	7888	8227	8556	8874	9182	9478	9765	10041
Aug	7132	7410	7690	7970	8253	8536	8821	9107
Sep	6466	6811	7172	7550	7944	8355	8782	9226
Oct	6008	6337	6681	7042	7418	7810	8218	8643
Nov	5216	5488	5772	6069	6378	6700	7035	7382
Dec	5966	6203	6450	6706	6972	7247	7531	7825

9.8.5 Load Forecasting Using Quadratic Regression

Model excluding the effect of weather. Figure 9.9 is nearly identical to the quadratic curves. Therefore, the general function y(t) for the quadratic curves is

$$y(t) = a_1 + a_2 t + a_3 t^2 \tag{9.61}$$

where t is the month number ((t =1.........,22)

$$f_1(t) = 1, f_2(t) = t_1 \text{ and } f_3(t) = t^2$$

The five steps of the regression model described in Section 9.8.1 are used to forecast the monthly peak load.

Table 9.6 gives the forecasted result obtained from using the quadratic regression model to forecast the monthly peak load. Figure 9.10 shows the forecasted annual load curves for the eight years.

The MATLAB® command is used to plot the absolute compression between actual and forecasted demands of years 2022–2029 by using the additive seasonal., as shown in Figure 9.7, and the forecasting demands of years 2022–2029 by using the seasonal additive model as shown in Figure 9.8. Figure 9.9 shows the real and last forecasted demands.

The following is a MATLAB program of the seasonal additive model for load forecasting. This program for eight years last and eight years next.

```
clc
clear
%THE ADDITIVE SEASONAL MODEL FOR LOAD FORECASTING &THIS
%PROGRAM FOR EIGHT YEARS LAST AND EIGHT YEARS NEXT
x = [4997 5148 4743 4896 5501 5656 6034 5830 4990 4601 4044 4967
   5202 5474 4985 4980 5799 5943 6421 5888 5256 4853 4255 5025
   5690 5445 5273 5301 6100 6226 6809 6382 5535 5118 4477 5331
   5914 5649 5554 5745 6399 6510 7169 6541 5820 5399 4711 5545
   6195 6315 5816 6064 6700 6793 7580 6934 6140 5695 4957 5718
```

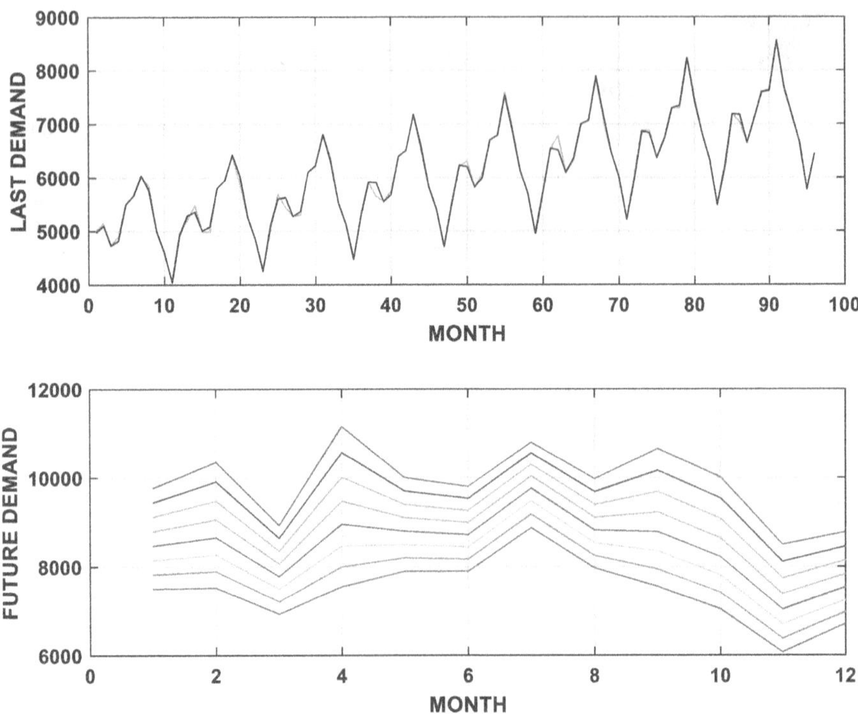

FIGURE 9.10 Compression between actual and forecasted demands of years 2002–2021 and the forecast demands of years 2022–2029 using the regression model.

```
 6539 6780 6103 6309 6999 7077 7850 7049 6469 6007 5215 6024
 6880 6879 6374 6744 7299 7300 8230 7485 6811 6336 5487 6152
 7180 7034 6648 7118 7599 7644 8563 7656 7172 6683 5774 6461];
X(1)=sum (x(1,:)/12);
X (8)=sum (x (8,:)/12);
b= (x (8) -X(1))/ (96-12);
a=X (1) - ((12+1)/2) *b;
q=0;
for i=1:8;
for t=1:12; q=q+1; g(q)=x (i, t); A(q) =a+ (b*q);
d(q) =g(q)-A (q); end; end; q=0; for i=1:8;
for t=1:12; q=2+1;
f(i,t)=d (q);
end;
end;
for i=1:12;
    D (i)=sum (f(:,i)/12);
end;
dd = sum (D)/12;
for i=1:12;
```

```
C(i) =D(i)-dd;
end;
U=0.006; k=0.06;
L=1; N(1) =U* (g(1) -C(1)) +(1-U) * (a+b);
P(1) = (k* (N(1)-a)) + ((1-k) *b);
C(13) = (L* (g(1) -N(1)))+((1-L) *C(1));
for t=2:96;
N(t) = (U* (x(t) -C(t))) +(1-0) * (N(t-1) +P (t-1));
P(t) = (k* (N(t)-N(t-1)))+((1-k) *P(t-1));
C(t+12) = (L* (g(t)-N(t)))+((1-1) *C(t));
end;
for t=1:96;
    R(t) = N (t) +P (t) +C (t+12);
end; s=0;
for t=1:12;
s=s+C (t+96);
end;
ad=s/12;
for t=1:12
d(96+t) =C (96+t)-ad;
end;
e=0;
for t=1:96;
e=e+ (g(t)-R(t))^2;
end;
%Mean Square Error = MSE
MSE=e/96; t=1:12;i=1:96;
%TO PLOT REAL & LAST FORECASTED DEMANDS
plot(i,R(i),'r-',i,g(i), 'b'),
grid,
xlabel("MONTH"),
ylabel("LAST DEMANDS"),
figure;
F1 (t)=N (60) + (P (60) * (t))+d (60+t);
F2 (t) =N (60) + (P(60) * (12+t))+d(60+t);
F3 (t) =N (60) + (P (60) * (24+t)) +d (60+t);
F4 (t) =N (60) + (P (60) * (36+t))+d(60+t);
F5 (t) =N (60) + (P (60) * (48+t))+d(60+t);
F6(t) =N (60) + (P(60) * (60+t))+d (60+t);
F7 (t) =N (60) + (P (60) * (72+t)) +d (60+t);
F8 (t) =N (60) + (P (60) * (84+t)) +d (60+t);
%TO PLOT THE FUTUREB DEMAND
plot(t, F1,t,F2, t,F3, t, F4,t,F5, t,F6,t, F7,t,F8),
grid,
xlabel('MONTH'),
ylabel('FUTURE DEMANDS'),
```

```
figure;
x=[4024 5202 5690 5914 5195 6539 6880 7180 ];
%f(t)=a(1)+a(2)*t+a(3)*t^2;
%g(1)=0;g(2)=0;g(3)=0;
for i=1:8;
g(1)=g(1)+x(i);
g(2)=g(2)+x(i)*i;
g(3)=g(3)+x(i)*i^2;
end;
for t=1:8;
q(t)=t^2;
p(t)=t^3;
h(t)=t^4;
A=sum(q);
B=sum(p);
C=sum(h);
end;
f=[8 sum(1:8) A;36 A B;A B C];
g=[g(1);g(2);g(3)];
a=inv(f)*g;
for t=1:8;
X(t)=a(1)+(a(2)*t)+(a(3)*t^2);
end;
e=0;
for i=1:8;
e=e+(x(i)-X(i))^2;
end;
MSE=e/8;
for t=1:8;
    W(t)=a(1)+(a(2)*(t+5))+(a(3)*(t+5)^2);
end;
t=1:5;

subplot(211),
plot(t,x(t), 'r',t,X(t),'b'),
grid,
xlabel('MONTH'),
ylabel('LAST DEMAND');
subplot(212),
plot(t, W(t),'m'),
grid,
xlabel('MONTH'),
ylabel('FUTURE DEMAND');
```

The MATLAB command is used to plot the compression between actual and forecasted demands of years 2002–2021 and the forecasting demands of years 2022–2029 using the regression model, as shown in Figure 9.10.

Following the regression model for load forecasting, this demand for years (2002–2021):

```
clc
clear
%THE REGRESSION MODEL FOR LOAD FORECASTING %THIS DEMAND FOR
YEARS (2002-2021)
X = [4997 5148 4743 4896 5501 5656 6034 5830 4990 4601 4044 4967
    5202 5474 4985 4980 5799 5943 6421 5888 5256 4853 4255 5025
    5690 5445 5273 5301 6100 6226 6809 6382 5535 5118 4477 5331
    5914 5649 5554 5745 6399 6510 7169 6541 5820 5399 4711 5545
    6195 6315 5816 6064 6700 6793 7580 6934 6140 5695 4957 5713
    6539 6780 6103 6309 6999 7077 7850 7049 6469 6007 5215 6024
    6880 6879 6374 6744 7299 7300 8230 7485 6811 6336 5487 6152
    7180 7034 6648 7118 7599 7644 8563 7656 7172 6683 5774 6461];
e=0;z=0;
for i=1:8;
 for p=1:12;z=z+1;w(z)=X(i,p);
 end
end;
for q=1:12;
    s=0; %TO EVALUATE G-MATREX
g(1)=0;g(2)=0;g(3)=0;
for i=q:12:96;
s=s+1;
g(1)= g(1)+w(i); g(2)=g(2)+w(i)*s; g(3)=g(3)+w(i)*s^2;
u(s)= s^2;p(s)=s^3;h(s)=s^4;
end;
A=sum(u);B=sum(p);C=sum(h); %TO EVALUATE F-MATREX
f=[8 sum(1:8) A;36 A B;A B C]; g=[g(1);g(2);g(3)]; %TO
EVALUATE A-MATREX
a=inv(f)*g;
s=0;
%TO EVALUTE LAST&FUTURE FORECASTED DEMANDS(X)-W %TO EVALUATE
  MEAN SQUARE ERROR(MSE)
for i=q:12:96;s=s+1;
X(i)=a(1)+(a(2)*s)+(a(3)*s^2); e=e+(w(i)-X(i))^2;
W(i)=a(1)+(a(2)*(s+8))+(a(3)*(s+8)^2);
end;
end;
MSE=e/96;
t=1:96;i=1:12;
%TO PLOT REAL& LAST FORECASTED DEMANDS
subplot(211),
plot(t,w(t),'r',t,X(t),'b'),
grid
```

```
xlabel('MONTH'),
ylabel('LAST DEMAND'),
%TO PLOT THE FUTURE DEMANDS FOR 5-YEARS 2005-2009
F1=W(i);
F2=W(i+12);
F3=W(i+24);
F4=W(i+36);
F5=W(i+48);
F6=W(i+60);
F7=W(i+72);
F8=W(i+84);
subplot(212),
plot(i,F1,i,F2,i,F3, i,F4, i,F5, i,F6, i,F7, i,F8),
grid,
xlabel('MONTH')
ylabel('FUTURE DEMAND')
```

Depending on the comparison, the best method was additive seasonal, then quadratic regression model excluding weather. Additive seasonal is the most simple and accurate model in the estimation. This is because, at each sample instant (t), there are values for a,b,d where

$$x_t = at + \mathrm{d}t + I \tag{9.62}$$

and estimation of x_t depends on x_t itself and the past demand patterns. Despite this, there is a disadvantage in the prediction where the swing in demands that are due to the seasonal influences are the same from year to year, but there is a shift in the magnitude of the curve (the prediction in this model does not depend on the past demand).

The regression model represented the linear relation between demands of months and produced demands. Thus, the additive seasonal is more practical than the regression model because it is much nearer to the actual value, and we can vary the forecasted demand (by varying the values of parameters U, L, K) to approach the nearest curve of demands.

PROBLEMS

9.1 A load having the data in Table 9.7:

TABLE 9.7
Load Data of Problem 1.4

Time(year)	Load1(MW)	Load2(MW)	Load3(MW)	Load4(MW)
0–0.1	100	150	50	50
0.1–0.25	150	100	100	100
0.25–0.75	150	150	150	100
0.75–1.0	100	50	50	50

Find the following:

1. The load factor of the total load on the network.
2. The diversity factor.
3. If it is required to supply the load by energy from a station of many 100 MW generating units, find the minimum number of units needed to supply the load, draw the operating schedule and find the use factor.

9.2 Draw the important components of a steam power station.
9.3 What factors are considered to select the site of the steam power station.
9.4 State the main parts of a thermal power station and discuss the function of each part.
9.5 Draw the important components of a hydroelectric power station.
9.6 What factors are considered to select the site of the hydroelectric power station.
9.7 Draw the important components of a nuclear power station.
9.8 What factors are considered to select the site of the nuclear power station.
9.9 What do you understand from the load curve, and what information is converged by a load curve.
9.10 Discuss the important points to consider while selecting the size and number of units generation.
9.11 Explain the terms connected load, load factor, plant use factor, diversity factor.
9.12 A power station is to supply three loads. The daily cycle of loads are given below (Tables 9.8A–9.8C):

TABLE 9.8A
Load A Table of Problem 9.12

Time in Hours	0–6	6–10	10–12	12–18	18–22	22–24
Loads A in MW	10	15	20	30	15	5

TABLE 9.8B
Load B Table of Problem 9.12

Time in Hours	0–10	10–20	20–24
Loads A in MW	0	15	0

TABLE 9.8C
Load C Table of Problem 9.12

Time in Hours	0–4	4–10	10–14	14–16	16–20	20–24
Loads A in MW	5	15	50	10	20	10

1. Draw the load and duration curves of the power station.
2. Calculate
   i. Load factor of each load.
   ii. Load factor of the power station.
   iii. Diversity factor of the power station.
   iv. The capacity factor when the station has reserve set 20% of the max. demand.

9.13 A power station is to supply two loads. The daily load curves of the two loads are shown in Figures 9.11.

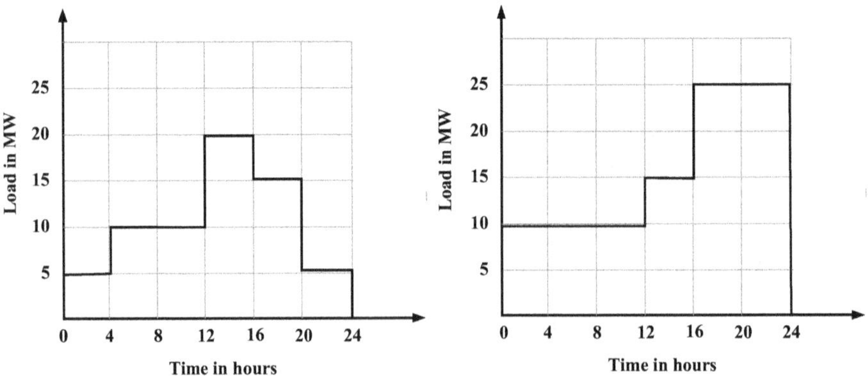

FIGURE 9.11 Load curves for Problem 9.13.

Load curve A Load curve B

1. Plot the load and duration curves of the power station.
2. Find:
   i. Load factor of each load.
   ii. Load factor and diversity factor of the power station.
   iii. The reserve capacity and capacity factor of the power station. If the power station consists of three units, the installed capacity of each unit is 20 MW.

9.14 A power station is to supply two loads. The daily loads are given below in Tables 9.9A–9.9B.

TABLE 9.9A
Load A Table of Problem 9.14

Time in Hours	0–6	6–8	8–16	16–22	22–24
Loads A in MW	4	12	18	30	35

TABLE 9.9B
Load B Table of Problem 9.14

Time in Hours	0–2	2–10	10–16	16–20	20–24
Loads A in MW	5	15	50	10	10

Draw the load and duration curves of the power station and find:

1. Load factor of each load.
2. Load factor of the power station.
3. Diversity factor of the power station.
4. The capacity factor when the station has a reserve set 20% of the max. demand.

Appendix A

MATLAB®, which stands for Matrix Laboratory, is a useful program for performing numerical and symbolic calculations. It is widely used in science and Engineering and in mathematics. The Basics of the Technical Language of the MATLAB is a technical language to ease scientific computations, and computational tool that employs matrices and vectors/arrays to carry out numerical analysis, and scientific visualization tasks.

MATLAB is an abbreviation for MATrix LABoratory. With MATLAB being opretional on Macintosh, Unix, and Windows operating systems.

This appendix introduces the reader to programming with the software package MATLAB. It is assumed that the reader has not had previous experience with a high-level programming language and is familiar with the techniques of writing loops, branching using logical relations, calling subroutines, and editing; therefore, this appendix.

A.1 MATLAB WINDOWS

This essential and present computational tool employs matrices and vectors/arrays to carry out numerical analysis, signal processing, and scientific visualization tasks.

A brief introduction to MATLAB presented in this appendix is sufficient for solving problems in this book.

MATLAB has become a powerful tool for technical professionals worldwide. A student version of MATLAB is available for PCs. A copy of MATLAB can be obtained from:

The Mathworks, Inc.
3 Apple Hill Drive
Natick, MA 01760-2098
Phone:(508) 647-7000
Website: http://www.Mathworks.com

To begin to use MATLAB, we use these operators. Type commands to MATLAB prompt ">>" in the Command window. To get help, type

```
>> help
```

MATLAB works with matrices. Everything MATLAB understands is in a matrix (from text to large cell arrays and structure arrays). The MATLAB Environment is shown in Figure A.1.

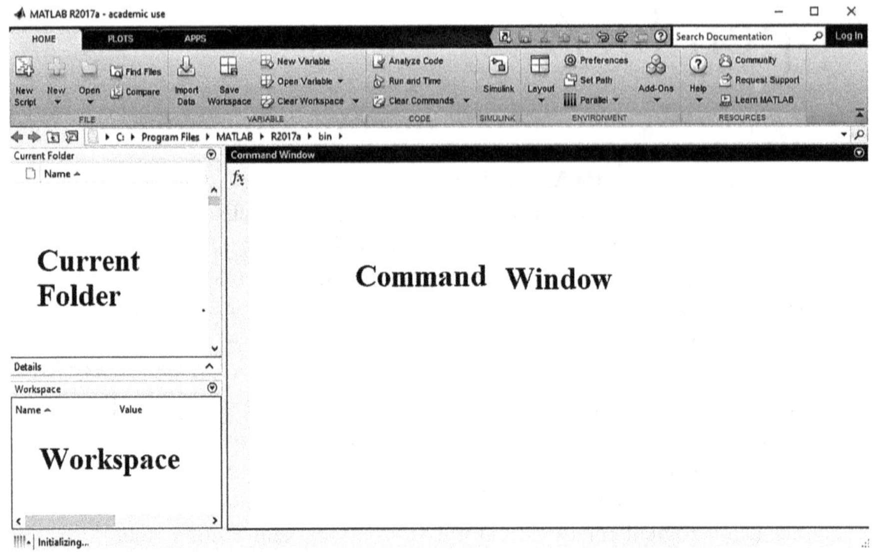

FIGURE A.1 The MATLAB environment.

MATLAB works through three basic windows.

Command Window: This is the main window. It is characterized by MATLAB command prompt >> when you launch the application program MATLAB puts you in this window. All commands, including those for user-written programs, are typed in this window at the MATLAB prompt.

Graphics window: The output of all graphics commands typed in the command window is flushed to the graphics or figure window. It is a separate gray window with a white background color. The user can create as many windows as the system memory will allow.

Edit window: This is where you write, edit, create, and save your programs in files called M files.

Input-output: MATLAB supports interactive computation by taking the input from the screen and flushing the output to the screen. Also, it can read input files and write output files.

Data Type: The fundamental data type in MATLAB is the array. It encompasses several distinct data objects – integers, real numbers, matrices, character strings, structures, and cells. There is no need to declare variables as real or complex; MATLAB automatically sets the variable to be real.

Dimensioning: Dimensioning is automatic in MATLAB. No dimension statements are required for vectors or arrays. We can find the dimensions of an existing matrix or a vector with the size and length commands.

All programs and commands can be entered either in the Command Window or the M file using MATLAB editor, then Save all M files in the folder 'work' in the current directory.

A.2 MATLAB BASICS

To find additional information about commands, options, and examples, the reader is urged to use the on-line help facility and the Reference and User's Guides that accompany the software.

TABLE A.1
Basic Operations

Operation	MATLAB Formula
Addition	a + b
Subtraction	a – b
Division (right)	a/b (means $a \div b$)
Division (left)	a\b (means $b \div a$)
Multiplication	a*b
Power	a^b

TABLE A.2
Basic Functions

Function	Remark
abs(x)	Absolute value or complex magnitude of x
acos, acosh(x)	Inverse cosine and inverse hyperbolic cosine of x in radians
acot, acoth (x)	Inverse cotangent and inverse hyperbolic cotangent of x in radians
angle(x)	Phase angle (in radian) of a complex number x
asin, asinh(x)	Inverse sine and inverse hyperbolic sine of x in radians
atan, atanh(x)	Inverse tangent and inverse hyperbolic tangent of x in radians
conj(x)	The complex conjugate of x
cos, cosh(x)	Cosine and hyperbolic cosine of x in radian
cot, coth(x)	Cotangent and hyperbolic cotangent of x in radian
exp(x)	Exponential of x
fix	Round toward zero
imag(x)	The imaginary part of a complex number x
log (x)	Natural logarithm of x
log2(x)	The logarithm of x to base 2
log10(x)	Common logarithms (base 10) of x
real (x)	A real part of a complex number x
sin, sinh(x)	Sine and hyperbolic sine of x in radian
sqrt (x)	The square root of x
tan, tanh	Tangent and hyperbolic tangent of x in radian

TABLE A.3
Matrix Operations

Operation	Remark
A'	Finds the transpose of matrix A
det(A)	Evaluates the determinant of matrix A
inv(A)	Calculates the inverse of matrix A
eig(A)	Determines the eigenvalues of matrix A
diag(A)	Finds the diagonal elements of matrix A
expm(A)	Exponential of matrix A

TABLE A.4
Special Matrices, Variables, and Constants

Matrix/Variable/Constant	Remark
eye	Identity matrix
ones	An array of ones
zeros	An array of zeros
i or j	Imaginary unit or sqrt(-1)
pi	3.142
NaN	Not a number
inf	Infinity
eps	A very small number, 3.2e-16
rand	Random element

TABLE A.5
Vector Operations

Vector	Remark
sum(a)	Sum of vector elements
mean(a)	Mean of vector elements
std(a)	Standard deviation
max(a)	Maximum
min(a)	Minimum

TABLE A.6
Complex Numbers

Real(A)	The Real Part of A
imag(A)	The imaginary part of A
conj(A)	The complex conjugate of A
abs(x)	The modulus of A.
angle(A)	The phase angle of A

TABLE A.7
Matrix Operations

Operation	MATLAB Formula
Matrix Addition	+
Matrix Subtraction	-
Matrix Multiplication	*
Right Matrix Division	/
Left Matrix Division	\
Raise to a Power	^
Transpose Matrix	‘

TABLE A.8
Array Operations

Operation	MATLAB Formula
Array Multiplication	.*
Right Array Division	./
Left Array Division	.\
Raise to a power	.^

TABLE A.9
Matrices and Functions

Operation	Remark
sqrtm(x)	The matrix square root.
expm(x)	The matrix exponential base e.
logm(x)	The matrix natural logarithm.

TABLE A.10
Utility Matrices

Operation	Remark
zeros(n)	n * n matrix where each element is zero.
zeros(m,n)	m * n matrix where each element is zero.
ones(n)	n * n matrix where each element is one.
ones(m,n)	m * n matrix where each element is one.
rand(n)	n * n matrix of random numbers.
rand(m,n)	m * n matrix of random numbers.
eye(n)	n * n identity matrix.

TABLE A.11
Relational and Logical Operations

Operation	MATLAB Formula
Less than	<
Less than or equal	<=
Greater than	>
Greater than or equal	>=
Equal	= =
Not equal	~=

TABLE A.12
Boolean Logic Operations

Operation	MATLAB Formula
AND	&
OR	|
NOT	~

TABLE A.13
Variable Control Commands

Operation	Remark
who	List all the variables in memory
whos	List all the variables in memory with more information
clear	Remove all variables from memory
clear <variable>	Remove specified variables from memory

TABLE A.14
File Control Commands

Operation	Remark
dir	List the contents of the current directory
ls	List the contents of the current directory
what	List the MATLAB files in the current directory
cd <directory>	Change the current directory
type <filename>	Display the contents of a text or .m file
delete <filename>	Delete a file
diary <filename>	Record all commands and results to a file
Diary	Off Stop above

TABLE A.15
Saving, Exporting, and Importing Data

Operation	Remark
Save	Save all variables to the file Matlab.mat
load	Load in variables from the file Matlab.mat
save <filename>	Save all the variables to the file filename.mat
load <filename>	Load in the variables from the file filename.mat
save <filename> <variable>	Save only the variable to the file filename.mat
load <filename> <variable>	Load in only the variable from the file filename.mat
save <filename> <variable> -ascii	Save the variable to the text file filename
load <filename>.<ext>	Load from the text file to the variable called filename
save <filename> <variable> -ascii -	double Save variable to the text file filename using double precision
csvwrite(<filename>.cvs, M)	Write matrix M to CSV file
M = csvread(<filename>.csv)	Load from the CSV file into variable M

A.3 USING MATLAB TO PLOT

To plot using MATLAB is easy. For a two-dimensional plot, use the plot command with *two* arguments as

```
>> plot(xdata,ydata)
```

where *xdata* and *ydata* are vectors of the same length containing the data to be plotted.

TABLE A.16
Various Color and Line Types

y	yellow	.	point
m	magenta	o	circle
c	cyan	x	x-mark
r	Red	+	plus
g	green	-	solid
b	blue	*	star
w	white	:	dotted
k	black	-.	dashdot
		--	dashed

TABLE A.17
Plotting Commands

Command	Comments
bar(x,y)	a bar graph
contour(z)	a contour plot
errorbar (x, y, l, u)	a plot with error bars
hist(x)	a histogram of the data
plot3(x, y, z)	a three-dimensional version of plot()
polar(r, angle)	a polar coordinate plot
stairs(x, y)	a stairstep plot
stem(n, x)	plots the data sequence as stems
subplot(m, n, p)	multiple (m-by-n) plots per window
surf(x, y, x, c)	a plot of 3-D colored surface.
hold on	Hold the plot on the screen.
hold off	Release the plot on the screen
hold	Toggle the hold state.

MATLAB will let you graph multiple plots together and distinguish them from different colors.

It is obtained with the command plot (xdata, ydata, 'color'), where the color is indicated by using a character string from the options listed in Table A.16.

A.4 BASIC COMMANDS OF MATLAB IN DISCRETE SYSTEMS

TABLE A.18
Basic Command in Discrete Systems

No.	Commands	Description
1	Y = 0:2:20	This instruction indicates a vector Y with an initial value 0 and a final value 20 with an increment of 3. Therefore Y = [0 2 4 6 8 10 12 14 16 18 20].
2	M= 40:5:100	M = [40 45 50 55 60 65 70 75 80 85 90 95 100]
3	N= 0: 1/pi: 1	N= [0, 0.3183, 0.6366, 0.9549]
4	zeros (1, 5)	Creates a vector of one row and five columns whose values are zero Output= [0 0 0 0 0]
5	ones (2,6)	Creates a vector of two rows and six columns Output = 1 1 1 1 1 1 1 1 1 1 1 1
6	a = [2 2 -5] b = [3 5 4]	a.*b = [6 10 -20]

TABLE A.18 (Continued)

No.	Commands	Description
7	plot (t, x) If x = [6 7 8 9] t = [1 2 3 4]	This instruction will display a figure window that indicates the plot of x versus t. 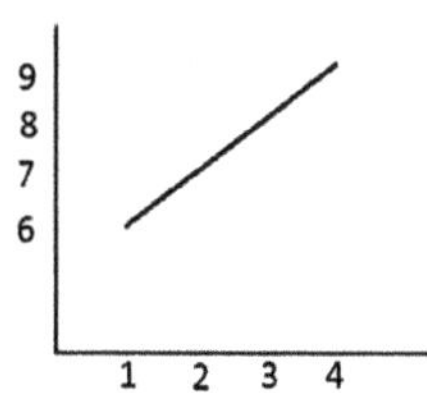
8	stem (t, x)	This instruction will display a figure window as shown. 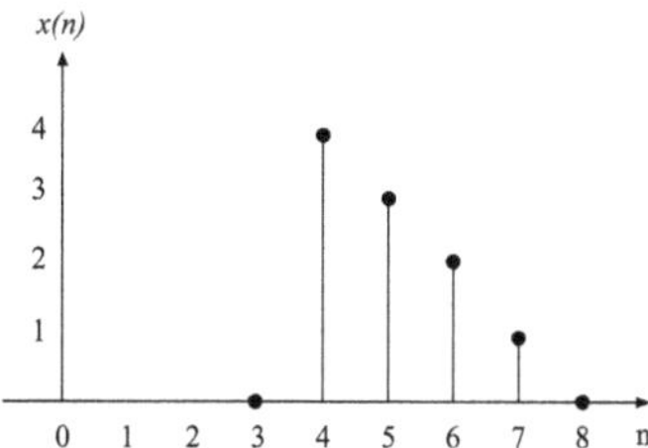
9	Subplot	This function divides the figure window into rows and columns. Subplot (2 2 1) divides the figure window into three rows and three columns 1, 2, 3, 4,... represent the number of the figure. (3, 3, 1) | (3, 3, 2) | (3, 3, 3) (3, 3, 4) | (3, 3, 5) | (3, 3, 6) (3, 3, 7) | (3, 3, 8) | (3, 3, 9)
10	Conv	Syntax: y = conv(a, b) Description:y=conv(a, b) convolves vectors a and b.
11	Disp	Syntax: disp(X) Description: disp(X) displays an array without printing the array name. If X contains a text string, the string is displayed.
12	FFT	FFT(X) is the discrete Fast Fourier transform (FFT) of vector X. For matrices, the FFT operation is applied to each column. For N-D arrays, the FFT operation operates on the first non-singleton dimension. FFT(X, N) is the N-point FFT, padded with zeros if X has less than N points and truncated if it has more.
13	ABS	Absolute value. ABS(X) is the absolute value of the elements of X. When X is complex, ABS(X) is the complex modulus (magnitude) of the elements of X.
14	ANGLE	Phase angle. ANGLE(H): the phase angles, in radians, of a matrix with complex elements.

(Continued)

TABLE A.18 (Continued)

No.	Commands	Description
15	INTERP	Y = INTERP(X, L) re-samples the sequence in vector X at L times the original sample rate. The resulting re-sampled vector Y is L times longer, LENGTH(Y) =L*LENGTH(X). Resample data at a higher rate using low-pass interpolation.
16	DECIMATE	Y = DECIMATE(X, M) re-samples the vector X sequence at 1/M times the original sample rate. The resulting re-sampled vector Y is M times shorter, LENGTH(Y) = CEIL(LENGTH(X)/M). By default, DECIMATE filters the data with an 8th order Chebyshev Type I low-pass filter with cut-off frequency. 8*(Fs/2)/R, before re-sampling. Resample data at a lower rate after low-pass filtering.
17	xlabel	Syntax: xlabel('string') Description: xlabel('string') labels the x-axis of the current axes.
18	ylabel	Syntax: ylabel('string') Description: ylabel('string') labels the y-axis of the current axes.
19	Title	Syntax: title('string') Description: title('string') outputs the string at the top and in the center of the current axes.
20	grid on	Syntax: grid on. Description: grid on adds major grid lines to the current axes.
21	*Help*	List topics on which support is available
22	*Help command name*	Provides help on the topic selected
23	*Demo*	Runs the demo program
24	*Who*	Lists variables currently in the workspace
25	*Whos*	Lists variables currently in the workspace with their size
26	*Clear*	Clears the workspace, all the variables are removed
32	*Clear x,y,z*	Clears only variables x,y,z.
33	*Quit*	Quits MATLAB
34	*fir, delay, cas, sos, cas2can*	FIR and IIR filtering
35	*cfir2, cdelay2, wrap2*	circular FIR filtering
36	*dtft*	DTFT computation
37	*sigav, sg, sgfilt, ecg*	Signal averaging, SG smoothing
38	*kwind, I0, kparm, kparm2*	Kaiser window
39	*klh, dlh, kbp, dbp, kdiff, ddiff, khilb, dhilb*	FIR filter design
40	*lhbutt, bpsbutt, lhcheb1, lhcheb2, bpcheb2, bscheb2*	IIR filter design

A.5 MATLAB/SIMULINK

The MATLAB program has been provided by Simulink library as shown in Figure A.2.

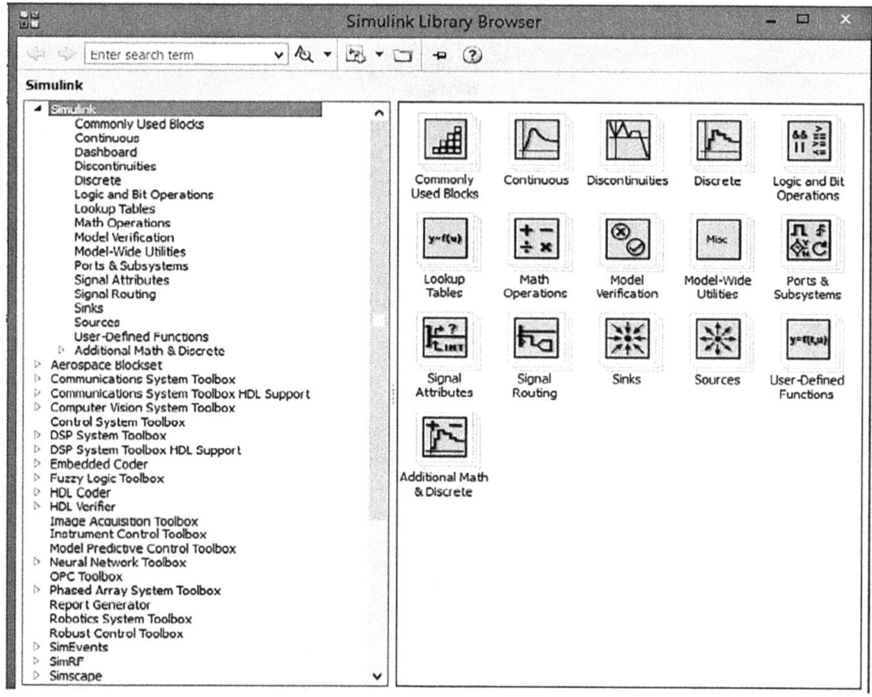

FIGURE A.2 Simulink library.

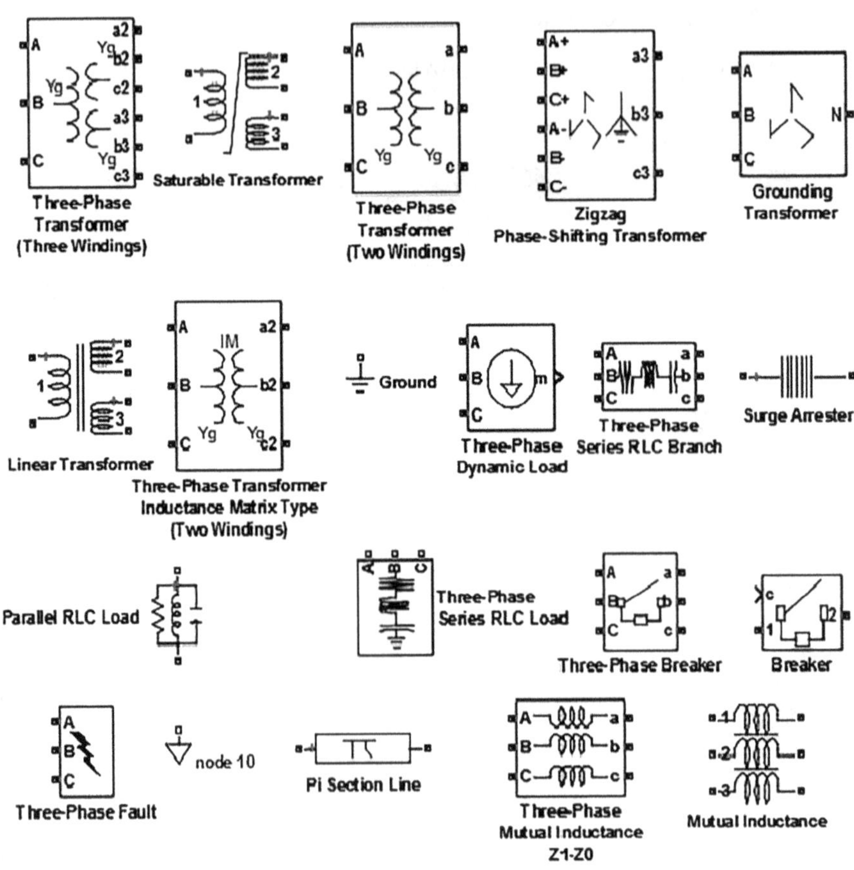

FIGURE A.3 Some specialized technology elements.

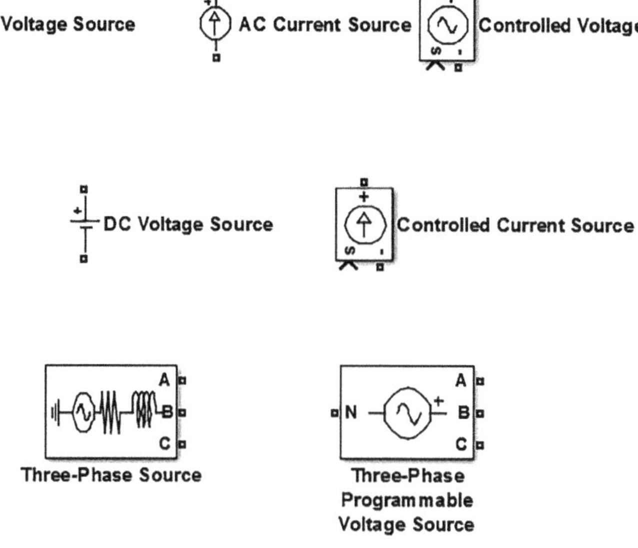

FIGURE A.4 Power system sources.

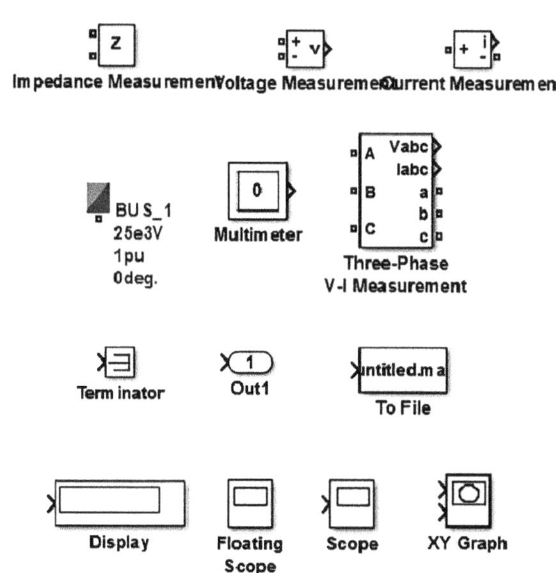

FIGURE A.5 Measurement elements.

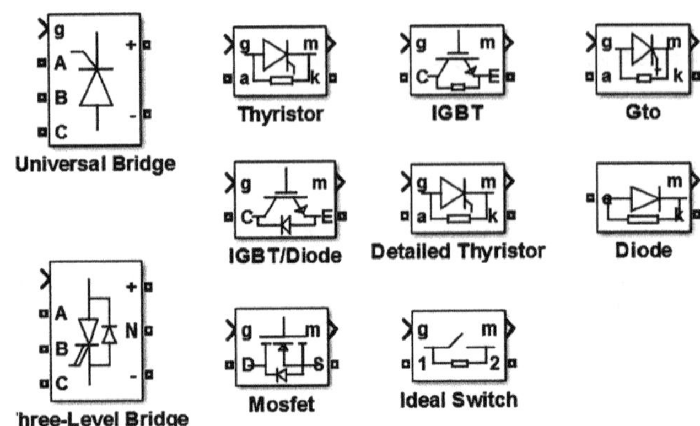

FIGURE A.6 Power electronics elements.

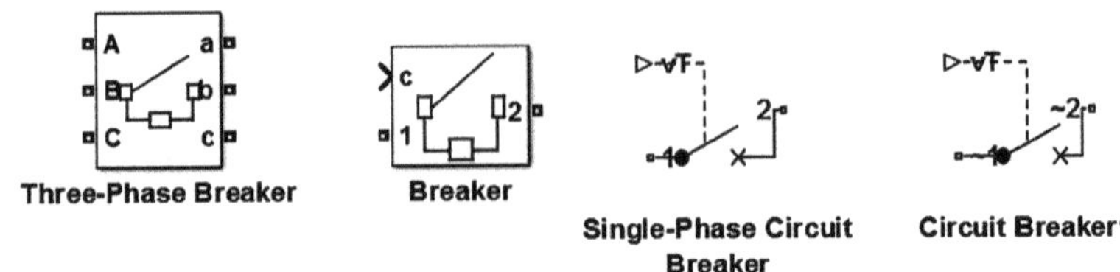

FIGURE A.7 Circuit breaker configuration using Simulink.

Bibliography

Abdel-Azim, M., Salah, H. E. D., and Eissa, M. E. (2018). IDS against Black-Hole Attack for MANET. IJ Network Security, 20(3), 585–592.

Abur, Ali, and G. E. Antonio, Power System State Estimation, NewYork: Marcel Dekker, 2004.

Adamson, C. and Hingorani, N. G., High Voltage Direct Current Power Transmission, Garaway Ltd., 1960.

Amara Korba, A., Nafaa, M., and Ghanemi, S. (2016). An efficient intrusion detection and prevention framework for ad hoc networks. Information & Computer Security, 24(4), 298–325.

Anderson, P. M., Analysis of Faulted Power Systems. Iowa State Univ. Press, Ames, IA, 1973a.

Anderson, P. M., Analysis of Faulted Power Systems. Iowa State Univ. Press, Ames, IA, 1973b.

Anwar, S., Jasni, M. Z., Mohamad, F. Z., Inayat, Z., Khan, S., Anthony, B., & Chang, V. (2017). From intrusion detection to an intrusion response system: fundamentals, requirements, and future directions. Algorithms, 10(2), 39. doi:10.3390/a10020039

Arrillaga, J., D. A. Bradley, and P. S. Bodger, Power System Harmonics, John Wiley and Sons, NewYork, 1989.

Assad, N., Elbhiri, B., Moulay, A. F., Ouadou, M., and Aboutajdine, D. (2015). Analysis of the deployment quality for intrusion detection in wireless sensor networks. Journal of Computer Networks and Communications, 5(2), 20. doi: 10.1155/2015/812613

Baggili, I., and Rogers, M. (2009). Self-reported cyber-crime: An analysis of the effects of anonymity and pre-employment. International Journal of Cyber Criminology, 3, 550–565. Retrieved from http://www.cybercrimejournal.com

Bahrami, M., and Bahrami, M. (2014). An overview of software architecture in the intrusion detection system. Journal of Computer Networks and Communications, 1(1), 1–8. doi: 10.7321/jscse.v1.n1.1

Bailetti, T., Gad, M., and Shah, A. (2016). Intrusion learning: an overview of an emergent discipline. Technology Innovation Management Review, 6(2): 15–20. doi:10.22215/timreview/964

Barot, V., Sameer, S. C., and Patel, B. (2014). Feature selection for modeling intrusion detection. International Journal of Computer Network and Information Security, 6(7), 56–62. doi:10.5815/ijcnis.2014.07.08

Ben-Asher, N., and Gonzalez, C. (2015). Effects of cybersecurity knowledge on attack detection. Computers in Human Behavior, 48, 51–61. doi: 10.1016/j.chb.2015.01.039

Bergen, A. R., and V. Vittal Power System Analysis, 2nd edition, Tom Robbins, Prentice-Hall Inc, 2000.

Berrie, T. W., Power System Economics, IEEE, London, 1983.

Bilal, M. B. (2014). A new classification scheme for intrusion detection systems. International Journal of Computer Network and Information Security, 6(8), 56–70. doi:10.5815/ijcnis.2014.08.08

Billinton, R., Power System Reliability Evaluation, Gqrdon and Breach, NewYork, 1970.

Billinton, R., and R. N. Allan, Reliability Evaluation of Power Systems, Plenum Press, NewYork, 1984.

Billirrton, R., R. J. Ringlee, and AJ Wood. Power System Reliabitity Calculations, The MIT Press, Boston, Mass, 1973.

Blanche, G. A. (2018). The cybersecurity workforce: Profession or not? (Order No. 13841062). Available from ProQuest Dissertations & Theses Global. (2179183753).

Boehle, O. B., Switchgear Manual, 8th edition, Trans. By David Stone, Published by Asea Brown Boveri, (A Book), 1988.

Boldea, I., and S. A. Nasar, Electric Machine Dynamics, CRC Press, NewYork, 1986.

Bowler, C. E. J., Concordia, C., and Tice, J. B., Subsynchronous torques on generating units feeding series-capacitor compensated lines, Proc. Am. Power Conf. 35, 1129–1136 (1973).

Brown, Homer E., Solution of Large Networks By Matrix Methods, 2nd edition, John Wiley & Sons, NewYork, 1985.

Canadian Electrical Association, Static Compensators for Reactive Power Control, Cantext Publications, Canada, 1984.

Carter, N., Bryant-Lukosius, D., DiCenso, A., Blythe, J., and Neville, A. J. (2014). The use of triangulation in qualitative research. Oncol Nurs Forum, 41, 545–547. doi: 10.1188/14.onf.545-547

Chebrolua, S., Abraham, B., and Thomas, J. (2005). Feature deduction and ensemble design of intrusion detection systems. Computers and Security, 24(4): 295–307. doi:10.1016/j.cose.2004.09.008

Chourasiya, R., Patel, V., and Shrivastava, A. (2018). Classification of cyber-attack using machine learning technique at Microsoft azure cloud. International Journal of Advanced Research in Computer Science. 6(1), 4–8.

Clair, M. L., Annual statistical report. Electr. World 193 (6) 49–80 (1980).

Concordia, C., and Rusteback, E., Self-excited oscillations in a transmission system using series capacitors. IEEE Trans. Power Appar. Syst. PAS-89 (no. 7) 1504–1512 (1970).

Cory, B. J., High Voltage Direct Current Converters and Systems, Macdonald, London, 1965.

Creswell, J. (2013). Research design: qualitative, quantitative, and mixed methods approach, 4th edition. Thousand Oaks, CA: SAGE Publications Inc.

Da Silva Cardoso, A. M., Lopes, R. F., Teles, A. S., and Magalhães, F. B. V. (2018, April). Real time DDoS detection based on complex event processing for IoT. In *Internet-of-Things Design and Implementation (IoTDI), 2018 IEEE/ACM Third International Conference on* (pp. 273–274). IEEE.

Debs, A. S., Modern Power Systems Contror and Operation, Kluwer Academic, Boston, 1988.

Denzin, N. K., and Lincoln, Y. S. (2005). Introduction: The discipline and practice of qualitative research. Thousand Oaks, CA: Sage.

Desai, A., Prajapati, H., and Bhatti, D. (2011). Problems and challenges in wireless network intrusion detection. National Journal of System and Information Technology, 4(2), 175–189.

DOEIEP-0005. USDOE, Office of Emergency Operations, Washington, DC, 1981.

Duan, T. (2016). Analysis focusing on intrusion detection technology when an outside party breaks into computer database. RISTI (Revista Iberica de Sistemas e Tecnologias de Informacao), 17B, 180–193.

Electric Power Research Institute, Transmission Line Reference Book: 345 kV and Above. EPRI, Palo Alto, CA, 1979a.

Electric Power Research Institute, Transmission Line Reference Book: 345 kV and Above. EPRI, Palo Alto, CA, 1979b.

Elgerd, Olle I., Electric Energy Systems Theory, 2nd edition, McGraw-Hill, NewYork, (A Book), 1982.

Elliott, L. C., Kilgore, L. A., and Taylor, E. R., The prediction and control of self-excited oscillations due to series capacitors in power systems. IEEE Trans. Power Appar. Syst. PAS-90 (no. 3) 1305–1311 (1971).

Endernyi, J., Reliability Modeling in Electric Power System, John Wiley and Sons, 1978.

Energy Information Administration, Energy Data Reports—Statistics of Privately-Owned Electric Utilities in the United States. US Dept. of Energy, Washington, DC, 1975–1978.

Fink, D. G., and Beaty, H. W., Standard Handbook for Electrical Engineers, 11th edition. McGraw-Hill, NewYork, 1978.

Fink, D. G., and Carroll, J. M. (eds.), Standard Handbook for Electrical Engineers, 10th edition, McGraw-Hill, NewYork, 1969.

Fuse and Types of Fuses – Construction, Operation & Applications.

Goldberg, D. E., Genetic Algorithm in Search, Optimization, and Machine Learning, Addison-Wiley, 1989.

Gonen, T., Electrical Power Distribution System Engineering. McGraw-Hill, NewYork, 1986.

Gonen, T., Electrical Power Distribution System Engineering, 2nd edition. CRC Press, Boca Raton, FL, 2008.

Gonen, T., Electrical Power Transmission System Engineering: Analysis and Design, 2nd edition. CRC Press, Boca Raton, FL, 2009.

Grigsby, Leonard (2007). Electric Power Generation, Transmission, and Distribution. CRC Press. ISBN 978-0-8493-9292-4.

Gross, C. A., Power System Analysis, Wiley, NewYork, 1979.

Grunbaum, R., M. Noroozian, and B. Thorvaldsson, FACTS Powerful systems for flexible power transmission, ABB Rev, May 1999, 4–17.

Guile, A. E., and W. Paterson, Electric Power Systems, Volume 1, Oliver & Boyd, Edinburgh, 1969.

Gupta, B. R., Power System Analysis and Design, Wheeles Bombay, 1985.

Haupt, Randy L., and Sue Ellen Haupt, Practical Genetic Algorithms, 2nd edition, John Wiley & Sons, 2004.

Hileman, A., Insulation coordination for power systems, Marcel Dekker Inc., NewYork, 1999

Horowitz, Stanley H., and Phadke, Arun G. *Power system relaying*, 3rd edition, John Wiley & Sons Ltd, 2008.

https://www.electricaltechnology.org/2014/11/fuse-types-of-fuses.html

IEEE Committee Report, Proposed definitions of terms for reporting and analyzing outages of electrical transmission and distribution facilities and interruptions, IEEE Trans. Power Appar. Syst. PAS-87 (5) 1318–1323 (1968).

IEEE Committee Report, Guidelines for use in developing a specific underground distribution system design standard, IEEE Trans. Power Appar. Syst. PAS-97 (3) 810–827 (1978).

IEEE Standard Definitions in Power Operations Terminology. IEEE Standard 346-1973, Nov. 2, 1973.

IEEE Standard Dictionary of Electrical and Electronics Terms. IEEE, NewYork, 1972.

Institute of Electrical and Electronics Engineers, Graphic Symbols for Electrical and Electronics Diagrams, IEEE Std. 315-1971 [or American National Standards Institute (ANSI) Y32.2-1971]. IEEE, NewYork, 1971.

Institute of Electrical and Electronics Engineers Committee Report, The significance of assumptions implied in long-range electric utility planning studies. IEEE Trans. Power Appar. Syst. PAS-99, 1047–1056 (1980).

Kalas, P., Reliability for Technology, Engineering and Management Prentice-Hall, Inc., USA, 1998.

Kilgore, L., Taylor, E. R., Jr., Ramey, D. G., Farmer, R. G., and Schwalb, A. L, Solutions to the problems of subsynchronous resonance in power systems with series capacitors. Proc. Am. Power Conf. 35, 1120–1128 (1973).

Kimbark, E. W., "Power system stability" volume 1, Elements of Stability Calculations, John Wiley & Sons, Inc., NewYork, printed in the USA, copyright, 1976.

Kisielewicz, T., G. B. Lo Piparo, F. Fiamingo, C. Mazzetti, B. Kuca, Z. Flisowski: Factors affecting selection, installation, and coordination of surge protective devices for low voltage systems, Electric Power Systems Research, ISSN: 0378-7796, Vol. 113, August 2014.

Kothari, D. P., Modern Power System Analysis McGraw-Hill, New Delhi, 2004.

Kron, G., Tensor Analysis of Networks, John Wiley & Sons, NewYork, 1939.

Kusic, G. L., Computer Aided Power System Analysis, Prentice-Hall, New Jersey, 1986.

Lakevi, E., Electricity Distribution Network Design, 1998.

Lewis, E. E., Introduction to Reliability Engineering, 2nd edition, John Wiley and Sons, Inc., 1996.

Mahalanabis, A. K., D. P. Kothari, and S. I. Ahson Computer Aided Power System Anlysis and Control, New Delhi: Tata McGraw-Hill, 1988.

Mamoh, James A.. Electric Power System Applications of Optimization, CRC Press Publisher, 2nd edition. 2008.

Meliopoulos, A. P. (1988). Power System Grounding and Transients: An Introduction, Marcel Dekker, Inc., NewYork and Basel.

Miller, T. J. E., Reactive Power Control in Electric Systems, John Wiley & Sons 1982.

Milligan, M. M. "Modeling Utility –Scale wind Power Plants" part 2: capacity credit. National Renewable Energy laberatory, May 2002.

Mohan, N., T. M. Undeland, and W. P. Robbins, Power Electronics: Converters, Applications, and Design, John Wiley and Sons, NewYork 1989.

Murphy, J. M. D., and F. G. Turnbull, Power Electronic Control of AC Motors, Pergamon Press, 1988.

National Electric Reliability Council, Tenth Annual Review of Overall Reliability and Adequacy of the North American Bulk Power Systems. NERC, Princeton, NJ, 1980.

National Electrical Safety Code, 1977 ed., ANSI C2, IEEE, NewYork, November 1977.

Neuenswander, J. R., Modern Power Systems. International Textbook Company, Scranton, PA, 1971.

NSK. Jaysekara Menik "Computer simulation of transient stability analysis of power system," University of Windsor, Canada, 2004.

O'Coonnor, P. D. T., Newton D., and Bromley R., Practical Reliability Engineering, 4th edition, John Wiley and Sons, Ltd., 2002.

Pack, S., J. Plesch: EMTP–RV Course, Graz, Austria, 18–20 October. 2010.

Padiyar, K. P., "Power System Dynamics Stability and Control", John Wiley & Sons (Asia), 1996.

Pender, H., and Del Mar, W. A., Electrical Engineers' Handbook—Electrical Power, 4th edition. Wiley, NewYork, 1962.

Proposed Standard Definitions of General Electrical and Electronics Terms. IEEE Standard 270, 1966.

Rashid, M. H., Power Electronics, Circuits Devices and Applications, Prentice-Hall International Editions, London, 1992.

Robert Eaton, J. and Edwin Cohen, Electric Power Transmission Systems, Prentice-Hall, Inc. 1991.

Saadat, Hadi, "Power System Analysis, McGraw-Hill Inc., NewYork, 1999a. Milwaukee School of Engineering Web site: http://www.home.att.net/~saadat, http://www.msoe.edu/~saadat

Saadat, Hadi Power system analysis, McGraw-Hill, NewYork, 1999b.

Schifreen, C. S., and Marble, W. C., Changing current limitations in the operation of high-voltage cable lines. Trans. Am. Inst. Electr. Eng. 26, 803–817 (1956).

Shooman, M. L., Probabilistic Reliability: An Engineering Approach, 2nd edition, McGraw-Hill, NewYork, 1990.

Simon, D.. Optimate State Estimation: Kalman, H infinity, and Non-linear Approaches, New Jersey: John Wiley & Sons, Inc. 2006a.

Simon, Dan. Optimal State Estimation, John Wiley & Sons Inc., 2006b.

Song, Yong-Hua (Ed.), Modern Optimization Techniquesin Power Systems, Kluwer Academic Publishers, London, 1999.

Stagg, G. W. and El-Abiad, A. H., Computer Methods in Power System Analysis, McGraw-Hill Book Company, NewYork, 1968.
Stark, P. A., Introduction to Numerical Methods, MacMillan Pub. Co, 1970.
Stevenson, W. D., Elements of Power System Analysis, 4th edition, Mc Graw-Hill, NewYork, 1989.
Stott, B., Review of Load Flow Calculation Methods, Proc. IEEE, July 1974.
Syed, Nasser A., Electric Energy System, Macmillan Publishing Company, NewYork, 1996.
Taylor, C. W., Power System Voltage Stability, McGraw-Hill Book Company, NewYork, 1994a.
Taylor, C. W., Power System Voltage Stability, McGraw-Hill Book Company, NewYork, 1994b.
Thorborg, K., Power Electronics, Prentice-Hall International (UK) Ltd., London, 1988.
US Department of Energy, The National Electric Reliability Study: Technical Study Reports,
Vadhera, S. S., Power System Analysis & Stability Delhi, 1987.
Van Cutsem, T. and Coustas Vournas, Voltage Stability of Electrical Power Systems, Kluwer Academic Publishers, Boston, USA., 1998.
Van Joolingen, Wouter R. and Ton De Jon, "Modeling Domain Knowledge for Intelligent Simulation Learning Environments," Computer Educ., 18, 1–3, 29–37, 1992.
Venkataraman, P., Applied Optimization with MATLAB Programming, McGraw-Hill Book Company, NewYork, 2001.
Vennard, E., Management of the Electric Energy Business. McGraw-Hill, NewYork, 1979.
Wagner, C. F., and Evans, R. D., Symmetrical Components. McGraw-Hill, NewYork, 1933.
Weedy, B. M., Electric Power Systems, 2nd edition. Wiley, NewYork, 1972.
Weedy, B. M., and B. J. Cory, Electric Power Systems, 4th edition. NewYork, John Wiley & Sons, 1998.
Wiseman, R. T., Discussions to charging current limitations in the operation of high-voltage cable lines, Trans. Am. Inst. Electr. Eng. 26, 803–817 (1956).
Wood, J., and Wollenberg, B. F., Power Generation Operation and Control, John Wiley & Sons Book Company, NewYork, 1996.
Woodford, D. A. (1998 March). HVDC Transmission, Professional Report from Manitoba HVDC Research Center, Winnipeg, Manitoba.

Index

For Product Safety Concerns and Information please contact our EU representative GPSR@taylorandfrancis.com
Taylor & Francis Verlag GmbH, Kaufingerstraße 24, 80331 München, Germany

www.ingramcontent.com/pod-product-compliance
Lightning Source LLC
Chambersburg PA
CBHW060809220326
41598CB00022B/2571

* 9 7 8 1 0 3 2 2 7 7 6 4 6 *